# The Human Aging & The Telomere

Edited by Paul F. Kisak

# Contents

# Chapter 1

# Genetics of aging

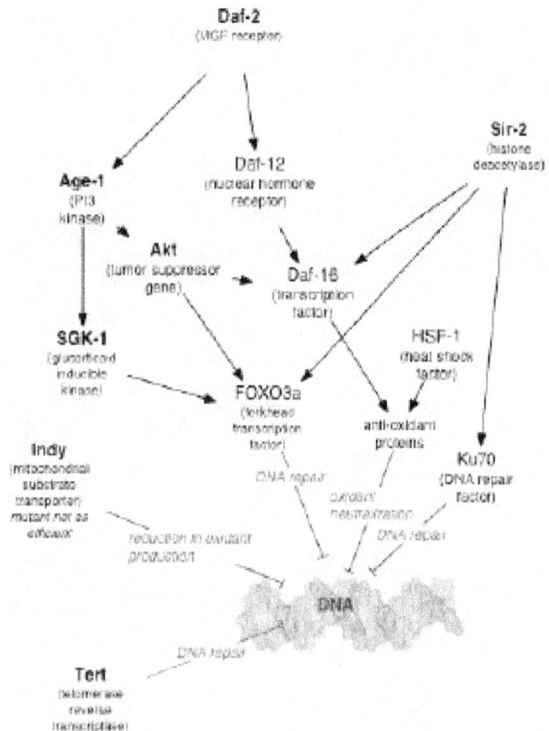

*Many life span influencing genes affect the rate of DNA damage or DNA repair*

**Genetics of aging** is generally concerned with life extension associated with genetic alterations, rather than with accelerated aging diseases leading to reduction in lifespan.

The first mutation found to increase longevity in an animal was the *age-1* gene in *Caenorhabditis elegans*. Michael Klass discovered that lifespan of *C. elegans* could be altered by mutations, but Klass believed that the effect was due to reduced food consumption (calorie restriction).[1] Thomas Johnson later showed that life extension of up to 65% was due to the mutation itself rather than due to calorie restriction,[2] and he named the gene *age-1* in the expectation that other genes that control aging would be found. The *age-1* gene encodes the catalytic subunit of class-I phosphatidylinositol 3-kinase (PI3K).

A decade after Johnson's discovery *daf-2*, one of the two genes that are essential for dauer larva formation,[3] was shown by Cynthia Kenyon to double *C. elegans* lifespan.[4] Kenyon showed that the *daf-2* mutants, which would form dauers above 25 °C (298 K; 77 °F) would bypass the dauer state below 20 °C (293 K; 68 °F) with a doubling of lifespan.[4] Prior to Kenyon's study it was commonly believed that lifespan could only be increased at the cost of a loss of reproductive capacity, but Kenyon's nematodes maintained youthful reproductive capacity as well as extended youth in general. Subsequent genetic modification (PI3K-null mutation) to *C. elegans* was shown to extend maximum life span tenfold.[5][6]

According to the GenAge database of aging-related genes, there are over 1800 genes altering lifespan in model organisms: 838 in the soil roundworm (*Caenorhabditis elegans*), 883 in the bakers' yeast (*Saccharomyces cerevisiae*), 170 in the fruit fly (*Drosophila melanogaster*) and 126 in the mouse (*Mus musculus*).[7]

Genetic modifications in other species have not achieved as great a lifespan extension as have been seen for *C. elegans*. *Drosophila melanogaster* lifespan has been doubled.[8] Genetic mutations in mice can increase maximum lifespan to 1.5 times normal, and up to 1.7 times normal when combined with calorie restriction.[9]

## 1.1 See also

- Strategies for Engineered Negligible Senescence

## 1.2 References

[1] Klass MR (1983). "A method for the isolation of longevity mutants in the nematode Caenorhabditis elegans and initial results". *MECHANISMS OF AGEING AND DEVELOPMENT.* **22** (3-4): 279–286. doi:10.1016/0047-6374(83)90082-9. PMID 6632998.

[2] Friedman DB, Johnson TE (1988). "A mutation in the age-1 gene in Caenorhabditis elegans lengthens life and reduces hermaphrodite fertility" (PDF). *Genetics (journal)*. **118** (1): 75–86. PMC 1203268⊝. PMID 8608934.

[3] Gottlieb S, Ruvkun G (1994). "daf-2, daf-16 and daf-23: genetically interacting genes controlling Dauer formation in Caenorhabditis elegans". *Genetics (journal)*. **137** (1): 107–120. PMC 1205929⊝. PMID 8056303.

[4] Kenyon C, Chang J, Gensch E, Rudner A, Tabtiang R (1993). "A C. elegans mutant that lives twice as long as wild type". *Nature*. **366** (6454): 461–464. doi:10.1038/366461a0. PMID 8247153.

[5] Ayyadevara S, Alla R, Thaden JJ, Shmookler Reis RJ (2008). "Remarkable longevity and stress resistance of nematode PI3K-null mutants". *Aging Cell*. **7** (1): 13–22. doi:10.1111/j.1474-9726.2007.00348.x. PMID 17996009.

[6] Shmookler Reis RJ, Bharill P, Tazearslan C, Ayyadevara S (2009). "Extreme-longevity mutations orchestrate silencing of multiple signaling pathways". *Biochimica et Biophysica Acta*. **1790** (10): 1075–1083. doi:10.1016/j.bbagen.2009.05.011. PMC 2885961⊝. PMID 19465083.

[7] "GenAge database". Retrieved 2016-02-11.

[8] Tatar M, Kopelman A, Epstein D, Tu MP, Yin CM, Garofalo RS (1988). "A mutant Drosophila insulin receptor homolog that extends life-span and impairs neuroendocrine function". *Science*. **292** (5514): 107–110. doi:10.1126/science.1057987. PMID 11292875.

[9] Bartke A, Wright JC, Mattison JA, Ingram DK, Miller RA, Roth GS (2001). "Extending the lifespan of long-lived mice". *Nature*. **414** (6862): 412. doi:10.1038/35106646. PMID 11719795.

## 1.3   External links

- Human Ageing Genomic Resources, a collection of databases and tools designed to help researchers study the genetics of human aging

- The NetAge Database, an online database and network analysis tools for biogerontological research

# Chapter 2

# Maximum life span

**Maximum life span** (or, for humans, **maximum reported age at death** or *MRAD*) is a measure of the maximum amount of time one or more members of a population have been observed to survive between birth and death. The term can also denote an estimate of the maximum amount of time that a member of a given species could survive between birth and death, provided circumstances that are optimal to that member's longevity.

Most living species have at least one upper limit on the number of times the cells of a member can divide. This is called the Hayflick limit, although number of cell divisions does not strictly control lifespan.

## 2.1 Definition

In animal studies, maximum span is often taken to be the mean life span of the most long-lived 10% of a given cohort. By another definition, however, maximum life span corresponds to the age at which the oldest known member of a species or experimental group has died. Calculation of the maximum life span in the latter sense depends upon initial sample size.[1]

*Maximum life span* contrasts with *mean life span (average life span, life expectancy)*, and *longevity*. Mean life span varies with susceptibility to disease, accident, suicide and homicide, whereas maximum life span is determined by "rate of aging".[2] Longevity refers only to the characteristics of the especially long lived members of a population, such as infirmities as they age or *compression of morbidity*, and not the specific life span of an individual.

## 2.2 In humans

Main articles: Oldest people and List of the verified oldest people

The longest-living person whose dates of birth and death were verified to the modern norms of *Guinness World Records* and the Gerontology Research Group was Jeanne Calment, a French woman who lived to 122. Reduction of infant mortality has accounted for most of the increased average life span longevity, but since the 1960s mortality rates among those over 80 years have decreased by about 1.5% per year. "The progress being made in lengthening lifespans and postponing senescence is entirely due to medical and public-health efforts, rising standards of living, better education, healthier nutrition and more salubrious lifestyles."[3] Animal studies suggest that further lengthening of human lifespan could be achieved through "calorie restriction mimetic" drugs or by directly reducing food consumption. Although calorie restriction has not been proven to extend the maximum human life span, as of 2014, results in ongoing primate studies have demonstrated that the assumptions derived from rodents are valid in primates as well [Reference: Nature 01.04.2014].[4]

No fixed theoretical limit to human longevity is apparent today.[5] "A fundamental question in aging research is whether humans and other species possess an immutable life-span limit."[6] "The assumption that the maximum human life span is fixed has been justified, [but] is invalid in a number of animal models and ... may become invalid for humans as well."[7] Studies in the biodemography of human longevity indicate a *late-life mortality deceleration law*: that death rates level off at advanced ages to a late-life mortality plateau. That is, there is no fixed upper limit to human longevity, or fixed maximal human lifespan.[8] This law was first quantified in 1939, when researchers found that the one-year probability of death at advanced age asymptotically approaches a limit of 44% for women and 54% for men.[9]

It has also been observed that the $VO_2$max value (a measure of the volume of oxygen flow to the cardiac muscle) decreases as a function of age. Therefore, the maximum lifespan of an individual can be determined by calculating when their $VO_2$max value drops below the basal metabolic rate necessary to sustain life - approximately 3 ml per kg per minute.[10] Noakes (p. 84) notes that, on the basis of this hypothesis, athletes with a $VO_2$max value between 50 and 60 at age 20 can be expected "to live for 100 to 125 years, provided they maintained their physical activity so that their rate of decline in $VO_2$max remained constant".

A theoretical study suggested the maximum human lifespan to be around 125 years using a modified stretched exponential function for human survival curves.[11] In another study, researchers claimed that there exists a maximum lifespan for humans, and that the human maximal lifespan has been declining since the 1990s.[12] A theoretical study also suggested that the maximum human life expectancy at birth is limited by the human life characteristic value δ, which is around 104 years.[13]

## 2.3  In other animals

Main article: List of long-living organisms

Small animals such as birds and squirrels rarely live to their maximum life span, usually dying of accidents, disease or predation. Grazing animals accumulate wear and tear to their teeth to the point where they can no longer eat, and they die of starvation.

The maximum life span of most species has not been accurately determined, because the data collection has been minimal and the number of species studied in captivity (or by monitoring in the wild) has been small.

Maximum life span is usually longer for species that are larger or have effective defenses against predation, such as bird flight, tortoise shells, porcupine quills, or large primate brains.

The differences in life span between species demonstrate the role of genetics in determining maximum life span ("rate of aging"). The records (in years) are these:

- for common house mouse, 4[14]

- for Norway rat, 7

- for dogs, 29 (See List of oldest dogs)[15]

- for cats, 38[16]

- for polar bears, 42[17] (Debby)

- for horses, 62[18]

- for Asian elephants, 86[19]

The longest-lived vertebrates have been variously described as

- Macaws (A parrot that can live up to 80–100 years in captivity)

- Koi (A Japanese species of fish, allegedly living up to 200 years, though generally not exceeding 50 – A specimen named Hanako was reportedly 226 years old upon her death)[20][21]

- Tortoises (Galápagos tortoise) (190 years)[22]

- Tuataras (a New Zealand reptile species, 100–200+ years[23])

- Eels, the so-called Brantevik eel (Swedish: Branteviksålen) is thought to have lived in a water well in southern Sweden since 1859, which makes it over 150 years old.[24] It was reported that it had died in August 2014 at an age of 155[25]

- Whales (Bowhead Whale) (*Balaena mysticetus* about 200 years)

    Although this idea was unproven for a time, recent research has indicated that bowhead whales recently killed still had harpoons in their bodies from about 1890,[26] which, along with analysis of amino acids, has indicated a maximum life span, stated as "the 211 year-old bowhead could have been from 177 to 245 years old".[27][28][29]

- Greenland Sharks are currently the vertebrate species with the longest known lifespan.[30] An examination of 28 specimens in one study published in 2016 determined by radiocarbon dating that the oldest of the animals that they sampled had lived for about 392 ± 120 years (a minimum of 272 years and a maximum of 512 years). The authors further concluded that the species reaches sexual maturity at about 150 years of age.[30]

With the possible exception of the Bowhead whale, the claims of lifespans >100 year rely on conjecture (e.g. counting otoliths) rather than empirical, continuous documentation.

Invertebrate species which continue to grow as long as they live (*e.g.*, certain clams, some coral species) can on occasion live hundreds of years:

- A bivalve mollusc (*Arctica islandica*) (aka "Ming", lived 507±2 years.[31][32])

### 2.3.1  Exceptions

- Some jellyfish species, including *Turritopsis dohrnii*, *Laodicea undulata*,[33] and *Aurelia* sp.1.[34] are able to revert to the polyp stage even after reproducing (so called life cycle reversal), rather than dying as in other jellyfish. Consequently, these species are considered biologically immortal and have no maximum lifespan.[35]

- There may be no natural limit to the Hydra's life span, but it is not yet clear how to estimate the age of a specimen.

- Flatworms, or Platyhelminthes, are known to be "almost immortal" as they have a great regeneration capacity, continuous growth and binary fission type cellular division.[36]

- Lobsters are sometimes said to be biologically immortal because they don't seem to slow down, weaken, or lose fertility with age. However, due to the energy needed for moulting, they don't live indefinitely.[37]

## 2.4   In plants

Main article: List of oldest trees

Plants are referred to as annuals which live only one year, biennials which live two years, and perennials which live longer than that. The longest-lived perennials, woody-stemmed plants such as trees and bushes, often live for hundreds and even thousands of years (one may question whether or not they may die of old age). A giant sequoia, General Sherman is alive and well in its third millennium. A Great Basin Bristlecone Pine called Methuselah is 4,845 years old (as of 2014) and the Bristlecone Pine called Prometheus was a little older still, at least 4,844 years (and possibly as old as 5,000 years), when it was cut down in 1964. The oldest known plant (possibly oldest living thing) is a clonal Quaking Aspen (*Populus tremuloides*) tree colony in the Fishlake National Forest in Utah called Pando at about 80,000 years.

## 2.5   Increasing maximum life span

Main article: Life extension

"Maximum life span" here means the mean life span of the most long-lived 10% of a given cohort. Caloric restriction has not yet been shown to break mammalian world records for longevity. Rats, mice, and hamsters experience maximum life-span extension from a diet that contains all of the nutrients but only 40–60% of the calories that the animals consume when they can eat as much as they want.

Mean life span is increased 65% and maximum life span is increased 50%, when caloric restriction is begun just before puberty.[38] For fruit flies the life extending benefits of calorie restriction are gained immediately at any age upon beginning calorie restriction and ended immediately at any age upon resuming full feeding.[39]

A few transgenic strains of mice have been created that have maximum life spans greater than that of wild-type or laboratory mice. The Ames and Snell mice, which have mutations in pituitary transcription factors and hence are deficient in Gh, LH, TSH, and secondarily IGF1, have extensions in maximal lifespan of up to 65%. To date, both in absolute and relative terms, these Ames and Snell mice have the maximum lifespan of any mouse not on caloric restriction (see below on GhR). Mutations/knockout of other genes affecting the GH/IGF1 axis, such as Lit, Ghr and Irs1 have also shown extension in lifespan, but much more modest both in relative and absolute terms. The longest lived laboratory mouse ever was a Ghr knockout mouse, which lived to ~1800 days in the lab of Andrzej Bartke at Southern Illinois University. The maximum for normal B6 mice under ideal conditions is 1200 days.

Most biomedical gerontologists believe that biomedical molecular engineering will eventually extend maximum lifespan and even bring about rejuvenation. Anti-aging drugs are a potential tool for extending life.[40]

Aubrey de Grey, a theoretical gerontologist, has proposed that aging can be reversed by Strategies for Engineered Negligible Senescence. De Grey has established The Methuselah Mouse Prize to award money to researchers who can extend the maximum life span of mice. So far, three Mouse Prizes have been awarded: one for breaking longevity records to Dr. Andrzej Bartke of Southern Illinois University (using GhR knockout mice); one for late-onset rejuvenation strategies to Dr. Stephen Spindler of the University of California (using caloric restriction initiated late in life); and one to Dr. Z. Dave Sharp for his work with the pharmaceutical rapamycin.[41]

## 2.6   Correlation with DNA repair capacity

Main article: DNA damage theory of aging

Accumulated DNA damage appears to be a limiting factor in the determination of maximum life span. The theory that DNA damage is the primary cause of aging, and thus a principal determinant of maximum life span, has attracted increased interest in recent years. This is based, in part, on evidence in human and mouse that inherited deficiencies in DNA repair genes often cause accelerated aging.[42][43][44] There is also substantial evidence that DNA damage accumulates with age in mammalian tissues, such as those of the brain, muscle, liver and kidney (reviewed by Bernstein et al.[45] and see DNA damage theory of aging and DNA damage (naturally occurring)). One expectation of the theory (that DNA damage is the primary cause of aging) is that among species with differing maximum life spans, the capacity to repair DNA damage should correlate with lifespan. The first experimental test of this idea was by Hart and Setlow[46] who measured the capacity of cells from seven different mammalian species to carry out DNA repair. They found that nucleotide excision repair capability increased systematically with species longevity. This correlation was striking and stimulated a series of 11 additional experiments in different laboratories over succeeding years on the relationship of nucleotide excision repair and life span in mammalian species (reviewed by Bernstein and Bernstein[47]). In general, the findings of these studies indicated a good correlation between nucleotide excision repair capacity and life span. The association between nucleotide excision repair capability and longevity is strengthened by the evidence that defects in nucleotide excision repair proteins in humans and rodents cause features of premature aging, as reviewed by Diderich.[43]

Further support for the theory that DNA damage is the primary cause of aging comes from study of Poly ADP ribose polymerases (PARPs). PARPs are enzymes that are activated by DNA strand breaks and play a role in DNA base excision repair. Burkle et al. reviewed evidence that PARPs, and especially PARP-1, are involved in maintaining mammalian longevity.[48] The life span of 13 mammalian species correlated with poly(ADP ribosyl)ation capability measured in mononuclear cells. Furthermore, lymphoblastoid cell lines from peripheral blood lymphocytes of humans over age 100 had a significantly higher poly(ADP-ribosyl)ation capability than control cell lines from younger individuals.

## 2.7   Research data

- A comparison of the heart mitochondria in rats (7-year maximum life span) and pigeons (35-year maximum life span) showed that pigeon mitochondria leak fewer free-radicals than rat mitochondria, despite the fact that both animals have similar metabolic rate and cardiac output[49]

- For mammals there is a direct relationship between mitochondrial membrane fatty acid saturation and maximum life span[50]

- Studies of the liver lipids of mammals and a bird (pigeon) show an inverse relationship between maximum life span and number of double bonds[51]

- Selected species of birds and mammals show an inverse relationship between telomere rate of change (shortening) and maximum life span[52]

- Maximum life span correlates negatively with antioxidant enzyme levels and free-radicals production and positively with rate of DNA repair[53]

- Female mammals express more Mn–SOD and glutathione peroxidase antioxidant enzymes than males. This has been hypothesized as the reason they live longer[54] However, mice entirely lacking in glutathione peroxidase 1 do not show a reduction in lifespan.

- The maximum life span of transgenic mice has been extended about 20% by overexpression of human catalase targeted to mitochondria[55]

- A comparison of 7 non-primate mammals (mouse, hamster, rat, guinea-pig, rabbit, pig and cow) showed that the rate of mitochondrial superoxide and hydrogen peroxide production in heart and kidney were inversely correlated with maximum life span[56]

- A study of 8 non-primate mammals showed an inverse correlation between maximum life span and oxidative damage to mtDNA (Mitochondrial DNA) in heart & brain[57]

- A study of several species of mammals and a bird (pigeon) indicated a linear relationship between oxidative damage to protein and maximum life span[58]

- There is a direct correlation between DNA repair and maximum life span for mammalian species[59]

- Drosophila (fruit-flies) bred for 15 generations by only using eggs that were laid toward the end of reproductive life achieved maximum life spans 30% greater than that of controls[60]

- Overexpression of the enzyme which synthesizes glutathione in long-lived transgenic Drosophila (fruit-flies) extended maximum lifespan by nearly 50%[61]

- A mutation in the **age–1** gene of the nematode worm *Caenorhabditis elegans* increased mean life span 65% and maximum life span 110%.[62] However, the degree of lifespan extension in relative terms by both the age-1 and daf-2 mutations is strongly dependent on ambient temperature, with ~10% extension at 16 °C and 65% extension at 27 °C.

- Fat-specific Insulin Receptor KnockOut (FIRKO) mice have reduced fat mass, normal calorie intake and an increased maximum life span of 18%.[63]

- The capacity of mammalian species to detoxify the carcinogenic chemical benzo(a)pyrene to a water-soluble form also correlates well with maximum life span.[64]

- Short-term induction of oxidative stress due to calorie restriction increases life span in *Caenorhabditis elegans* by promoting stress defense, specifically by inducing an enzyme called catalase. As shown by Michael Ristow and co-workers nutritive antioxidants completely abolish this extension of life span by inhibiting a process called mitohormesis.[65]

## 2.8 See also

- Ageing

- Aging brain

- American Aging Association

- Aubrey de Grey

- Biodemography

- Biological immortality

- Calorie restriction

- Compression of morbidity

- DNA damage theory of aging

- Extreme longevity tracking

- Genetics of aging

- Gerontology

- Hayflick limit

- Indefinite lifespan

- Life expectancy

- Life extension

- List of long-living organisms

- Longevity

- Methuselah Mouse Prize

- Michael Ristow

- Mitohormesis

- Oldest people

- Senescence

- Strategies for Engineered Negligible Senescence (SENS)

## 2.9 References

[1] Gavrilov, Leonid A.; Gavrilova, Natalia S. (1991). *The Biology of Life Span: A Quantitative Approach*. New York: Harwood Academic. ISBN 978-3-7186-4983-9.

[2] Brody, Jane E. (August 25, 2008). "Living Longer, in Good Health to the End". *The New York Times*. p. D7.

[3] Vaupel, James W. (2010). "Biodemography of human ageing". *Nature*. **464** (7288): 536–42. doi:10.1038/nature08984. PMID 20336136.

[4] Ingram, Donald K.; Roth, George S.; Lane, Mark A.; Ottinger, Mary Ann; Zou, Sige; Cabo, Rafael; Mattison, Julie A. (2006). "The potential for dietary restriction to increase longevity in humans: Extrapolation from monkey studies". *Biogerontology*. **7** (3): 143–8. doi:10.1007/s10522-006-9013-2. PMID 16732404.

[5] Gavrilov, L. A.; Gavrilova, N. S. (1991). *The Biology of Life Span: A Quantitative Approach*. New York City: Starwood Academic Publishers. In Gavrilov, Leonid A.; Gavrilova, Natalia S.; Center on Aging, NORC/University of Chicago (June 2000). "Book Reviews: *Validation of Exceptional Longevity*" (PDF). *Population Dev Rev*. **26** (2): 403–4. Retrieved 2009-05-18.

[6] Wilmoth, J. R.; Deegan, LJ; Lundström, H; Horiuchi, S (2000). "Increase of Maximum Life-Span in Sweden, 1861–1999". *Science*. **289** (5488): 2366–8. doi:10.1126/science.289.5488.2366. PMID 11009426.

[7] Banks, D. A. (1997). "Telomeres, cancer, and aging. Altering the *human life* span". *JAMA: The Journal of the American Medical Association*. **278** (16): 1345–8. doi:10.1001/jama.278.16.1345.

[8] Gavrilov, Leonid A.; Center on Aging, NORC/University of Chicago (2004-03-05). "Biodemography of Human Longevity (Keynote Lecture)". International Conference on Longevity. Retrieved 2009-05-18.

[9] Greenwood, M.; Irwin, J. O. (1939). "The biostatics of senility" (PDF). *Human Biology*. **11**: 1–23. Retrieved 2009-05-18.

[10] Noakes, T. (1985). *The Lore of Running*. Oxford University Press.

[11] B. M. Weon & J. H. Je (2009). "Theoretical estimation of maximum human lifespan". *Biogerontology*. **10**: 65–71. doi:10.1007/s10522-008-9156-4.

[12] X. Dong; B. Milholland & J. Vijg (2016). "Evidence for a limit to human lifespan". *Nature*. **538**: 257–259. doi:10.1038/nature19793.

[13] X. Liu (2015). "Life equations for the senescence process". *Biochemistry and Biophysics Reports*. **4**: 228–233. doi:10.1016/j.bbrep.2015.09.020.

[14] "Longevity, Aging, and Life History of *Mus musculus*". Retrieved 2009-08-13.

[15] http://www.iberianet.com/news/max-misses-world-s-oldest-dog-title/article_a432c462-c251-11e2-96b5-001a4bcf887a.html

[16] *Guinness World Records 2010*. Bantam. 2010. p. 320. ISBN 978-0-553-59337-2. The oldest cat ever was Creme Puff, who was born on August 3, 1967, and lived until August 6, 2005—38 years and 3 days in total.

[17] "World's oldest polar bear". Retrieved 2008-11-19.

[18] Ensminger, M. E. (1990). *Horses and Horsemanship: Animal Agricultural Series* (Sixth ed.). Danville, Indiana: Interstate Publishers. ISBN 0-8134-2883-1. OCLC 21977751., pp. 46–50

[19] "Lin Wang, an Asian elephant (*Elephas maximus*) at Taipei Zoo". Retrieved 2009-08-13.

[20] "International Nishikigoi Promotion Center-Genealogy". Japan-nishikigoi.org. Retrieved 2009-04-11.

[21] Barton, Laura (2007-04-12). "Will you still feed me ... ?". *The Guardian*. London. Retrieved 2009-04-11.

[22] Seed: *Week In Science*: 6/23 - 6/29 Archived October 31, 2007, at the Wayback Machine.

[23] Tuatara#cite note-43

[24] "Brantevik Eels may be the world's oldest". 2008-04-11.

[25] "The world's oldest Eek dead - Lived 155 years in a well (Article in Swedish )". 2014-08-08.

[26] "125-Year-old New Bedford Bomb Fragment Found Embedded in Alaskan Bowhead Whale".

[27] "Bowhead Whales May Be the World's Oldest Mammals". 2001.

[28] "Bowhead Whales May Be the World's Oldest Mammals". 2007 [2001].

[29] John C. George; Jeffrey Bada; Judith Zeh; Laura Scott; Stephen E. Brown; Todd O'Hara & Robert Suydam (1999). "Age and growth estimates of bowhead whales (*Balaena mysticetus*) via aspartic acid racemization". *Canadian Journal of Zoology*. **77** (4): 571–580. doi:10.1139/cjz-77-4-571.

[30] Nielsen, Julius; Hedeholm, Rasmus B.; Heinemeier, Jan; Bushnell, Peter G.; Christiansen, Jørgen S.; Olsen, Jesper; Ramsey, Christopher Bronk; Brill, Richard W.; Simon, Malene; Steffensen, Kirstine F.; Steffensen, John F. (2016). "Eye lens radiocarbon reveals centuries of longevity in the Greenland shark (*Somniosus microcephalus*)". *Science*. **353** (6300): 702–4. doi:10.1126/science.aaf1703. PMID 27516602. Lay summary – *Science News* (August 12, 2016).

[31] Butler, Paul G.; Wanamaker, Alan D.; Scourse, James D.; Richardson, Christopher A.; Reynolds, David J. (2013). "Variability of marine climate on the North Icelandic Shelf in a 1357-year proxy archive based on growth increments in the bivalve *Arctica islandica*". *Palaeogeography, Palaeoclimatology, Palaeoecology*. **373**: 141–151. doi:10.1016/j.palaeo.2012.01.016.

[32] Lise Brix (2013-11-06). "New record: World's oldest animal is 507 years old". *Sciencenordic*. Archived from the original on 2013-11-15. Retrieved 2013-11-14.

[33] De Vito; et al. (2006). "Evidence of reverse development in Leptomedusae (Cnidaria, Hydrozoa): the case of *Laodicea undulata* (Forbes and Goodsir 1851)". *Marine Biology*. **149**: 339–346. doi:10.1007/s00227-005-0182-3. Retrieved 2015-12-31.

[34] He; et al. (2015-12-21). "Life Cycle Reversal in *Aurelia* sp.1 (Cnidaria, Scyphozoa)". *PLOS ONE*. **10**: e0145314. doi:10.1371/journal.pone.0145314. PMC 4687044. PMID 26690755. Retrieved 2015-12-31.

[35] Piraino, Stefano; F. Boero; B. Aeschbach; V. Schmid (1996). "Reversing the life cycle: medusae transforming into polyps and cell transdifferentiation in 'Turritopsis nutricula (Cnidaria, Hydrozoa)". Biological Bulletin. *Biological Bulletin, vol. 190, no. 3.* **190** (3): 302–312. doi:10.2307/1543022. JSTOR 1543022.

[36] Saló E. (2006). "The power of regeneration and the stem-cell kingdom: freshwater planarians (Platyhelminthes)". *BioEssays*. **28** (5): 546–559. doi:10.1002/bies.20416. PMID 16615086.

[37] Marina Koren (June 3, 2013). "Don't Listen to the Buzz: Lobsters Aren't Actually Immortal". Smithsonian.com.

[38] Koubova J, Guarente L (2003). "How does calorie restriction work?". *Genes & Development*. **17** (3): 313–321. doi:10.1101/gad.1052903. PMID 12569120.

[39] Mair W, Goymer P, Pletcher SD, Partridge L (2003). "Demography of dietary restriction and death in *Drosophila*". *Science*. **301** (5640): 1731–1733. doi:10.1126/science.1086016. PMID 14500985.

[40] Kaeberlein, Matt (2010). "Resveratrol and rapamycin: are they anti-aging drugs?". *BioEssays*. **32** (2): 96–99. doi:10.1002/bies.200900171. PMID 20091754.

[41] "Work". Methuselah Foundation.

[42] Hoeijmakers, JH (2009). "DNA damage, aging, and cancer". *New England Journal of Medicine*. **361** (15): 1475–1485. doi:10.1056/NEJMra0804615. PMID 19812404.

[43] Diderich K, Alanazi M. Hoeijmakers JH (2011). Premature aging and cancer in nucleotide excision repair-disorders. DNA Repair (Amst) 10(7):772-780. doi: 10.1016/j.dnarep.2011.04.025. Review. PMID 21680258

[44] Freitas AA, de Magalhães JP (2011). A review and appraisal of the DNA damage theory of ageing. *Mutation Research* 728 (1–2): 12–22. Review. doi:10.1016/j.mrrev.2011.05.001 PMID 21600302

[45] Bernstein H, Payne CM, Bernstein C, Garewal H, Dvorak K (2008). Cancer and aging as consequences of un-repaired DNA damage. In: *New Research on DNA Damages* (Editors: Honoka Kimura and Aoi Suzuki) Nova Science Publishers, Inc., New York, Chapter 1, pp. 1-47. ISBN 1604565810 ISBN 978-1604565812

[46] Hart, RW; Setlow, RB (1974). "Correlation between deoxyribonucleic acid excision-repair and life-span in a number of mammalian species". *Proceedings of the National Academy of Sciences of the United States of America*. **71** (6): 2169–2173. doi:10.1073/pnas.71.6.2169. PMID 4526202.

[47] Bernstein C, Bernstein H. (1991) *Aging, Sex, and DNA Repair*. Academic Press, San Diego. ISBN 0120928604 ISBN 978-0120928606

[48] Bürkle A, Brabeck C, Diefenbach J, Beneke S (2005). The emerging role of poly(ADP-ribose) polymerase-1 in longevity. *International Journal of Biochemistry and Cell Biology* 37(5):1043–1053. Review. doi:10.1016/j.biocel.2004.10.006 PMID 15743677

[49] Herrero A, Barja G (1997). "Sites and mechanisms responsible for the low rate of free radical production of heart mitochondria in the long-lived pigeon". *Mechanisms of Aging and Development*. **98** (2): 95–111. doi:10.1016/S0047-6374(97)00076-6. PMID 9379714

[50] Pamplona R, Portero-Otin M, Riba D, Ruiz C, Prat J, Bellmunt MJ, Barja G (1 October 1998). "Mitochondrial membrane peroxidizability index is inversely related to maximum life span in mammals". *Journal of Lipid Research*. **39** (2): 1989–1994. PMID 9788245.

[51] Pamplona R, Portero-Otin M, Riba D, Requena JR, Thorpe SR, Lopez-Torres M, Barja G (2000). "Low fatty acid unsaturation: a mechanism for lowered lipoperoxidative modification of tissue proteins in mammalian species with long life spans". *Journals of Gerontology Series A Biological Sciences and Medical Sciences*. **55A** (6): B286–B291. PMID 10843345.

[52] Haussmann MF, Winkler DW, O'Reilly KM, Huntington CE, Nisbet IC, Vleck CM (2003). "Telomeres shorten more slowly in long-lived birds and mammals than in short-lived ones". *Proceedings of the Royal Society B: Biological Sciences*. **270** (1522): 1387–1392.

doi:10.1098/rspb.2003.2385. PMC 1691385. PMID 12965030.

[53] Perez-Campo R, Lopez-Torres M, Cadenas S, Rojas C, Barja G (1998). "The rate of free radical production as a determinant of the rate of aging: evidence from the comparative approach". *Journal of Comparative Physiology B*. **168** (3): 149–158. doi:10.1007/s003600050131. PMID 9591361.

[54] Vina J, Borras C, Gambini J, Sastre J, Pallardo FV (2005). "Why females live longer than males? Importance of the upregulation of longevity-associated genes by oestrogenic compounds". *FEBS Letters*. **579** (12): 2541–2545. doi:10.1016/j.febslet.2005.03.090. PMID 15862287.

[55] Schriner SE, Linford NJ, Martin GM, Treuting P, Ogburn CE, Emond M, Coskun PE, Ladiges W, Wolf N, Van Remmen H, Wallace DC, Rabinovitch PS (2005). "Extension of murine life span by overexpression of catalase targeted to mitochondria". *Science*. **308** (5730): 1909–1911. doi:10.1126/science.1106653. PMID 15879174.

[56] Ku HH, Brunk UT, Sohal RS (1993). "Relationship between mitochondrial superoxide and hydrogen peroxide production and longevity of mammalian species". *Free Radical Biology & Medicine*. **15** (6): 621–627. doi:10.1016/0891-5849(93)90165-Q. PMID 8138188.

[57] Barja G, Herrero A (1 February 2000). "Oxidative damage to mitochondrial DNA is inversely related to maximum life span in the heart and brain of mammals". *FASEB Journal*. **14** (2): 312–318. PMID 10657987.

[58] Agarwal S, Sohal RS (1996). "Relationship between susceptibility to protein oxidation, aging, and maximum life span potential of different species". *Experimental Gerontology*. **31** (3): 365–372. doi:10.1016/0531-5565(95)02039-X. PMID 9415119.

[59] Cortopassi GA, Wang E (1996). "There is substantial agreement among interspecies estimates of DNA repair activity". *Mechanisms of Aging and Development*. **91** (3): 211–218. doi:10.1016/S0047-6374(96)01788-5. PMID 9055244.

[60] Kurapati R, Passananti HB, Rose MR, Tower J (2000). "Increased hsp22 RNA levels in *Drosophila* lines genetically selected for increased longevity". *Journals of Gerontology Series A Biological Sciences and Medical Sciences*. **55A** (11): B552–B559. PMID 11078089.

[61] Orr WC, Radyuk SN, Prabhudesai L, Toroser D, Benes JJ, Luchak JM, Mockett RJ, Rebrin I, Hubbard JG, Sohal RS (2005). "Overexpression of glutamate-cysteine ligase extends life span in *Drosophila melanogaster*". *The Journal of Biological Chemistry*. **280** (45): 37331–37338. doi:10.1074/jbc.M508272200. PMID 16148000.

[62] Friedman DB, Johnson TE (1 January 1988). "A mutation in the age-1 gene in *Caenorhabditis elegans* lengthens life and reduces hermaphrodite fertility". *Genetics*. **118** (1): 75–86. PMC 1203268. PMID 8608934.

[63] Bluher M, Kahn BB, Kahn CR (2003). "Extended longevity in mice lacking the insulin receptor in adipose tissue". *Science*. **299** (5606): 572–574. doi:10.1126/science.1078223. PMID 12543978.

[64] Moore CJ, Schwartz AG (1978). "Inverse correlation between species lifespan and capacity of cultured fibroblasts to convert benzo(a)pyrene to water-soluble metabolites". *Experimental Cell Research*. **116** (2): 359–364. doi:10.1016/0014-4827(78)90459-7. PMID 101383.

[65] "Publication demonstrating that oxidative stress is promoting life span". Cellmetabolism.org. Retrieved 2010-11-04.

## 2.10   External links

- AnAge Database

- Informational website on the biology of aging

- Mechanisms of Aging

# Chapter 3

# Epigenetic clock

An **epigenetic clock** is a type of **DNA clock** based on measuring natural DNA methylation levels to estimate the biological age of a tissue, cell type or organ. A pre-eminent example for an epigenetic clock is **Horvath's clock**, which is based on 353 epigenetic markers on the human genome.[1][2][3][4]

## 3.1 History

The strong effect of age on DNA methylation levels has been known since the late 1960s.[5] A vast literature describes sets of CpGs whose DNA methylation levels correlate with age, e.g.[6][7][8][9][10] Two publications describe age estimators based on DNA methylation levels in either saliva [11] or blood.[12]

Horvath's epigenetic clock was developed by Steve Horvath, a professor of human genetics at the David Geffen School of Medicine at UCLA and of biostatistics at the UCLA Fielding School of Public Health. The scientific article was first published on Oct 21, 2013, in Genome Biology.[1][3] Horvath spent over 4 years collecting publicly available Illumina DNA methylation data and identifying suitable statistical methods.[13] The personal story behind the discovery was featured in Nature.[14] The age estimator was developed using 8,000 samples from 82 Illumina DNA methylation array datasets, encompassing 51 healthy tissues and cell types. The major innovation of Horvath's epigenetic clock lies in its wide applicability: the same set of 353 CpGs and the same prediction algorithm is used irrespective of the DNA source within the organism, i.e. it does not require any adjustments or offsets.[1] This property allows one to compare the ages of different areas of the human body using the same aging clock.

## 3.2 Relationship to a cause of biological aging

It is not yet known what exactly is measured by DNA methylation age. Horvath hypothesized that DNA methylation age measures the cumulative effect of an epigenetic maintenance system but details are unknown. The fact that DNA methylation age of blood predicts all-cause mortality in later life [15][16][17][18] strongly suggests that it relates to a process that causes aging. However, it is unlikely that the 353 clock CpGs are special or play a direct causal role in the aging process.[1] Rather, the epigenetic clock captures an emergent property of the epigenome.

## 3.3 Motivation for biological clocks

In general, biological aging clocks and biomarkers of aging are expected to find many uses in biological research since age is a fundamental characteristic of most organisms. Accurate measures of biological age (**biological aging clocks**) could be useful for

- testing the validity of various theories of biological aging,

- diagnosing various age related diseases and for defining cancer subtypes,

- predicting/prognosticating the onset of various diseases,

- serving as surrogate markers for evaluating therapeutic interventions including rejuvenation approaches,

- studying developmental biology and cell differentiation,

- forensic applications, for example to estimate the age of a suspect based on blood left on a crime scene.

Overall, biological clocks are expected to be useful for studying what causes aging and what can be done against it.

## 3.4 Properties of Horvath's clock

The clock is defined as an age estimation method based on 353 epigenetic markers on the DNA. The 353 markers measure DNA methylation of CpG dinucleotides. Estimated age ("predicted age" in mathematical usage), also referred to as DNA methylation age, has the following properties: first, it is close to zero for embryonic and induced pluripotent stem cells; second, it correlates with cell passage number; third, it gives rise to a highly heritable measure of age acceleration; and, fourth, it is applicable to chimpanzee tissues (which are used as human analogs for biological testing purposes). Organismal growth (and concomitant cell division) leads to a high ticking rate of the epigenetic clock that slows down to a constant ticking rate (linear dependence) after adulthood (age 20).[1] The fact that DNA methylation age of blood predicts all-cause mortality in later life even after adjusting for known risk factors [15][16] suggests that it relates to a process that causes aging. Similarly, markers of physical and mental fitness are associated with the epigenetic clock (lower abilities associated with age acceleration).[19]

Salient features of Horvath's epigenetic clock include its high accuracy and its applicability to a broad spectrum of tissues and cell types. Since it allows one to contrast the ages of different tissues from the same subject, it can be used to identify tissues that show evidence of accelerated age due to disease.

### 3.4.1  Statistical approach

The basic approach is to form a weighted average of the 353 clock CpGs, which is then transformed to DNAm age using a calibration function. The calibration function reveals that the epigenetic clock has a high ticking rate until adulthood, after which it slows to a constant ticking rate. Using the training data sets, Horvath used a penalized regression model (Elastic net regularization) to regress a calibrated version of chronological age on 21,369 CpG probes that were present both on the Illumina 450K and 27K platform and had fewer than 10 missing values. DNAm age is defined as estimated ("predicted") age. The elastic net predictor automatically selected 353 CpGs. 193 of the 353 CpGs correlate positively with age while the remaining 160 CpGs correlate negatively with age. R software and a freely available web-based tool can be found at the following webpage.[20]

### 3.4.2  Accuracy

The median error of estimated age is 3.6 years across a wide spectrum of tissues and cell types.[1] The epigenetic clock performs well in heterogeneous tissues (for example, whole blood, peripheral blood mononuclear cells, cerebellar samples, occipital cortex, buccal epithelium, colon, adipose, kidney, liver, lung, saliva, uterine cervix, epidermis, muscle) as well as in individual cell types such as CD4 T cells, CD14 monocytes, glial cells, neurons, immortalized B cells, mesenchymal stromal cells.[1] However, accuracy depends to some extent on the source of the DNA. In particular, DNAm age is higher than chronological age in female breast tissue that is adjacent to breast cancer tissue.[1] Since normal tissue, which is adjacent to other cancer types, does not exhibit a similar age acceleration effect, this finding suggests that normal female breast tissue ages faster than other parts of the body.[1]

### 3.4.3  Comparison with other biological clocks

The epigenetic clock leads to a chronological age prediction that has a Pearson correlation coefficient of r=0.96 with chronological age (Figure 2 in [1]). Thus the age correlation is close to its maximum possible correlation value of 1. Other biological clocks are based on a) telomere length, b) p16INK4a expression levels (also known as INK4a/ARF locus),[21] and c) microsatellite mutations.[22] The correlation between chronological age and telomere length is r=−0.51 in women and r=−0.55 in men.[23] The correlation between chronological age and expression levels of p16INK4a in T cells is r=0.56.[24] p16INK4a expression levels only relate to age in T cells, a type of white blood cells. The microsatellite clock measures not chronological age but age in terms of elapsed cell divisions within a tissue.

## 3.5 Applications of Horvath's clock

By contrasting DNA methylation age (estimated age) with chronological age, one can define measures of age acceleration. Age acceleration can be defined as the difference between DNA methylation age and chronological age. Alternatively, it can be defined as the residual that results from regressing DNAm age on chronological age. The latter measure is attractive because it does not correlate with chronological age. A positive/negative value of epigenetic age acceleration suggests that the underlying tissue ages faster/slower than expected.

### 3.5.1 Genetic studies of epigenetic age acceleration

The broad sense heritability (defined via Falconer's formula) of age acceleration of blood from older subjects is around 40% but it appears to be much higher in newborns.[1] Similarly, the age acceleration of brain tissue (prefrontal cortex) was found to be 41% in older subjects.[25] Genome-wide association studies of cerebellar age acceleration have identified several SNPs at a genomewide significance level.[26] Gene and SNP sets found by genome-wide association analysis of epigenetic age acceleration exhibit significant overlap with those of Alzheimer's disease, age-related macular degeneration, and Parkinson's disease.[26]

### 3.5.2 Cancer tissue

Cancer tissues show both positive and negative age acceleration effects. For most tumor types, no significant relationship can be observed between age acceleration and tumor morphology (grade/stage).[1][2] On average, cancer tissues with mutated TP53 have a lower age acceleration than those without it.[1] Further, cancer tissues with high age acceleration tend to have fewer somatic mutations than those with low age acceleration.[1][2] Age acceleration is highly related to various genomic aberrations in cancer tissues. Somatic mutations in estrogen receptors or progesterone receptors are associated with accelerated DNAm age in breast cancer.[1] Colorectal cancer samples with a BRAF (V600E) mutation or promoter hypermethylation of the mismatch repair gene MLH1 are associated with an increased age acceleration.[1] Age acceleration in glioblastoma multiforme samples is highly significantly associated with certain mutations in H3F3A.[1] One study suggests that the epigenetic age of blood tissue may be prognostic of lung cancer incidence.[27]

### 3.5.3 Obesity

The epigenetic clock was used to study the relationship between high body mass index (BMI) and the DNA methylation ages of human blood, liver, muscle and adipose tissue.[28] A significant correlation (r=0.42) between BMI and epigenetic age acceleration could only be observed for the liver.

### 3.5.4 Trisomy 21 (Down syndrome)

Down Syndrome (DS) entails an increased risk of many chronic diseases that are typically associated with older age. The clinical manifestations of accelerated aging suggest that trisomy 21 increases the biological age of tissues, but molecular evidence for this hypothesis has been sparse. According to the epigenetic clock, trisomy 21 significantly increases the age of blood and brain tissue (on average by 6.6 years).[29]

### 3.5.5 Alzheimer's disease related neuropathology

Epigenetic age acceleration of the human prefrontal cortex was found to be correlated with several neuropathological measurements that play a role in Alzheimer's disease [25] Further, it was found to be associated with a decline in global cognitive functioning, and memory functioning among individuals with Alzheimer's disease.[25] The epigenetic age of **blood** relates to cognitive functioning in the elderly.[19] Overall, these results strongly suggest that the epigenetic clock lends itself for measuring the biological age of the brain.

### 3.5.6 Cerebellum ages slowly

It has been difficult to identify tissues that seem to evade aging due to the lack of biomarkers of tissue age that allow one to contrast compare the ages of different tissues. An application of epigenetic clock to 30 anatomic sites from six centenarians and younger subjects revealed that the cerebellum ages slowly: it is about 15 years younger than expected in a centenarian.[30] This finding might explain why the cerebellum exhibits fewer neuropathological hallmarks of age related dementias compared to other brain regions. In younger subjects (e.g. younger than 70), brain regions and brain cells appear to have roughly the same age.[1][30] Several SNPs and genes have been identified that relate to the epigenetic age of the cerebellum [26]

### 3.5.7 Huntington's disease

Huntington's disease has been found to increase the epigenetic aging rates of several human brain regions.[31]

### 3.5.8 Centenarians age slowly

The offspring of semi-supercentenarians (subjects who reached an age of 105–109 years) have a lower epigenetic age than age-matched controls (age difference=5.1 years in blood) and centenarians are younger (8.6 years) than expected based on their chronological age.[18]

### 3.5.9 HIV infection

Infection with the Human Immunodeficiency Virus-1 (HIV) is associated with clinical symptoms of accelerated aging, as evidenced by increased incidence and diversity of age-related illnesses at relatively young ages. But it has been difficult to detect an accelerated aging effect on a molecular level. An epigenetic clock analysis of human DNA from HIV+ subjects and controls detected a significant age acceleration effect in brain (7.4 years) and blood (5.2 years) tissue due to HIV-1 infection.[32]

### 3.5.10 Parkinson's disease

A large-scale study suggests that the blood of Parkinson's disease subjects exhibits (relatively weak) accelerated aging effects.[33]

### 3.5.11 Developmental disorder: syndrome X

Children with a very rare disorder known as syndrome X maintain the façade of persistent toddler-like features while aging from birth to adulthood. Since the physical development of these children is dramatically delayed, these children appear to be a toddler or at best a preschooler. According to an epigenetic clock analysis, blood tissue from syndrome X cases is not younger than expected.[34]

### 3.5.12 Menopause accelerates epigenetic aging

The following results strongly suggest that the loss of female hormones resulting from menopause accelerates the epigenetic aging rate of blood and possibly that of other tissues.[35] First, early menopause has been found to be associated with an increased epigenetic age acceleration of blood.[35] Second, surgical menopause (due to bilateral oophorectomy) is associated with epigenetic age acceleration in blood and saliva. Third, menopausal hormone therapy, which mitigates hormonal loss, is associated with a negative age acceleration of buccal cells (but not of blood cells).[35] Fourth, genetic markers that are associated with early menopause are also associated with increased epigenetic age acceleration in blood.[35]

### 3.5.13 Cellular senescence versus epigenetic aging

A confounding aspect of biological aging is the nature and role of senescent cells. It is unclear whether the three major types of cellular senescence, namely replicative senescence,

oncogene-induced senescence and DNA damage-induced senescence are descriptions of the same phenomenon instigated by different sources, or if each of these is distinct, and how they are associated with epigenetic aging. Induction of replicative senescence (RS) and oncogene-induced senescence (OIS) were found to be accompanied by epigenetic aging of primary cells but senescence induced by DNA damage was not, even though RS and OIS activate the cellular DNA damage response pathway.[36] These results highlight the independence of cellular senescence from epigenetic aging. Consistent with this, telomerase-immortalised cells continued to age (according to the epigenetic clock) without having been treated with any senescence inducers or DNA-damaging agents, re-affirming the independence of the process of epigenetic ageing from telomeres, cellular senescence, and the DNA damage response pathway. Although the uncoupling of senescence from cellular aging appears at first sight to be inconsistent with the fact that senescent cells contribute to the physical manifestation of organism ageing, as demonstrated by Baker et al., where removal of senescent cells slowed down aging.[37] However, the epigenetic clock analysis of senescence suggests that cellular senescence is a state that cells are forced into as a result of external pressures such as DNA damage, ectopic oncogene expression and exhaustive proliferation of cells to replenish those eliminated by external/environmental factors.[36] These senescent cells, in sufficient numbers, will probably cause the deterioration of tissues, which is interpreted as organism ageing. However, at the cellular level, aging, as measured by the epigenetic clock, is distinct from senescence. It is an intrinsic mechanism that exists from the birth of the cell and continues. This implies that if cells are not shunted into senescence by the external pressures described above, they would still continue to age. This is consistent with the fact that mice with naturally long telomeres still age and eventually die even though their telomere lengths are far longer than the critical limit, and they age prematurely when their telomeres are forcibly shortened, due to replicative senescence. Therefore, cellular senescence is a route by which cells exit prematurely from the natural course of cellular aging.[36]

### 3.5.14 Effect of sex and race/ethnicity

Men age faster than women according to epigenetic age acceleration in blood, brain, saliva, and many other tissues.[38] The epigenetic clock method applies to all examined racial/ethnic groups in the sense that DNAm age is highly correlated with chronological age. But ethnicity can be associated with epigenetic age acceleration.[38] For example, the blood of Hispanics and the Tsimané ages more slowly than that of other populations which might explain the Hispanic mortality paradox.[38]

## 3.6 Biological mechanism behind the epigenetic clock

The precise biological mechanism behind the epigenetic clock is currently unknown. However, the following explanations have been proposed in the literature.

### 3.6.1 Possible explanation 1: Epigenomic maintenance system

Horvath hypothesized that his clock arises from a methylation footprint left by an epigenomic maintenance system.[1]

### 3.6.2 Possible explanation 2: Unrepaired DNA damages

Endogenous DNA damages occur frequently including about 50 double-strand DNA breaks per cell cycle[39] and about 10,000 oxidative damages per day (see DNA damage (naturally occurring)). During repair of double-strand breaks many epigenetic alterations are introduced, and in a percentage of cases epigenetic alterations remain after repair is completed, including increased methylation of CpG island promoters.[40][41][42] Similar, but usually transient epigenetic alterations were recently found during repair of oxidative damages caused by $H_2O_2$, and it was suggested that occasionally these epigenetic alterations may also remain after repair.[43] These accumulated epigenetic alterations may contribute to the epigenetic clock. Accumulation of epigenetic alterations may parallel the accumulation of un-repaired DNA damages that are proposed to cause aging (see DNA damage theory of aging).

## 3.7 Other age estimators based on DNA methylation levels

Several other age estimators have been described in the literature.

1) Weidner et al. (2014) describe an age estimator for DNA from blood that uses only three CpG sites of genes hardly affected by aging (cg25809905 in integrin, alpha 2b (ITGA2B); cg02228185 in aspartoacylase (ASPA) and cg17861230 in phosphodiesterase 4C, cAMP specific (PDE4C)).[44] The age estimator by Weidener et al. (2014) applies only to blood. Even in blood this sparse estimator is far less accurate than Horvath's epigenetic clock (Horvath 2014) when applied to data generated by the Illumina 27K or 450K platforms. [45] But the sparse estimator was developed for pyrosequencing data and is highly cost effective.

[46]

2) Hannum et al. (2013) [12] report several age estimators: one for each tissue type. Each of these estimators requires covariate information (e.g. gender, body mass index, batch). The authors mention that each tissue led to a clear linear offset (intercept and slope). Therefore, the authors had to adjust the blood-based age estimator for each tissue type using a linear model. When the Hannum estimator is applied to other tissues, it leads to a high error (due to poor calibration) as can be seen from Figure 4A in Hannum et al. (2013). Hannum et al. adjusted their blood-based age estimator (by adjusting the slope and the intercept term) in order to apply it to other tissue types. Since this adjustment step removes differences between tissue, the blood-based estimator from Hannum et al. cannot be used to compare the ages of different tissues/organs. In contrast, a salient characteristic of the epigenetic clock is that one does not have to carry out such a calibration step:[1] it always uses the same CpGs and the same coefficient values. Therefore, Horvath's epigenetic clock can be used to compare the ages of different tissues/cells/organs from the same individual. While the age estimators from Hannum et al. cannot be used to compare the ages of different normal tissues, they can be used to compare the age of a cancerous tissue with that of a corresponding normal (non-cancerous) tissue. Hannum et al. reported pronounced age acceleration effects in all cancers. In contrast, Horvath's epigenetic clock [12][47] reveals that some cancer types (e.g. triple negative breast cancers or uterine corpus endometrial carcinoma) exhibit negative age acceleration, i.e. cancer tissue can be much younger than expected. An important difference relates to additional covariates. Hannum's age estimators make use of covariates such as gender, body mass index, diabetes status, ethnicity, and batch. Since new data involve different batches, one cannot apply it directly to new data. However, the authors present coefficient values for their CpGs in Supplementary Tables which can be used to define an aggregate measure that tends to be strongly correlated with chronological age but may be poorly calibrated (i.e. lead to high errors).

3.) Giuliani et al. identify genomic regions whose DNA methylation level correlates with age in human teeth. They propose the evaluation of DNA methylation at ELOVL2, FHL2, and PENK genes in DNA recovered from both cementum and pulp of the same modern teeth.[50] They wish to apply this method also to historical and relatively ancient human teeth.

In a multicenter benchmarking study 18 research groups from three continents compared all promising methods for analyzing DNA methylation in the clinic and identified the most accurate methods, having concluded that epigenetic tests based on DNA methylation are a mature technology ready for broad clinical use.[51]

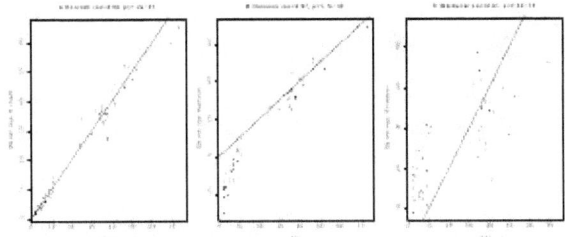

*Comparison of the 3 age predictors described in A) Horvath (2013),[1] B) Hannum (2013),[45] and C) Weidener (2014),[49] respectively. The x-axis depicts the chronological age in years whereas the y-axis shows the predicted age. The solid black line corresponds to y=x. These results were generated in an independent blood methylation data set that was not used in the construction of these predictors (data generated in Nov 2014).*

## 3.8   References

[1] Horvath, S (2013). "DNA methylation age of human tissues and cell types". *Genome Biology*. **14**: R115. doi:10.1186/gb-2013-14-10-r115. PMC 4015143. PMID 24138928.

[2] Horvath, S (2015). "Erratum to: DNA methylation age of human tissues and cell types". *Genome Biology*. **16** (1): 96. doi:10.1186/s13059-015-0649-6. PMC 4427927. PMID 25968125.

[3] University of California, Los Angeles (UCLA), Health Sciences (20 October 2013). "Scientist uncovers internal clock able to measure age of most human tissues; Women's breast tissue ages faster than rest of body". ScienceDaily. Retrieved 22 October 2013.

[4] University of California, Los Angeles (UCLA), Health Sciences (21 October 2013). "Scientists discover new biological clock with age-measuring potential". Forbes. Retrieved 21 October 2013.

[5] Berdyshev, G; Korotaev, G; Boiarskikh, G; Vaniushin, B (1967). "Nucleotide composition of DNA and RNA from somatic tissues of humpback and its changes during spawning". *Biokhimiia*. **31**: 88–993.

[6] Rakyan, VK; Down, TA; Maslau, S; Andrew, T; Yang, TP; Beyan, H; Whittaker, P; McCann, OT; Finer, S; Valdes, AM; Leslie, RD; Deloukas, P; Spector, TD (2010). "Human aging-associated DNA hypermethylation occurs preferentially at bivalent chromatin domains". *Genome Res*. **20**: 434–439. doi:10.1101/gr.103101.109.

[7] Teschendorff, AE; Menon, U; Gentry-Maharaj, A; Ramus, SJ; Weisenberger, DJ; Shen, H; Campan, M; Noushmehr, H; Bell, CG; Maxwell, AP; Savage, DA; Mueller-Holzner, E; Marth, C; Kocjan, G; Gayther, SA; Jones, A; Beck, S; Wagner, W; Laird, PW; Jacobs, IJ; Widschwendter, M (2010).

"Age-dependent DNA methylation of genes that are suppressed in stem cells is a hallmark of cancer". *Genome Res*. **20**: 440–446. doi:10.1101/gr.103606.109.

[8] Koch, CM; Wagner, W (Oct 2011). "Epigenetic-aging-signature to determine age in different tissues". *Aging*. **3** (10): 1018–27.

[9] Horvath, S; Zhang, Y; Langfelder, P; Kahn, R; Boks, M; van Eijk, K; van den Berg, L; Ophoff, RA (2012). "Aging effects on DNA methylation modules in human brain and blood tissue". *Genome Biol*. **13** (10): R97. doi:10.1186/gb-2012-13-10-r97. PMC 4053733. PMID 23034122.

[10] Bell, JT; Tsai, PC; Yang, TP; Pidsley, R; Nisbet, J; Glass, D; Mangino, M; Zhai, G; Zhang, F; Valdes, A; Shin, SY; Dempster, EL; Murray, RM; Grundberg, E; Hedman, AK; Nica, A; Small, KS; Dermitzakis, ET; McCarthy, MI; Mill, J; Spector, TD; Deloukas, P (2012). "Epigenome-wide scans identify differentially methylated regions for age and age-related phenotypes in a healthy ageing population". *PLoS Genet*. **8**: e1002629. doi:10.1371/journal.pgen.1002629.

[11] Bocklandt, S; Lin, W; Sehl, ME; Sánchez, FJ; Sinsheimer, JS; Horvath, S; Vilain, E (2011). "Epigenetic Predictor of Age". *PLoS ONE*. **6** (6): e14821. doi:10.1371/journal.pone.0014821. PMC 3120753. PMID 21731603.

[12] Hannum, G; Guinney, J; Zhao, L; Zhang, L; Hughes, G; Sadda, S; Klotzle, B; Bibikova, M; Fan, JB; Gao, Y; Deconde, R; Chen, M; Rajapakse, I; Friend, S; Ideker, T; Zhang, K (2013). "Genome-wide methylation profiles reveal quantitative views of human aging rates". *Mol Cell*. **49**: 359–367. doi:10.1016/j.molcel.2012.10.016. PMC 3780611. PMID 23177740.

[13] Biome on 21st October 2013 Novel epigenetic clock predicts tissue age

[14] Gibbs, WT (2014). "Biomarkers and ageing: The clock-watcher". *Nature*. **508**: 168–170. doi:10.1038/508168a.

[15] Chen, B; Marioni, ME (2016). "DNA methylation-based measures of biological age: meta-analysis predicting time to death.". *Aging (US Albany)*. **8** (9): 1844–1865. doi:10.18632/aging.101020. PMC 5076441. PMID 27690265.

[16] Marioni, R; Shah, S; McRae, A; Chen, B; Colicino, E; Harris, S; Gibson, J; Henders, A; Redmond, P; Cox, S; Pattie, A; Corley, J; Murphy, L; Martin, N; Montgomery, G; Feinberg, A; Fallin, M; Multhaup, M; Jaffe, A; Joehanes, R; Schwartz, J; Just, A; Lunetta, K; Murabito, JM; Starr, J; Horvath, S; Baccarelli, A; Levy, D; Visscher, P; Wray, N; Deary, I (2015). "DNA methylation age of blood predicts all-cause mortality in later life". *Genome Biology*. **16** (1): 25. doi:10.1186/s13059-015-0584-6.

[17] Christiansen, L (2015). "DNA methylation age is associated with mortality in a longitudinal Danish twin study". *Aging Cell*. **15** (1): 149–154. doi:10.1111/acel.12421. PMC 4717264. PMID 26594032.

[18] Horvath, S (2015). "Decreased epigenetic age of PBMCs from Italian semi-supercentenarians and their offspring.". *Aging (Albany NY)* (Dec).

[19] Marioni, R; Shah, S; McRae, A; Ritchie, S; Muniz-Terrera, GH; SE; Gibson, J; Redmond, P; SR, C; Pattie, A; Corley, J; Taylor, A; Murphy, L; Starr, J; Horvath, S; Visscher, P; Wray, N; Deary, I (2015). "The epigenetic clock is correlated with physical and cognitive fitness in the Lothian Birth Cohort 1936". *Int J of Epidemiology*. **44**: 1388–1396. doi:10.1093/ije/dyu277. PMC 4588858. PMID 25617346.

[20] DNA methylation age software

[21] Collado, M; Blasco, MA; Serrano, M (Jul 2007). "Cellular senescence in cancer and aging". *Cell*. **130** (2): 223–33. doi:10.1016/j.cell.2007.07.003.

[22] Forster, P; Hohoff, C; Dunkelmann, B; Schürenkamp, M; Pfeiffer, H; Neuhuber, F; Brinkmann, B (2015). "Elevated germline mutation rate in teenage fathers". *Proc Biol Sci*. **282**: 20142898. doi:10.1098/rspb.2014.2898.

[23] Nordfjäll, K; Svenson, U; Norrback, KF; Adolfsson, R; Roos, G (Mar 2010). "Large-scale parent-child comparison confirms a strong paternal influence on telomere length". *Eur J Hum Genet*. **18** (3): 385–9. doi:10.1038/ejhg.2009.178. PMC 2987222. PMID 19826452.

[24] Wang, Y; Zang, X; Wang, Y; Chen, P (2012). "High expression of p16INK4a and low expression of Bmi1 are associated with endothelial cellular senescence in the human cornea" (PDF). *Molecular Vision*. **18**: 803–815.

[25] Levine, M (2015). "Epigenetic age of the pre-frontal cortex is associated with neuritic plaques, amyloid load, and Alzheimer's disease related cognitive functioning". *Aging (Albany NY)*. **7** (Dec): 1198–211. PMC 4712342. PMID 26684672.

[26] Lu, A (2016). "Genetic variants near MLST8 and DHX57 affect the epigenetic age of the cerebellum.". *Nature Communications*. **7**: 10561. doi:10.1038/ncomms10561.

[27] Levine, M (2015). "DNA methylation age of blood predicts future onset of lung cancer in the women's health initiative.". *Aging (Albany NY)*. **7**: 690–700. doi:10.18632/aging.100809. PMC 4600626. PMID 26411804.

[28] Horvath, S; Erhart, W; Brosch, M; Ammerpohl, O; von Schoenfels, W; Ahrens, M; Heits, N; Bell, JT; Tsai, PC; Spector, TD; Deloukas, P; Siebert, R; Sipos, B; Becker, T; Roecken, C; Schafmayer, C; Hampe, J (2014). "Obesity accelerates epigenetic aging of human liver". *Proc Natl Acad Sci U S A*. **111**: 15538–43. doi:10.1073/pnas.1412759111. PMC 4217403. PMID 25313081.

[29] Horvath, S; Garagnani, P; Bacalini, MG; Pirazzini, C; Salvioli, S; Gentilini, D; Di Blasio, AM; Giuliani, C; Tung, S; Vinters, HV; Franceschi, C (Feb 2015). "Accelerated epigenetic aging in Down syndrome". *Aging Cell*. **14**: 491–5. doi:10.1111/acel.12325. PMC 4406678. PMID 25678027.

[30] Horvath, S; Mah, V; Lu, AT; Woo, JS; Choi, OW; Jasinska, AJ; Riancho, JA; Tung, S; Coles, NS; Braun, J; Vinters, HV; Coles, LS (2015). "The cerebellum ages slowly according to the epigenetic clock" (PDF). *Aging (Albany NY)*. **7** (5): 294–306. doi:10.18632/aging.100742. PMC 4468311. PMID 26000617.

[31] Horvath, S (2016). "Huntington's disease accelerates epigenetic aging of human brain and disrupts DNA methylation levels". *Aging (Albany NY)*. **8** (7): 1485–512. doi:10.18632/aging.101005. PMC 4993344. PMID 27479945.

[32] Horvath, S; Levine, AJ (2015). "HIV-1 infection accelerates age according to the epigenetic clock". *J Infect Dis*. **212**: 1563–73. doi:10.1093/infdis/jiv277. PMC 4621253. PMID 25969563.

[33] Horvath, S (2015). "Increased epigenetic age and granulocyte counts in the blood of Parkinson's disease patients.". *Aging (Albany NY)*. **7**: 1130–42. doi:10.18632/aging.100859. PMC 4712337. PMID 26655927.

[34] Walker, RF; Liu, JS; Peters, BA; Ritz, BR; Wu, T; Ophoff, RA; Horvath, S (2015). "Epigenetic age analysis of children who seem to evade aging". *Aging (Albany NY)*. **7** (5): 334–9. doi:10.18632/aging.100744. PMC 4468314. PMID 25991677.

[35] Levine, M (2016). "Menopause accelerates biological aging". *Proc Natl Acad Sci USA*. **113**: 201604558. doi:10.1073/pnas.1604558113. PMID 27457926.

[36] Lowe, D (2016). "Epigenetic clock analyses of cellular senescence and ageing.". *Oncotarget*. **7** (8): 8524–8531. doi:10.18632/oncotarget.7383. PMID 26885756.

[37] Baker, DJ (2011). "Clearance of p16Ink4a-positive senescent cells delays ageing-associated disorders.". *Nature*. **479** (7372): 232–6. doi:10.1038/nature10600. PMC 3468323. PMID 22048312.

[38] Horvath S, Gurven M, Levine ME, Trumble BC, Kaplan H, Allayee H, Ritz BR, Chen B, Lu AT, Rickabaugh TM, Jamieson BD, Sun D, Li S, Chen W, Quintana-Murci L, Fagny M, Kobor MS, Tsao PS, Reiner AP, Edlefsen KL, Absher D, Assimes TL (2016). "An epigenetic clock analysis of race/ethnicity, sex, and coronary heart disease.". *Genome Biol*. **17** (1): 171. doi:10.1186/s13059-016-1030-0. PMC 4980791. PMID 27511193.

[39] Vilenchik MM, Knudson AG (2003). "Endogenous DNA double-strand breaks: production, fidelity of repair, and induction of cancer". *Proc. Natl. Acad. Sci. U.S.A.* **100** (22): 12871–6. doi:10.1073/pnas.2135498100. PMC 240711⊙. PMID 14566050.

[40] Cuozzo C, Porcellini A, Angrisano T, Morano A, Lee B, Di Pardo A, Messina S, Iuliano R, Fusco A, Santillo MR, Muller MT, Chiariotti L, Gottesman ME, Avvedimento EV (2007). "DNA damage, homology-directed repair, and DNA methylation". *PLoS Genet.* **3** (7): e110. doi:10.1371/journal.pgen.0030110. PMC 1913100⊙. PMID 17616978.

[41] O'Hagan HM, Mohammad HP, Baylin SB (2008). "Double strand breaks can initiate gene silencing and SIRT1-dependent onset of DNA methylation in an exogenous promoter CpG island". *PLoS Genet.* **4** (8): e1000155. doi:10.1371/journal.pgen.1000155. PMC 2491723⊙. PMID 18704159.

[42] Morano A, Angrisano T, Russo G, Landi R, Pezone A, Bartollino S, Zuchegna C, Babbio F, Bonapace IM, Allen B, Muller MT, Chiariotti L, Gottesman ME, Porcellini A, Avvedimento EV (2014). "Targeted DNA methylation by homology-directed repair in mammalian cells. Transcription reshapes methylation on the repaired gene". *Nucleic Acids Res.* **42** (2): 804–21. doi:10.1093/nar/gkt920. PMC 3902918⊙. PMID 24137009.

[43] Ding N, Bonham EM, Hannon BE, Amick TR, Baylin SB, O'Hagan HM (2016). "Mismatch repair proteins recruit DNA methyltransferase 1 to sites of oxidative DNA damage". *J Mol Cell Biol.* **8** (3): 244–54. doi:10.1093/jmcb/mjv050. PMID 26186941. free version at http://jmcb.oxfordjournals.org/content/early/2015/08/06/jmcb.mjv050

[44] Weidner, C. I.; Lin, Q.; Koch, C. M.; Eisele, L.; Beier, F.; Ziegler, P.; Wagner, W. (2014). "Aging of blood can be tracked by DNA methylation changes at just three CpG sites". *Genome Biology.* **15** (2): R24. doi:10.1186/gb-2014-15-2-r24. PMC 4053864⊙. PMID 24490752.

[45] Horvath S (2014-02-18 16:34) Comparison with the epigenetic clock (2014). Reader Comment.

[46] Wagner W (2014) Response to comment "comparison with the epigenetic clock by Horvath 2013"

[47] Horvath S (2013-11-04 11:00) Erratum in cancer tissues Reader Comment

[48] Hannum, G; Guinney, J; Zhao, L; Zhang, L; Hughes, G; Sadda, S; Klotzle, B; Bibikova, M; Fan, JB; Gao, Y; Deconde, R; Chen, M; Rajapakse, I; Friend, S; Ideker, T; Zhang, K (Jan 2013). "Genome-wide methylation profiles reveal quantitative views of human aging rates.". *Mol Cell.* **49**: 359–67. doi:10.1016/j.molcel.2012.10.016. PMC 3780611⊙. PMID 23177740.

[49] Weidner, CI; Lin, Q; Koch, CM; Eisele, L; Beier, F; Ziegler, P; Bauerschlag, DO; Jöckel, KH; Erbel, R; Mühleisen, TW; Zenke, M; Brümmendorf, TH; Wagner, W (2014). "Aging of blood can be tracked by DNA methylation changes at just three CpG sites.". *Genome Biol.* **15**: R24. doi:10.1186/gb-2014-15-2-r24. PMC 4053864⊙. PMID 24490752.

[50] Giuliani, C.; Cilli, E.; Bacalini, M. G.; Pirazzini, C.; Sazzini, M.; Gruppioni, G.; Franceschi, C.; Garagnani, P.; Luiselli, D. (2016). "Inferring chronological age from DNA methylation patterns of human teeth". *Am. J. Phys. Anthropol.* **159**: 585–595. doi:10.1002/ajpa.22921.

[51] Christoph Bock et al. (2016). Quantitative comparison of DNA methylation assays for biomarker development and clinical applications. Nature Biotechnology. doi:10.1038/nbt.3605

# Chapter 4

# Senescence

This article is about the aging of living things. For aging specifically in humans, see aging. For the study of aging in humans, see gerontology. For the science of the care of the elderly, see geriatrics. For experimental gerontology, see life extension.

For the aging of plants specifically, see plant senescence.

For premature aging disorders, see Progeroid syndromes.

**Senescence** (/sɪˈnesəns/) (from Latin: *senescere*, meaning

*An elderly man at a nursing home in Norway*

"to grow old", from *senex*) or **biological aging** (also spelled **biological ageing**) is the gradual deterioration of function characteristic of most complex lifeforms, arguably found in all biological kingdoms, that on the level of the organism increases mortality after maturation. The word *senescence* can refer either to cellular senescence or to senescence of the whole organism. It is commonly believed that cellular senescence underlies organismal senescence. The science of biological aging is biogerontology.

Senescence is not the inevitable fate of all organisms and can be delayed. The discovery, in 1934, that calorie restriction can extend lifespan 50% in rats, and the existence of species having negligible senescence and potentially immortal species such as *Hydra*, have motivated research into delaying and preventing senescence and thus age-related diseases. Organisms of some taxonomic groups (*taxa*), including some animals, experience chronological decrease

in mortality, for all or part of their life cycle.[1] On the other extreme are accelerated aging diseases, rare in humans. There is also the extremely rare and poorly understood "Syndrome X," whereby a person remains physically and mentally an infant or child throughout one's life.[2][3]

Even if environmental factors do not cause aging, they may affect it; in such a way, for example, overexposure to ultraviolet radiation accelerates skin aging. Different parts of the body may age at different rates. Two organisms of the same species can also age at different rates, so that biological aging and chronological aging are quite distinct concepts.

Albeit indirectly, senescence is by far the leading cause of death (other than in the trivially accurate sense that cerebral hypoxia, *i.e.*, lack of oxygen to the brain, is the immediate cause of all human death). Of the roughly 150,000 people who die each day across the globe, about two thirds – 100,000 per day – die of age-related causes; in industrialized nations, moreover, the proportion is much higher, reaching 90%.[4]

There are a number of hypotheses as to why senescence occurs; for example, some posit it is programmed by gene expression changes, others that it is the cumulative damage caused by biological processes. Whether senescence as a biological process itself can be slowed down, halted or even reversed, is a subject of current scientific speculation and research.[5]

## 4.1 Cellular senescence or cellular aiging

*Cellular senescence* is the phenomenon by which normal diploid cells cease to divide. In culture, fibroblasts can reach a maximum of 50 cell divisions before becoming senescent. This phenomenon is known as "replicative senescence", or the Hayflick limit.[6] Replicative senescence is the result of telomere shortening that ultimately triggers a DNA damage response. Cells can also be induced to senesce via DNA

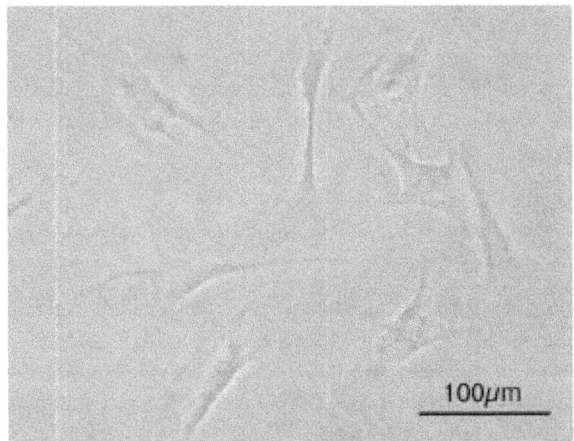

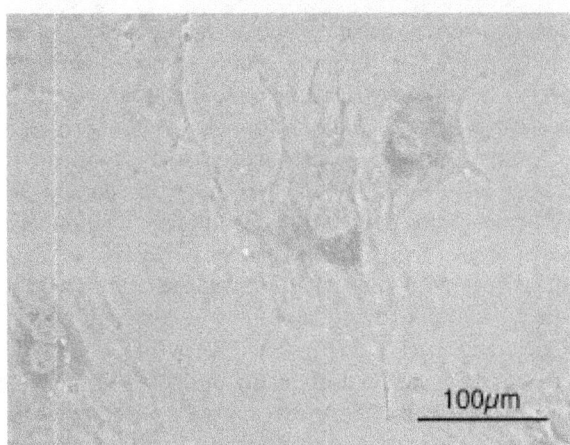

**Cellular senescence**
*(upper) Primary mouse embryonic fibroblast cells (MEFs) before senescence. Spindle-shaped. (lower) MEFs became senescent after passages. Cells grow larger, flatten shape and expressed senescence-associated β-galactosidase (SABG, blue areas), a marker of cellular senescence.*

damage in response to elevated reactive oxygen species (ROS), activation of oncogenes and cell-cell fusion, independent of telomere length. As such, cellular senescence represents a change in "cell state" rather than a cell becoming "aged" as the name confusingly suggests.

Although senescent cells can no longer replicate, they remain metabolically active and commonly adopt an immunogenic phenotype consisting of a pro-inflammatory secretome, the up-regulation of immune ligands, a pro-survival response, promiscuous gene expression (pGE) and stain positive for senescence-associated β-galactosidase activity.[7] The nucleus of senescent cells is characterized by senescence-associated heterochromatin foci (SAHF) and DNA segments with chromatin alterations reinforcing senescence (DNA-SCARS).[8] Senescent cells affect tumour suppression, wound healing and possibly embryonic/placental development and a pathological role in age-related diseases.[9]

The experimental elimination of senescent cells from transgenic progeroid mice[10] and non-progeroid, naturally-aged mice[11][12][13] led to greater resistance against aging-associated diseases.

### 4.1.1   Epigenetic clock analysis of cellular senescence

According to a molecular biomarker of aging known as epigenetic clock,[14] the three major types of cellular senescence, namely replicative senescence, oncogene-induced senescence and DNA damage-induced senescence are distinct because induction of replicative senescence (RS) and oncogene-induced senescence (OIS) were found to be accompanied by epigenetic aging of primary cells but senescence induced by DNA damage was not, even though RS and OIS activate the cellular DNA damage response pathway.[15] These results highlight the independence of cellular senescence from epigenetic aging. Consistent with this, telomerase-immortalised cells continued to age (according to the epigenetic clock) without having been treated with any senescence inducers or DNA-damaging agents, re-affirming the independence of the process of epigenetic ageing from telomeres, cellular senescence, and the DNA damage response pathway. Although the uncoupling of senescence from cellular aging appears at first sight to be inconsistent with the fact that senescent cells contribute to the physical manifestation of organism ageing, as demonstrated by Baker et al., where removal of senescent cells slowed down aging.[10] However, the epigenetic clock analysis of senescence suggests that cellular senescence is a state that cells are forced into as a result of external pressures such as DNA damage, ectopic oncogene expression and exhaustive proliferation of cells to replenish those eliminated by external/environmental factors.[15] These senescent cells, in sufficient numbers, will undoubtedly cause the deterioration of tissues, which is interpreted as organism ageing. However, at the cellular level, aging, as measured by the epigenetic clock, is distinct from senescence. It is an intrinsic mechanism that exists from the birth of the cell and continues. This implies that if cells are not shunted into senescence by the external pressures described above, they would still continue to age. This is consistent with the fact that mice with naturally long telomeres still age and eventually die even though their telomere lengths are far longer than the critical limit, and they age prematurely when their telomeres are forcibly shortened, due to replicative senescence. Hence senescence is a route by which cells exit prematurely from the natural course of cellular ageing.[15]

## 4.2 Aging of the whole organism

*Organismal senescence* is the aging of whole organisms. In general, aging is characterized by the declining ability to respond to stress, increased homeostatic imbalance, and increased risk of aging-associated diseases. Death is the ultimate consequence of aging, though "old age" is not a scientifically recognized cause of death because there is always a specific proximal cause, such as cancer, heart disease, or liver failure. Aging of whole organisms is therefore a complex process that can be defined as "a progressive deterioration of physiological function, an intrinsic age-related process of loss of viability and increase in vulnerability."[16]

Differences in maximum life span among species correspond to different "rates of aging." For example, inherited differences in the rate of aging make a mouse elderly at 3 years and a human elderly at 80 years.[17] These genetic differences affect a variety of physiological processes, including the efficiency of DNA repair, antioxidant enzymes, and rates of free radical production.

*Supercentenarian Ann Pouder (8 April 1807 – 10 July 1917) photographed on her 110th birthday. A heavily lined face is common in human senescence.*

Senescence of the organism gives rise to the Gompertz–Makeham law of mortality, which says that mortality rate accelerates rapidly with age.

Some animals, such as some reptiles and fish, age slowly (negligible senescence) and exhibit very long lifespans. Some even exhibit "negative senescence", in which mortality falls with age, in disagreement with the Gompertz–Makeham "law".[1]

Whether replicative senescence (Hayflick limit) plays a causative role in organismal aging is at present an active area of investigation.

The oft-quoted evolutionary theorist George Williams wrote, "It is remarkable that after a seemingly miraculous feat of morphogenesis, a complex metazoan should be unable to perform the much simpler task of merely maintaining what is already formed."[18]

There is a current debate as to whether or not the pursuit of longevity and the postponement of senescence are cost-effective health care goals given finite health care resources. Because of the accumulated infirmities of old age, bioethicist Ezekiel Emanuel, opines that the pursuit of longevity via the compression of morbidity hypothesis is a "fantasy" and that human life is not worth living after age 75; longevity then should not be a goal of health care policy.[19] This opinion has been refuted by neurosurgeon and medical ethicist Miguel Faria, who states that life can be worthwhile during old age, and that longevity should be pursued in association with the attainment of quality of life.[20] Faria claims that postponement of senescence as well as happiness and wisdom can be attained in old age in a large proportion of those who lead healthy lifestyles and remain intellectually active.[21]

## 4.3 Theories of aging

The exact etiology of senescence is still largely unclear and yet to be discovered. The process of senescence is complex, and may derive from a variety of different mechanisms and exist for a variety of different reasons. However, senescence is not universal. In a few simple species, such as those in the genus *Hydra*, senescence is negligible and cannot be detected.

All such species have no "post-mitotic" cells; they reduce the effect of damaging free radicals by cell division and dilution. Another related mechanism is that of the biologically immortal planarian flatworms, which have "apparently limitless [telomere] regenerative capacity fueled by a population of highly proliferative adult stem cells."[22] These organisms are biologically immortal but not immortal in the traditional sense as they are nonetheless susceptible to trauma and infectious and non-infectious disease. Moreover, average lifespans can vary greatly within and between species. This suggests that both genetic and environmental factors contribute to aging.

In general, theories that explain senescence have been divided between the programmed and stochastic theories of aging. Programmed theories imply that aging is regulated by biological clocks operating throughout the lifespan. This regulation would depend on changes in gene expression that affect the systems responsible for maintenance, repair, and defense responses. The reproductive-cell cycle theory suggests that aging is caused by changes in hormonal signaling over the lifespan.[23] Stochastic theories blame environmental impacts on living organisms that induce cumulative damage at various levels as the cause of aging, examples of which ranging from damage to DNA, damage to tissues and cells by oxygen radicals (widely known as free radicals countered by the even more well-known antioxidants), and cross-linking.

However, aging is seen as a progressive failure of homeodynamics–systemic preservation of homeostasis, involving genes for maintenance and repair, stochastic events leading to molecular damage and molecular heterogeneity, and chance events determining the probability of death. Since complex and interacting systems of maintenance and repair comprise the homeodynamic space of a biological system, aging is considered to be a progressive shrinkage of homeodynamic space mainly due to increased molecular heterogeneity. In 2013, a group of scientists defined nine hallmarks of aging that are common between organisms with emphasis on mammals: genomic instability, telomere attrition, epigenetic alterations, loss of proteostasis, deregulated nutrient sensing, mitochondrial dysfunction, cellular senescence, stem cell exhaustion, and altered intercellular communication.[24]

### 4.3.1  Evolutionary theories

Main article: Evolution of ageing

A gene can be expressed at various stages of life. Therefore, natural selection can support lethal and harmful alleles, if their expression occurs after reproduction. Senescence may be the product of such selection.[25][26][27] In addition, ageing is believed to have evolved because of the increasingly smaller probability of an organism still being alive at older age, due to predation and accidents, both of which may be random and age-invariant. The antagonistic plietropy theory states that strategies which result in a higher reproductive rate at a young age, but shorter overall lifespan, result in a higher lifetime reproductive success and are therefore favoured by natural selection. In essence, aging is, therefore, the result of investing resources in reproduction, rather than maintenance of the body (the "Disposable Soma" theory[28]), in light of the fact that accidents, predation, and disease kill organisms regardless of how much energy is devoted to repair of the body. Various other theories of aging exist, and are not necessarily mutually exclusive.

The geneticist J. B. S. Haldane wondered why the dominant mutation that causes Huntington's disease remained in the population, and why natural selection had not eliminated it. The onset of this neurological disease is (on average) at age 45 and is invariably fatal within 10–20 years. Haldane assumed that, in human prehistory, few survived until age 45. Since few were alive at older ages and their contribution to the next generation was therefore small relative to the large cohorts of younger age groups, the force of selection against such late-acting deleterious mutations was correspondingly small. However, if a mutation affected younger individuals, selection against it would be strong. Therefore, late-acting deleterious mutations could accumulate in populations over evolutionary time through genetic drift, which has been demonstrated experimentally. This concept of higher accumulation of deleterious mutations for older organisms came to be known as the selection shadow.[29]

Peter Medawar formalised this observation in his mutation accumulation theory of aging.[30][31] "The force of natural selection weakens with increasing age—even in a theoretically immortal population, provided only that it is exposed to real hazards of mortality. If a genetic disaster... happens late enough in individual life, its consequences may be completely unimportant". The 'real hazards of mortality' are, in typical circumstances, predation, disease, and accidents. So, even an immortal population, whose fertility does not decline with time, will have fewer individuals alive in older age groups. This is called 'extrinsic mortality'. Young cohorts, not depleted in numbers yet by extrinsic mortality, contribute far more to the next generation than the few remaining older cohorts, so the force of selection against late-acting deleterious mutations, which affect only these few older individuals, is very weak. The mutations may not be selected against, therefore, and may spread over evolutionary time into the population.

The major testable prediction made by this model is that species that have high extrinsic mortality in nature will age more quickly and have shorter intrinsic lifespans. This is borne out among mammals, the best-studied in terms of life history. There is a correlation among mammals between body size and lifespan, such that larger species live longer than smaller species under controlled/optimum conditions, but there are notable exceptions. For instance, many bats and rodents are of similar size, yet bats live much longer. For instance, the little brown bat, half the size of a mouse, can live 30 years in the wild. A mouse will only live 2–3 years even under optimum conditions. The explanation is that bats have fewer predators, and therefore low extrinsic mortality. More individuals survive to later ages, so the force of selection against late-acting deleterious mutations is stronger. Fewer late-acting deleterious mutations equates to slower aging and therefore a longer lifespan. Birds are also warm-blooded and are similar in size to many small mammals, yet often live 5–10 times as long. They have less predation pressure than ground-dwelling mammals. Seabirds, which, in general, have the fewest predators of all birds, live longest.

When examining the body-size vs. lifespan relationship, one also observes that predatory mammals tend to live longer than prey mammals in a controlled environment, such as a zoo or nature reserve. The explanation for the long lifespans of primates (such as humans, monkeys, and apes) relative to body size is that their intelligence, and often their sociality, help them avoid becoming prey. High position in the food chain, intelligence and cooperativeness all reduce extrinsic mortality in species.

Another evolutionary theory of aging was proposed by George C. Williams[132] and involves antagonistic pleiotropy. A single gene may affect multiple traits. Some traits that increase fitness early in life may also have negative effects later in life. But, because many more individuals are alive at young ages than at old ages, even small positive effects early can be strongly selected for, and large negative effects later may be very weakly selected against. Williams suggested the following example: Perhaps a gene codes for calcium deposition in bones, which promotes juvenile survival and will therefore be favored by natural selection; however, this same gene promotes calcium deposition in the arteries, causing negative atherosclerotic effects in old age. Thus, harmful biological changes in old age may result from selection for pleiotropic genes that are beneficial early in life but harmful later on. In this case, selection pressure is relatively high when Fisher's reproductive value is high and relatively low when Fisher's reproductive value is low.

## 4.3.2 Gene regulation

See also: Genetics of aging

A number of genetic components of aging have been identified using model organisms, ranging from the simple budding yeast *Saccharomyces cerevisiae* to worms such as *Caenorhabditis elegans* and fruit flies (*Drosophila melanogaster*). Study of these organisms has revealed the presence of at least two conserved aging pathways.

One of these pathways involves the gene *Sir2*, a NAD+-dependent histone deacetylase. In yeast, Sir2 is required for genomic silencing at three loci: the yeast mating loci, the telomeres and the ribosomal DNA (rDNA). In some species of yeast, replicative aging may be partially caused by homologous recombination between rDNA re-

peats; excision of rDNA repeats results in the formation of extrachromosomal rDNA circles (ERCs). These ERCs replicate and preferentially segregate to the mother cell during cell division, and are believed to result in cellular senescence by titrating away (competing for) essential nuclear factors. ERCs have not been observed in other species (nor even all strains of the same yeast species) of yeast (which also display replicative senescence), and ERCs are not believed to contribute to aging in higher organisms such as humans (they have not been shown to accumulate in mammals in a similar manner to yeast). Extrachromosomal circular DNA (eccDNA) has been found in worms, flies, and humans. The origin and role of eccDNA in aging, if any, is unknown.

Despite the lack of a connection between circular DNA and aging in higher organisms, extra copies of Sir2 are capable of extending the lifespan of both worms and flies (though, in flies, this finding has not been replicated by other investigators, and the activator of Sir2 resveratrol does not reproducibly increase lifespan in either species.[33]) Whether the Sir2 homologues in higher organisms have any role in lifespan is unclear, but the human SIRT1 protein has been demonstrated to deacetylate p53, Ku70, and the forkhead family of transcription factors. SIRT1 can also regulate acetylates such as CBP/p300, and has been shown to deacetylate specific histone residues.

RAS1 and RAS2 also affect aging in yeast and have a human homologue. RAS2 overexpression has been shown to extend lifespan in yeast.

Other genes regulate aging in yeast by increasing the resistance to oxidative stress. Superoxide dismutase, a protein that protects against the effects of mitochondrial free radicals, can extend yeast lifespan in stationary phase when overexpressed.

In higher organisms, aging is likely to be regulated in part through the insulin/IGF-1 pathway. Mutations that affect insulin-like signaling in worms, flies, and the growth hormone/IGF1 axis in mice are associated with extended lifespan. In yeast, Sir2 activity is regulated by the nicotinamidase PNC1. PNC1 is transcriptionally upregulated under stressful conditions such as caloric restriction, heat shock, and osmotic shock. By converting nicotinamide to niacin, nicotinamide is removed, inhibiting the activity of Sir2. A nicotinamidase found in humans, known as PBEF, may serve a similar function, and a secreted form of PBEF known as visfatin may help to regulate serum insulin levels. It is not known, however, whether these mechanisms also exist in humans, since there are obvious differences in biology between humans and model organisms.

Sir2 activity has been shown to increase under calorie restriction. Due to the lack of available glucose in the cells, more NAD+ is available and can activate Sir2. Resveratrol, a stilbenoid found in the skin of red grapes, was reported to extend the lifespan of yeast, worms, and flies (the lifespan extension in flies and worms have proved to be irreproducible by independent investigators[33]). It has been shown to activate Sir2 and therefore mimics the effects of calorie restriction, if one accepts that caloric restriction is indeed dependent on Sir2.

Gene expression is imperfectly controlled, and it is possible that random fluctuations in the expression levels of many genes contribute to the aging process as suggested by a study of such genes in yeast.[34] Individual cells, which are genetically identical, none-the-less can have substantially different responses to outside stimuli, and markedly different lifespans, indicating the epigenetic factors play an important role in gene expression and aging as well as genetic factors.

According to the GenAge database of aging-related genes there are over 700 genes associated with aging in model organisms: 555 in the soil roundworm (*Caenorhabditis elegans*), 87 in the bakers' yeast (*Saccharomyces cerevisiae*), 75 in the fruit fly (*Drosophila melanogaster*) and 68 in the mouse (*Mus musculus*).[35] The following is a list of genes connected to longevity through research on model organisms:

### 4.3.3   Cellular senescence

As noted above, senescence is not universal. It was once thought that senescence did not occur in single-celled organisms that reproduce through the process of cellular mitosis.[36] Recent investigation has unveiled a more complex picture. Single cells do accumulate age-related damage. On mitosis the debris is not evenly divided between the new cells. Instead it passes to one of the cells leaving the other cell pristine. With successive generations the cell population becomes a mosaic of cells with half ageless and the rest with varying degrees of senescence.[37]

Moreover, cellular senescence is not observed in several organisms, including perennial plants, sponges, corals, and lobsters. In those species where cellular senescence is observed, cells eventually become post-mitotic when they can no longer replicate themselves through the process of cellular mitosis; i.e., cells experience *replicative senescence*. How and why some cells become post-mitotic in some species has been the subject of much research and speculation, but (as noted above) it is sometimes suggested that cellular senescence evolved as a way to prevent the onset and spread of cancer. Somatic cells that have divided many times will have accumulated DNA mutations and would therefore be in danger of becoming cancerous if cell division continued. As such, it is becoming apparent that senescent cells undergo conversion to an immunogenic phe-notype that enables them to be eliminated by the immune system.[38]

Lately, the role of telomeres in cellular senescence has aroused general interest, especially with a view to the possible genetically adverse effects of cloning. The successive shortening of the chromosomal telomeres with each cell cycle is also believed to limit the number of divisions of the cell, thus contributing to aging. There have, on the other hand, also been reports that cloning could alter the shortening of telomeres. Some cells do not age and are, therefore, described as being "biologically immortal". It is theorized by some that when it is discovered exactly what allows these cells, whether it be the result of telomere lengthening or not, to divide without limit that it will be possible to genetically alter other cells to have the same capability. It is further theorized that it will eventually be possible to genetically engineer all cells in the human body to have this capability by employing gene therapy and, therefore, stop or reverse aging, effectively making the entire organism potentially immortal.

The length of the telomere strand has senescent effects: telomere shortening activates extensive alterations in alternative RNA splicing that produce senescent toxins such as progerin, which degrades the tissue and makes it more prone to failure.[39]

Cancer cells are usually immortal. In about 85% of tumors, this evasion of cellular senescence is the result of up-activation of their telomerase genes.[40] This simple observation suggests that reactivation of telomerase in healthy individuals could greatly increase their cancer risk.

Ned Sharpless and collaborators demonstrated the first in vivo link between p16-expressing senescent cells and lifespan.[41] They found delayed senescent cell accumulation in mice with mutations that extend lifespan, as well as in mice that had their lifespan extended by food restriction. Later, Jan van Deursen and Darren Baker in collaboration with Andre Terzic at the Mayo Clinic in Rochester, Minn., provided the first in vivo evidence for a causal link between cellular senescence and aging by preventing the accumulation of senescent cells in BubR1 progeroid mice.[42] In the absence of senescent cells, the mice's tissues showed a major improvement in the usual burden of age-related disorders. They did not develop cataracts, avoided the usual wasting of muscle with age. They retained the fat layers in the skin that usually thin out with age and, in people, cause wrinkling. Jan van Deursen, James Kirkland, Tamara Tchkonia, Nathan LeBrasseur, and Darren Baker at the Mayo Clinic in Rochester, Minn., provided the first direct in vivo evidence that cellular senescence causes signs of aging by eliminating senescent cells from progeroid mice by introducing a drug-inducible suicide gene and then treating the mice with the drug to kill senescent cells selectively, as opposed to decreasing whole body p16.[10] Another Mayo study led by James Kirkland in collaboration with Scripps and other groups demonstrated that senolytics, drugs that target senescent cells, enhance cardiac function and improve vascular reactivity in old mice, alleviate gait disturbance caused by radiation in mice, and delay frailty, neurological dysfunction, and osteoporosis in progeroid mice. Discovery of senolytic drugs was based on a hypothesis-driven approach: the investigators leveraged the observation that senescent cells are resistant to apoptosis to discover that pro-survival pathways are up-regulated in these cells. They demonstrated that these survival pathways are the "Achilles heel" of senescent cells using RNA interference approaches, including Bcl-2-, AKT-, p21-, and tyrosine kinase-related pathways. They then used drugs known to target the identified pathways and showed these drugs kill senescent cells by apoptosis in culture and decrease senescent cell burden in multiple tissues in vivo. Importantly, these drugs had long term effects after a single dose, consistent with removal of senescent cells, rather than a temporary effect requiring continued presence of the drugs. This was the first study to show that clearing senescent cells enhances function in chronologically aged mice.[43]

### 4.3.4 Chemical damage

*Elderly Klamath woman photographed by Edward S. Curtis in 1924*

One of the earliest aging theories was the *Rate of Living Hypothesis* described by Raymond Pearl in 1928[44] (based on earlier work by Max Rubner), which states that fast basal metabolic rate corresponds to short maximum life span.

While there may be some validity to the idea that for various types of specific damage detailed below that are by-products of metabolism, all other things being equal, a fast metabolism may reduce lifespan, in general this theory does not adequately explain the differences in lifespan either within, or between, species. Calorically restricted animals process as much, or more, calories per gram of body mass, as their *ad libitum* fed counterparts, yet exhibit substantially longer lifespans. Similarly, metabolic rate is a poor predictor of lifespan for birds, bats and other species that, it is presumed, have reduced mortality from predation, and therefore have evolved long lifespans even in the presence of very high metabolic rates.[45] In a 2007 analysis it was shown that, when modern statistical methods for correcting for the effects of body size and phylogeny are employed, metabolic rate does not correlate with longevity in mammals or birds.[46] (For a critique of the *Rate of Living Hypothesis* see *Living fast, dying when?*[47])

With respect to specific types of chemical damage caused by metabolism, it is suggested that damage to long-lived biopolymers, such as structural proteins or DNA, caused by ubiquitous chemical agents in the body such as oxygen and sugars, are in part responsible for aging. The damage can include breakage of biopolymer chains, cross-linking of biopolymers, or chemical attachment of unnatural substituents (haptens) to biopolymers.

Under normal aerobic conditions, approximately 4% of the oxygen metabolized by mitochondria is converted to superoxide ion, which can subsequently be converted to hydrogen peroxide, hydroxyl radical and eventually other reactive species including other peroxides and singlet oxygen, which can, in turn, generate free radicals capable of damaging structural proteins and DNA. Certain metal ions found in the body, such as copper and iron, may participate in the process. (In Wilson's disease, a hereditary defect that causes the body to retain copper, some of the symptoms resemble accelerated senescence.) These processes termed oxidative stress are linked to the potential benefits of dietary polyphenol antioxidants, for example in coffee,[48] red wine and tea.[49]

Sugars such as glucose and fructose can react with certain amino acids such as lysine and arginine and certain DNA bases such as guanine to produce sugar adducts, in a process called *glycation*. These adducts can further rearrange to form reactive species, which can then cross-link the structural proteins or DNA to similar biopolymers or other biomolecules such as non-structural proteins. People with diabetes, who have elevated blood sugar, develop senescence-associated disorders much earlier than the general population, but can delay such disorders by rigorous control of their blood sugar levels. There is evidence that sugar damage is linked to oxidant damage in a process termed *glycoxidation*.

Free radicals can damage proteins, lipids or DNA. Glycation mainly damages proteins. Damaged proteins and lipids accumulate in lysosomes as lipofuscin. Chemical damage to structural proteins can lead to loss of function; for example, damage to collagen of blood vessel walls can lead to vessel-wall stiffness and, thus, hypertension, and vessel wall thickening and reactive tissue formation (atherosclerosis); similar processes in the kidney can lead to renal failure. Damage to enzymes reduces cellular functionality. Lipid peroxidation of the inner mitochondrial membrane reduces the electric potential and the ability to generate energy. It is probably no accident that nearly all of the so-called "accelerated aging diseases" are due to defective DNA repair enzymes.

It is believed that the impact of alcohol on aging can be partly explained by alcohol's activation of the HPA axis, which stimulates glucocorticoid secretion, long-term exposure to which produces symptoms of aging.[50]

### 4.3.5   DNA damage theory

Main article: DNA damage theory of aging

Alexander[51] was the first to propose that DNA damage is the primary cause of aging. Early experimental evidence supporting this idea was reviewed by Gensler and Bernstein.[52] By the early 1990s experimental support for this proposal was substantial, and further indicated that DNA damage due to reactive oxygen species was a major source of the DNA damages causing aging.[53][54][55][56][57] The current state of evidence bearing on this theory is reviewed in DNA damage theory of aging and by Bernstein et al.[58]

### 4.3.6   Reliability theory

Main article: Reliability theory of aging and longevity

Reliability theory suggests that biological systems start their adult life with a high load of initial damage. Reliability theory is a general theory about systems failure. It allows researchers to predict the age-related failure kinetics for a system of given architecture (reliability structure) and given reliability of its components. Reliability theory predicts that even those systems which are composed entirely of non-aging elements (with a constant failure rate) will neverthe-

less deteriorate (fail more often) with age, if these systems are redundant in irreplaceable elements. Aging, therefore, is a direct consequence of systems.

Reliability theory also predicts the late-life mortality deceleration with subsequent leveling-off, as well as the late-life mortality plateaus, as an inevitable consequence of redundancy exhaustion at extreme old ages. The theory explains why mortality rates increase exponentially with age (the Gompertz law) in many species, by taking into account the initial flaws (defects) in newly formed systems. It also explains why organisms "prefer" to die according to the Gompertz law, while technical devices usually fail according to the Weibull (power) law. Reliability theory allows to specify conditions when organisms die according to the Weibull distribution: Organisms should be relatively free of initial flaws and defects. The theory makes it possible to find a general failure law applicable to all adult and extreme old ages, where the Gompertz and the Weibull laws are just special cases of this more general failure law. The theory explains why relative differences in mortality rates of compared populations (within a given species) vanish with age (compensation law of mortality), and mortality convergence is observed due to the exhaustion of initial differences in redundancy levels.

## 4.4   Miscellanea

Biological clocks, which objectively measure the biological age of cells and tissues, may become useful for testing different biological aging theories.[14]

A set of rare hereditary (genetic) disorders, each called progeria, has been known for some time. Sufferers exhibit symptoms resembling accelerated aging, including wrinkled skin. The cause of Hutchinson–Gilford progeria syndrome was reported in the journal *Nature* in May 2003.[59] This report suggests that DNA damage, not oxidative stress, is the cause of this form of accelerated aging.

Recently, a kind of early senescence has been alleged to be a possible unintended outcome of early cloning experiments. The issue was raised in the case of Dolly the sheep, following her death from a contagious lung disease. The claim that Dolly's early death involved premature senescence has been vigorously contested,[60] and Dolly's creator, Dr. Ian Wilmut has expressed the view that her illness and death were probably unrelated to the fact that she was a clone.

## 4.5   See also

• Accelerated aging disease

- Advanced adult

- Ageing

- Aging brain

- Aging-associated disease

- Anti-aging medicine

- Anti-aging movement

- Biogerontology

- DNA damage theory of aging

- DNA repair

- Free radicals

- Genetics of aging

- Homeostatic capacity

- Immortality

- Indefinite lifespan

- List of life extension-related topics

- Mitohormesis

- Oxidative stress

- Phenoptosis

- Plant senescence

- Programmed cell death

- Regenerative medicine

- Rejuvenation (aging)

- SAGE KE

- Strategies for Engineered Negligible Senescence (SENS)

- Sub-lethal damage

- Transgenerational design

- Stem cell theory of aging

- Timeline of senescence research

# 4.6 References

[1] Ainsworth, C; Lepage, M (2007). "Evolution's greatest mistakes". *New Scientist.* **195** (2616): 36–39. doi:10.1016/S0262-4079(07)62033-8.

[2] Walker, R.; Pakula, L.; Sutcliffe, M.; Kruk, P.; Graakjaer, J.; Shay, J. (2009). "A case study of "disorganized development" and its possible relevance to genetic determinants of aging". *Mechanisms of ageing and development.* **130** (5): 350–356. doi:10.1016/j.mad.2009.02.003. PMID 19428454.

[3] Brown, Bob (23 June 2006). "Doctors Baffled, Intrigued by Girl Who Doesn't Age". *Health.* ABC News. Retrieved 27 June 2009.

[4] Aubrey D.N.J. de Grey (2007). "Life Span Extension Research and Public Debate: Societal Considerations" (PDF). *Studies in Ethics, Law, and Technology.* **1** (1). doi:10.2202/1941-6008.1011. Article 5.

[5] "SENS Foundation".

[6] Hayflick L; Moorhead PS (December 1961). "The serial cultivation of human diploid cell strains". *Exp. Cell Res.* **25**: 585–621. doi:10.1016/0014-4827(61)90192-6. PMID 13905658.

[7] Campisi, Judith (February 2013). "Aging, Cellular Senescence, and Cancer". *Annual Review of Physiology.* **75**: 685–705. doi:10.1146/annurev-physiol-030212-183653. PMC 4166529. PMID 23140366.

[8] Rodier, F.; Campisi, J. (14 February 2011). "Four faces of cellular senescence". *The Journal of Cell Biology.* **192** (4): 547–556. doi:10.1083/jcb.201009094.

[9] Burton, Dominick G. A.; Krizhanovsky, Valery (31 July 2014). "Physiological and pathological consequences of cellular senescence". *Cellular and Molecular Life Sciences.* **71** (22): 4373–4386. doi:10.1007/s00018-014-1691-3.

[10] Baker, D.; Wijshake, T.; Tchkonia, T.; LeBrasseur, N.; Childs, B.; van de Sluis, B.; Kirkland, J.; van Deursen, J. (10 November 2011). "Clearance of p16Ink4a-positive senescent cells delays ageing-associated disorders". *Nature.* **479**: 232–6. doi:10.1038/nature10600. PMC 3468323. PMID 22048312.

[11] Xu, M; Palmer, AK; Ding, H; Weivoda, MM; Pirtskhalava, T; White, TA; Sepe, A; Johnson, KO; Stout, MB; Giorgadze, N; Jensen, MD; LeBrasseur, NK; Tchkonia, T; Kirkland, JL (2015). "Targeting senescent cells enhances adipogenesis and metabolic function in old age". *eLife.* **4**. doi:10.7554/eLife.12997. PMC 4758946. PMID 26687007.

[12] Quick, Darren (February 3, 2016). "Clearing out damaged cells in mice extends lifespan by up to 35 percent". *www.gizmag.com.* Retrieved 2016-02-04.

[13] Regalado, Antonio (February 3, 2016). "In New Anti-Aging Strategy, Clearing Out Old Cells Increases Life Span of Mice by 25 Percent". *MIT Technology Review*. Retrieved 2016-02-04.

[14] Horvath S (2013). "DNA methylation age of human tissues and cell types". *Genome Biology*. **14**: R115. doi:10.1186/gb-2013-14-10-r115. PMC 4015143. PMID 24138928.

[15] Lowe, D (2016). "Epigenetic clock analyses of cellular senescence and ageing.". *Oncotarget*. **7** (8): 8524–8531. doi:10.18632/oncotarget.7383. PMID 26885756.

[16] "Aging and Gerontology Glossary". Retrieved 26 February 2011.

[17] Austad, S (2009). "Comparative Biology of Aging". *J Gerontol a Biol Sci Med Sci*. **64** (2): 199–201. doi:10.1093/gerona/gln060. PMC 2655036. PMID 19223603.

[18] Williams, G.C. (1957). "Pleiotropy, natural selection, and the evolution of senescence". *Evolution*. **11**: 398–411. doi:10.2307/2406060.

[19] Emmanuel EJ. "Why I hope to die at 75: An argument that society and families – and you – will be better off if nature takes its course swiftly and promptly". The Atlantic. Retrieved 7 April 2015.

[20] Faria MA. "Bioethics and why I hope to live beyond age 75 attaining wisdom!: A rebuttal to Dr. Ezekiel Emanuel's 75 age limit.". Surg Neurol Int 2015;6:35. Retrieved 7 April 2015.

[21] Faria MA. "Longevity and compression of morbidity from a neuroscience perspective: Do we have a duty to die by a certain age?". Surg Neurol Int 2015;6:49. Retrieved 7 April 2015.

[22] Thomas C. J. Tan; Ruman Rahman; Farah Jaber-Hijazi; Daniel A. Felix; Chen Chen; Edward J. Louis & Aziz Aboobaker (February 2012). "Telomere maintenance and telomerase activity are differentially regulated in asexual and sexual worms" (PDF). *PNAS*. **109** (9): 4209–4214. doi:10.1073/pnas.1118885109.

[23] Bowen RL; Atwood CS (2011). "The reproductive-cell cycle theory of aging: an update.". *Experimental Gerontology*. **46** (2): 100–7. doi:10.1016/j.exger.2010.09.007. PMID 20851172.

[24] Lopez-Otin, C; et al. (2013). "The hallmarks of aging.". *Cell*. **153**: 1194–217. doi:10.1016/j.cell.2013.05.039. PMC 3836174. PMID 23746838.

[25] Medawar, P.B. (1952). *An Unsolved problem of biology; an inaugural lecture delivered at University College, London, 6 December, 1951*. London: H.K. Lewis. OCLC 8482093.

[26] Williams, G.C. (1957). "Pleiotropy, Natural Selection, and the Evolution of Senescence". *Evolution*. **11**: 398–411. doi:10.2307/2406060.

[27] Hamilton WD (September 1966). "The moulding of senescence by natural selection". *J. Theor. Biol.* **12** (1): 12–45. doi:10.1016/0022-5193(66)90184-6. PMID 6015424.

[28] Kirkwood TB (November 1977). "Evolution of ageing". *Nature*. **270** (5635): 301–4. doi:10.1038/270301a0. PMID 593350.

[29] Fabian, Daniel; Flatt, Thomas (2011). "The Evolution of Aging". *Scitable*. Nature Publishing Group. Retrieved December 9, 2014.

[30] Medawar PB (1946). "Old age and natural death". *Modern Quarterly*. **1**: 30–56.

[31] Medawar, Peter B. (1952). *An Unsolved Problem of Biology*. London: H. K. Lewis.

[32] Williams, George C. (December 1957). "Pleiotropy, Natural Selection, and the Evolution of Senescence" (PDF). *Evolution*. **11** (4): 398–411. doi:10.2307/2406060. JSTOR 2406060.

[33] Bass TM; Weinkove D; Houthoofd K; Gems D; Partridge L (October 2007). "Effects of resveratrol on lifespan in Drosophila melanogaster and Caenorhabditis elegans". *Mechanisms of Ageing and Development*. **128** (10): 546–52. doi:10.1016/j.mad.2007.07.007. PMID 17875315.

[34] Ryley J; Pereira-Smith OM (2006). "Microfluidics device for single cell gene expression analysis in Saccharomyces cerevisiae". *Yeast*. **23** (14–15): 1065–73. doi:10.1002/yea.1412. PMID 17083143.

[35] "GenAge database". Retrieved 26 February 2011.

[36] Gavrilov LA; Gavrilova NS (December 2001). "The reliability theory of aging and longevity". *Journal of Theoretical Biology*. **213** (4): 527–45. doi:10.1006/jtbi.2001.2430. PMID 11742523.

[37] Stephens C (April 2005). "Senescence: even bacteria get old". *Curr. Biol.* **15** (8): R308–10. doi:10.1016/j.cub.2005.04.006. PMID 15854899.

[38] Burton; Faragher (2015). "Cellular senescence: from growth arrest to immunogenic conversion". *AGE*. **37**. doi:10.1007/s11357-015-9764-2.

[39] Collins FS, et al. (13 June 2011). "Progerin and telomere dysfunction collaborate to trigger cellular senescence in normal human fibroblasts". *J Clin Invest*. **121** (7): 2833–44. doi:10.1172/JCI43578. PMC 3223819. PMID 21670498.

[40] Hanahan D; Weinberg RA (January 2000). "The hallmarks of cancer". *Cell*. **100** (1): 57–70. doi:10.1016/S0092-8674(00)81683-9. PMID 10647931.

[41] Krishnamurthy, J; Torrice, C; Ramsey, MR; Kovalev, GI; Al-Regaiey, K; Su, L; Sharpless, NE (2004). "Ink4a/Arf expression is a biomarker of aging". *J. Clin. Invest.* **114** (9): 1299–1307. doi:10.1172/JCI22475. PMC 524230. PMID 15520862.

[42] Baker DJ; Dawlaty MM; Wijshake T; Jeganathan KB; Malureanu L; van Ree JH; Crespo-Diaz R; Reyes S; Seaburg L; Shapiro V; Behfar A; Terzic A; van de Sluis B; van Deursen JM (Jan 2013). "Increased expression of BubR1 protects against aneuploidy and cancer and extends healthy lifespan". *Nat Cell Biol.* **15** (1): 96–102. doi:10.1038/ncb2643. PMID 23242215.

[43] Zhu, Y; Tchkonia, T; Pirtskhalava, T; Gower, AC; Ding, H; Giorgadze, N; Palmer, AK; Ikeno, Y; Hubbard, GB; Lenburg, M; O'Hara, SP; LaRusso, NF; Miller, JD; Roos, CM; Verzosa, GC; LeBrasseur, NK; Wren, JD; Farr, JN; Khosla, S; Stout, MB; McGowan, SJ; Fuhrmann-Stroissnigg, H; Gurkar, AU; Zhao, J; Colangelo, D; Dorronsoro, A; Ling, YY; Barghouthy, AS; Navarro, DC; Sano, T; Robbins, PD; Niedernhofer, LJ; Kirkland, JL (9 March 2015). "The Achilles' heel of senescent cells: from transcriptome to senolytic drugs.". *Aging Cell.* **14**: 644–58. doi:10.1111/acel.12344. PMID 25754370.

[44] Pearl, Raymond (1928). *The Rate of Living, Being an Account of Some Experimental Studies on the Biology of Life Duration.* New York: Alfred A. Knopf.

[45] Brunet-Rossinni AK; Austad SN (2004). "Ageing studies on bats: a review". *Biogerontology.* **5** (4): 211–22. doi:10.1023/B:BGEN.0000038022.65024.d8. PMID 15314271.

[46] de Magalhães JP; Costa J; Church GM (1 February 2007). "An Analysis of the Relationship Between Metabolism, Developmental Schedules, and Longevity Using Phylogenetic Independent Contrasts". *The Journals of Gerontology Series A: Biological Sciences and Medical Sciences.* **62** (2): 149–60. doi:10.1093/gerona/62.2.149. PMC 2288695. PMID 17339640. Archived from the original on 23 December 2014.

[47] Speakman JR; Selman C; McLaren JS; Harper EJ (1 June 2002). "Living fast, dying when? The link between aging and energetics". *The Journal of Nutrition.* **132** (6 Suppl 2): 1583S–97S. PMID 12042467.

[48] Freedman ND; Park Y; Abnet CC; Hollenbeck AR; Sinha R (May 2012). "Association of coffee drinking with total and cause-specific mortality". *N. Engl. J. Med.* **366** (20): 1891–904. doi:10.1056/NEJMoa1112010. PMC 3439152. PMID 22591295.

[49] Yang Y; Chan SW; Hu M; Walden R; Tomlinson B (2011). "Effects of some common food constituents on cardiovascular disease". *ISRN Cardiol.* **2011**: 397136. doi:10.5402/2011/397136. PMC 3262529. PMID 22347642.

[50] Spencer RL; Hutchison KE (1999). "Alcohol, aging, and the stress response" (PDF). *Alcohol Research & Health.* **23** (4): 272–83. PMID 10890824.

[51] Alexander P (1967). "The role of DNA lesions in the processes leading to aging in mice". *Symp. Soc. Exp. Biol.* **21**: 29–50. PMID 4860956.

[52] Gensler HL; Bernstein H (September 1981). "DNA damage as the primary cause of aging". *Q Rev Biol.* **56** (3): 279–303. doi:10.1086/412317. PMID 7031747.

[53] Bernstein C; Bernstein H (1991). *Aging, Sex, and DNA Repair.* San Diego CA: Academic Press. ISBN 0123960037.

[54] Ames BN; Gold LS (1991). "Endogenous mutagens and the causes of aging and cancer". *Mutat. Res.* **250** (1-2): 3–16. doi:10.1016/0027-5107(91)90157-j. PMID 1944345.

[55] Holmes GE; Bernstein C; Bernstein H (September 1992). "Oxidative and other DNA damages as the basis of aging: a review". *Mutat. Res.* **275** (3-6): 305–15. doi:10.1016/0921-8734(92)90034-M. PMID 1383772.

[56] Rao KS; Loeb LA (September 1992). "DNA damage and repair in brain: relationship to aging". *Mutat. Res.* **275** (3-6): 317–29. doi:10.1016/0921-8734(92)90035-N. PMID 1383773.

[57] Ames BN; Shigenaga MK; Hagen TM (September 1993). "Oxidants, antioxidants, and the degenerative diseases of aging". *Proc. Natl. Acad. Sci. U.S.A.* **90** (17): 7915–22. doi:10.1073/pnas.90.17.7915. PMC 47258. PMID 8367443.

[58] Bernstein, H; Payne, CM; Bernstein, C; Garewal, H; Dvorak, K (2008). "Cancer and aging as consequences of unrepaired DNA damage.". In Kimura, Honoka; Suzuki, Aoi. *New Research on DNA Damage.* Nova Science Publishers. pp. 1–47. ISBN 978-1604565812.

[59] Mounkes LC; Kozlov S (2003). "A progeroid syndrome in mice is caused by defects in A-type lamins" (PDF). *Nature.* **423** (6937): 298–301. doi:10.1038/nature01631.

[60] Macintosh, Kerry Lynn (2005). *Illegal Beings: Human Clones and the Law.* Cambridge: Cambridge University Press. ISBN 0-521-85328-1.

## 4.7 External links

- Senescence.info

- Fight Aging!

- SENS Foundation

- AgeLab (Massachusetts Institute of Technology)

- Senescence at DMOZ

- Aging Cell

- Telomere Shortening

- Jones, Owen R.; Scheuerlein, Alexander; Salguero-Gómez, Roberto; Camarda, Carlo Giovanni; Schaible, Ralf; Casper, Brenda B.; Dahlgren, Johan P.; Ehrlén, Johan; García, María B.; Menges, Eric S.; Quintana-Ascencio, Pedro F.; Caswell, Hal; Baudisch, Annette; Vaupel, James W. (2013). "Diversity of ageing across the tree of life". *Nature*. **505**: 169–173. doi:10.1038/nature12789. Lay summary – *National Geographic* (8 December 2013).

# Chapter 5

# Shelterin

**Shelterin** (also called **telosome**) is a protein complex known to protect telomeres in many eukaryotes from DNA repair mechanisms, as well as regulate telomerase activity. In mammals and other eukaryotes, telomeric DNA consists of double- and single-stranded TTAGGG repeats and a single-stranded, G-rich overhang. Subunits of shelterin bind to these regions and induce the formation of a t-loop, a cap structure that deters DNA-damage-sensing machinery from mistakenly repairing telomeres. The absence of shelterin causes telomere uncapping and thereby activates damage-signaling pathways that may lead to non-homologous end joining (NHEJ), homology directed repair (HDR),[1] senescence, or apoptosis.[2]

## 5.1 Subunits

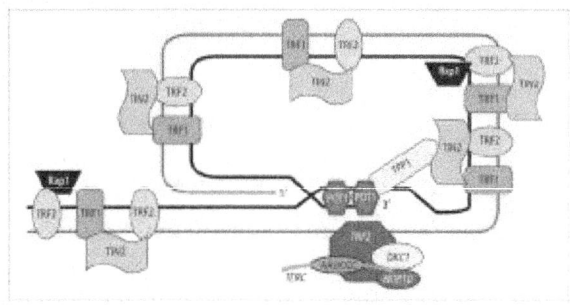

*Shelterin co-ordinates the T-loop formation of telomeres*

Shelterin has six subunits: TRF1, TRF2, POT1, RAP1, TIN2, and TPP1.[3] They can operate in smaller subsets to regulate the length of or protect telomeres.

- TRF1 (Telomere Repeat Factor 1): TRF1 is a homodimeric protein[4] that binds to the double-stranded TTAGGG region of the telomere. TRF1 along with TRF2 normally prevents telomerase from adding more telomere units to telomeres.[5] But when telomere lengthening is required, TRF1 recruits helicases to facilitate the process.[6] TRF1 is highly ex-

pressed in stem cells, and is essential for generation of induced pluripotent stem cells.[7] TRF1 may recruit PINX1 to inhibit telomere elongation by telomerase.[2]

- TRF2 (Telomere Repeat Factor 2) TRF2 is a homodimeric protein[4] that binds to the double-stranded TTAGGG region of the telomere and prevents the recognition of double-strand DNA breaks.[8]

- RAP1 (Repressor / Activator Protein 1): RAP1 is a stabilizing protein associated with TRF2.[9]

- POT1 (Protection of Telomere 1): POT1 contains OB-folds (oligonucleotide/oligosaccharide binding) that bind POT1 to single-stranded DNA,[10] which increase its affinity for single-stranded TTAGGG region of telomeric DNA. POT1 prevents the degradation of this single stranded DNA by nucleases and shelters the G-overhang.[3] Humans only have a single POT1, whereas mice have POT1a and POT1b.[11]

- TPP1 (ACD (gene)): TPP1 is a protein associated with POT1. The loss of TPP1 leads to impaired POT1 function.[2] When telomeres are to be lengthened, TPP1 is a central factor in recruiting telomerase to telomeres.[12] The gene which encodes for TPP1 (ACD) is distinct from the unrelated TPP1 gene on chromosome 11, which encodes tripeptidyl-peptidase I.[13]

- TIN2 (TRF1- and TRF2-Interacting Nuclear Protein 2) TIN2 is a stabilizing protein that binds to the TRF1, TRF2, and the TPP1-POT1 complex.[14] thereby bridging units attached to double-stranded DNA and units attached to single-stranded DNA.[2]

## 5.2 Repression of DNA repair mechanisms

There are two main DNA-damage-signaling pathways that shelterin represses: the ATR kinase pathway, blocked by

POT1, and the ATM kinase pathway, blocked by TRF2.[4] In the ATR kinase pathway, ATR and ATRIP sense the presence of single-stranded DNA and induce a phosphorylation cascade that leads to cell cycle arrest. To prevent this signal, POT1 "shelters" the single-stranded region of telomeric DNA. The ATM kinase pathway, which starts from ATM and other proteins sensing double strand breaks, similarly ends with cell cycle arrest. TRF2 may also hide the ends of telomeres, just as POT1 hides the single-stranded regions. Another theory proposes the blocking of the signal downstream. This will lead to a dynamic instability of the cells over time.

The structure of the t-loop may prevent NHEJ.[4] For NHEJ to occur, the Ku heterodimer must be able to bind to the ends of the chromosome. Another theory offers the mechanism proposed earlier: TRF2 hides the ends of telomeres.[2]

## 5.3   Species differences

At least four factors contribute to telomere maintenance in most eurkaryotes: telomerase, shelterin, TERRA and the CST Complex.[15] Fission yeast (Schizosaccharomyces pombe) has a shelterin complex for protection and maintenance of telomeres, but in budding yeast (Saccharomyces cerevisiae) this function is performed by the CST Complex.[16] For fission yeast, Rap1 and Pot1 are conserved, but Tpz1 is an ortholog of TPP1 and Taz1 is an ortholog of TRF1 and TRF2.[17]

Plants contain a variety of telomere-protecting proteins which can resemble either shelterin or the CST Complex.[18]

The fruit fly Drosophila melanogaster lacks both shelterin and telomerase, but instead uses retrotransposons to maintain telomeres.[19]

## 5.4   Non-telomeric functions of shelterin proteins

TIN2 can localize to mitochondria where it promotes glycolysis.[20]

RAP1 regulates transcription and affects NF-κB signaling.[6]

## 5.5   See also

- TERRA (biology)

- CST Complex

## 5.6   References

[1] Rodriguez, Raphaël; Müller, Sebastian; Yeoman, Justin A.; Trentesaux, Chantal; Riou, Jean-François; Balasubramanian, Shankar (2008). "A Novel Small Molecule That Alters Shelterin Integrity and Triggers a DNA-Damage Response at Telomeres". Journal of the American Chemical Society. 130: 15758–59. doi:10.1021/ja805615w.

[2] Palm, Wilhelm, and Titia de Lange. "How Shelterin Protects Mammalian Telomeres." Annual Reviews 42 (2008): 301-34. doi: 10.1146/annurev.genet.41.110306.130350.

[3] Xin, Huawei; Liu, Dan; Songyang, Zhou (2008). "The telomere/shelterin complex and its functions". Genome Biology. 9: 232. doi:10.1186/gb-2008-9-9-232.

[4] de Lange, Titia (2010). "How Shelterin Solves the Telomere End-Protection Problem". Cold Spring Harbor Symposia on Quantitative Biology. 75: 167–77. doi:10.1101/sqb.2010.75.017.

[5] Diotti R1, Loayza D (2011). "Shelterin complex and associated factors at human telomeres". NUCLEUS. 2 (2): 119–135. doi:10.4161/nucl.2.2.15135. PMC 3127094⊘. PMID 21738835.

[6] Sfeir A (2012). "Telomeres at a glance". Journal of Cell Science. 125 (Pt 18): 4173–4178. doi:10.1242/jcs.106831. PMID 23135002.

[7] Schneider RP, Garrobo I, Foronda M, Palacios JA, Marión RM, Flores I, Ortega S, Blasco MA (2013). "TRF1 is a stem cell marker and is essential for the generation of induced pluripotent stem cells". Nature Communications. 4: 1946. doi:10.1038/ncomms2946. PMID 23735977.

[8] Choi, Kyung H.; Farrell, Amy S.; Lakamp, Amanda S.; Ouellette, Michel M. (2011). "Characterization of the DNA binding specificity of Shelterin complexes". Nucleic Acids Research. 39: 9206–23. doi:10.1093/nar/gkr665.

[9] Nandakumar J1, Cech TR (2013). "Finding the end: recruitment of telomerase to telomeres". Nature Reviews Molecular Cell Biology. 14 (2): 69–82. doi:10.1038/nrm3505. PMC 3805138⊘. PMID 23299958.

[10] Flynn RL, Zou L (2010). "Oligonucleotide/oligosaccharide-binding fold proteins: a growing family of genome guardians". Critical Reviews in Biochemistry and Molecular Biology. 45 (4): 266–275. doi:10.3109/10409238.2010.488216. PMC 2906097⊘. PMID 20515430.

[11] Martínez P1, Blasco MA (2010). "Role of shelterin in cancer and aging". Aging Cell. 9 (5): 653–666. doi:10.1111/j.1474-9726.2010.00596.x. PMID 20569239.

[12] Abreu E1, Aritonovska E, Reichenbach P, Cristofari G, Culp B, Terns RM, Lingner J, Terns MP (2010). "TIN2-tethered TPP1 recruits human telomerase to telomeres in

vivo". *Molecular and Cellular Biology.* **30** (12): 2971–2982. doi:10.1128/MCB.00240-10. PMC 2876666ⓓ. PMID 20404094.

[13] "ACD ACD, shelterin complex subunit and telomerase recruitment factor [Homo sapiens (human)] - Gene - NCBI". *www.ncbi.nlm.nih.gov.* Retrieved 2017-02-03.

[14] Takai, Kaori K.; Hooper, Sarah; Blackwood, Stephanie; Gandhi, Rita; de Lange, Titia (2010). "In Vivo Stoichiometry of Shelterin Components". *Journal of Biological Chemistry.* **285**: 1457–67. doi:10.1074/jbc.M109.038026. PMC 2801271ⓓ. PMID 19864690.

[15] Giraud-Panis MJ, Teixeira MT, Géli V, Gilson E (2010). "CST meets shelterin to keep telomeres in check". *Molecular Cell.* **39** (5): 665–676. doi:10.1016/j.molcel.2010.08.024. PMID 20832719.

[16] Price CM, Boltz KA, Chaiken MF, Stewart JA, Beilstein MA, Shippen DE (2010). "Evolution of CST function in telomere maintenance". *Cell Cycle (journal).* **9** (16): 3157–3165. doi:10.4161/cc.9.16.12547. PMC 3041159ⓓ. PMID 20697207.

[17] Miyagawa K, Low RS, Santosa V, Tsuji H, Moser BA, Fujisawa S, Harland JL, Raguimova ON, Go A, Ueno M, Matsuyama A, Yoshida M, Nakamura TM, Tanaka K (2014). "SUMOylation regulates telomere length by targeting the shelterin subunit Tpz1(Tpp1) to modulate shelterin-Stn1 interaction in fission yeast" (PDF). *PNAS.* **111** (16): 5950–5955. doi:10.1073/pnas.1401359111. PMC 4000806ⓓ. PMID 24711392.

[18] Procházková Schrumpfová P, Schořová Š, Fajkus J (2016). "Telomere- and Telomerase-Associated Proteins and Their Functions in the Plant Cell". *FRONTIERS IN PLANT SCIENCE.* **7**: 851. doi:10.3389/fpls.2016.00851. PMC 4924339ⓓ. PMID 27446102.

[19] Pardue ML, DeBaryshe PG (2011). "Retrotransposons that maintain chromosome ends" (PDF). *PNAS.* **108** (51): 20317–20324. doi:10.1073/pnas.1100278108. PMC 3251079ⓓ. PMID 21821789.

[20] Chen LY, Zhang Y, Zhang Q, Li H, Luo Z, Fang H, Kim SH, Qin L, Yotnda P, Xu J, Tu BP, Bai Y, Songyang Z (2012). "Mitochondrial localization of telomeric protein TIN2 links telomere regulation to metabolic control". *Molecular Cell.* **47** (6): 839–850. doi:10.1016/j.molcel.2012.07.002. PMC 3462252ⓓ. PMID 22885005.

# Chapter 6

# Hayflick limit

The **Hayflick limit** or **Hayflick phenomenon** is the number of times a normal human cell population will divide until cell division stops. Empirical evidence shows that the telomeres associated with each cell's DNA will get slightly shorter with each new cell division until they shorten to a critical length.[1][2]

The concept of the Hayflick limit was advanced by American anatomist Leonard Hayflick in 1961,[1] at the Wistar Institute in Philadelphia, Pennsylvania. Hayflick demonstrated that a population of normal human fetal cells in a cell culture will divide between 40 and 60 times. The population will then enter a senescence phase, which refutes the contention by Nobel laureate Alexis Carrel that normal cells are immortal. Each mitosis slightly shortens each of the telomeres on the DNA of the cells. Telomere shortening in humans eventually makes cell division impossible, and this aging of the cell population appears to correlate with the overall physical aging of the human body.

Australian Nobel laureate Sir Macfarlane Burnet coined the name "Hayflick limit" in his book *Intrinsic Mutagenesis: A Genetic Approach to Ageing*, published in 1974.[3]

## 6.1 History

### 6.1.1 Belief of cell immortality

Prior to Hayflick's discovery, it was believed that vertebrate cells had an unlimited potential to replicate. Alexis Carrel, a Nobel prize-winning surgeon, had stated "that all cells explanted in culture are immortal, and that the lack of continuous cell replication was due to ignorance on how best to cultivate the cells".[3] He supported this hypothesis by claiming to have cultivated fibroblasts from chicken hearts and to have kept the culture growing for 34 years.[4] This indicated that cells of vertebrates could continue to divide indefinitely in a culture. However, other scientists have been unable to repeat Carrel's result.

Carrel's result is suspected to be due to an error in experimental procedure. To provide required nutrients, embryonic stem cells of chickens may have been readded to the culture daily. This would have easily allowed the cultivation of new, fresh cells in the culture, so there was not an infinite reproduction of the original cells.[1] If this is true, it has been speculated that Carrel knew about the error, but he never admitted it.[5][6]

Also, it has been theorized that the cells Carrel used were young enough to contain pluripotent stem cells, which, if supplied with a supporting telomerase-activation nutrient, would have been capable of staving off replicative senescence, or even possibly reversing it. Cultures not containing telomerase-active pluripotent stem cells would have been populated with telomerase-inactive cells, which would have been subject to the 50–60 mitosis cycles until apoptosis occurs as described in Leonard Hayflick's findings.

### 6.1.2 Experiment and discovery

Dr. Leonard Hayflick first became suspicious of Carrel's theory while working in a lab at the Wistar Institute. Hayflick was preparing normal human cells to be exposed to extracts of cancer cells when he noticed the normal cells had stopped proliferating. At first he thought that he had made a technical error in preparing the experiment, but later he began to think that the cell division processes had a counting mechanism. Working with Paul Moorhead, a cytogeneticist, he designed an experiment to test Carrel's theory of cell division.

The experiment proceeded as follows. Hayflick and Moorhead mixed equal numbers of normal human male fibroblasts that had divided many times (cells at the 40th population doubling) with female fibroblasts that had divided only a few times (cells at the 10th population doubling). Unmixed cell populations were kept as controls. When the male control culture stopped dividing, the mixed culture was examined and only female cells were found. This showed that the old male cells *remembered* they were old, even when surrounded by young cells, and that technical errors or contaminating viruses were unlikely explanations as to why only the male cell component had died.[1][3] The cells had stopped dividing and had become senescent based purely upon how many times the cell had divided.

These results disproved the immortality theory of Carrel and established the Hayflick limit as a credible biological theory that, unlike Carrel's experiment, has been repeated by other scientists.

## 6.2 Cell phases

Hayflick describes three phases in the life of a cell. At the start of his experiment he named the primary culture "phase one". Phase two is defined as the period when cells are proliferating – Hayflick called it the time of "luxuriant growth". After months of doubling the cells eventually reach phase three, a phenomenon of senescence – cell growth diminishes and then cell division stops altogether.

## 6.3 Telomere length

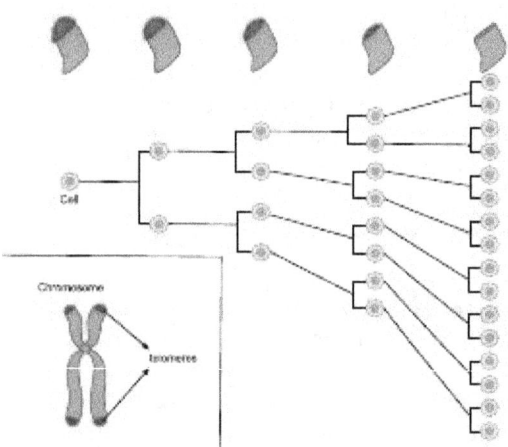

*The average cell will divide between 50-70 times before cell death. As the cell divides the telomeres on the end of the chromosome get smaller. The Hayflick Limit is the theory that due to the telomeres shortening through each division, the telomeres will eventually no longer be present on the chromosome. This end stage is known as senescence and proves the concept that links the deterioration of telomeres and aging.*

The Hayflick limit has been found to correlate with the length of the telomere region at the end of a strand of DNA. During the process of DNA replication, small segments of DNA at each end of the DNA strand (telomeres) are unable to be copied and are lost after each time DNA is duplicated.[7] The telomere region of DNA does not code for any protein; it is simply a repeated code on the end region of DNA that is lost. After many divisions, the telomeres become depleted and the cell begins apoptosis. This is a mechanism that prevents replication error that would cause mutations in DNA. Once the telomeres are depleted, due to the cell dividing many times, it will no longer divide. This is when the cell has reached its Hayflick limit.[8][9]

This process does not take place in most cancer cells due to an enzyme called telomerase. This enzyme maintains telomere length, which results in the telomeres of cancer cells never shortening. This gives these cells infinite replicative potential.[10] A proposed treatment for cancer is the usage of telomerase inhibitors that would prevent the restoration of the telomere, allowing the cell to die like other body cells.[11] On the other hand, telomerase activators might repair or extend the telomeres of healthy cells, thus extending their Hayflick limit. Telomerase activation might also lengthen the telomeres of immune system cells enough to prevent cancerous cells from developing from cells with very short telomeres.

## 6.4 See also

- Ageing
- Apoptosis
- Biological immortality
- HeLa cells
- Induced stem cells

## 6.5 References

[1] Hayflick L, Moorhead PS (1961). "The serial cultivation of human diploid cell strains". *Exp Cell Res.* **25** (3): 585–621. doi:10.1016/0014-4827(61)90192-6. PMID 13905658.

[2] Hayflick L. (1965). "The limited in vitro lifetime of human diploid cell strains". *Exp. Cell Res.* **37** (3): 614–636. doi:10.1016/0014-4827(65)90211-9. PMID 14315085.

[3] Shay JW, Wright WE; Wright (2000). "Hayflick, his limit, and cellular ageing" (PDF). *Nature Reviews Molecular Cell Biology.* **1** (1): 72–76. doi:10.1038/35036093. PMID 11413492.

[4] Carrel A, Ebeling AH (1921). "Age and multiplication of fibroblasts". *J. Exp. Med.* **34** (6): 599–606. doi:10.1084/jem.34.6.599.

[5] Witkowski JA (1985). "The myth of cell immortality". *Trends Biochem. Sci.* **10** (7): 258–260. doi:10.1016/0968-0004(85)90076-3.

[6] Witkowski JA (1980). "Dr. Carrel's immortal cells". *Med. Hist.* **24** (2): 129–142. doi:10.1017/S0025727300040126. PMC 1082700⊙. PMID 6990125.

[7] Watson JD (1972).  "Origin of concatemeric T7 DNA". *Nature New Biol.* **239** (94): 197–201. doi:10.1038/newbio239197a0. PMID 4507727.

[8] Olovnikov AM (1996). "Telomeres, telomerase and aging: Origin of the theory". *Exp. Gerontol.* **31** (4): 443–448. doi:10.1016/0531-5565(96)00005-8. PMID 9415101.

[9] Olovnikov, A. M. (1971).  "Принцип маргинотомии в матричном синтезе полинуклеотидов" [Principles of marginotomy in template synthesis of polynucleotides]. *Doklady Akademii Nauk SSSR.* **201**: 1496–1499.

[10] Feng F; et al.  (1995).  "The RNA component of human telomerase". *Science.* **269** (5228): 1236–1241. doi:10.1126/science.7544491. PMID 7544491.

[11] Wright WE, Shay JW (2000). "Telomere dynamics in cancer progression and prevention: Fundamental differences in human and mouse telomere biology". *Nature Medicine.* **6** (8): 849–851. doi:10.1038/78592. PMID 10932210.

## 6.6  Literature

- Watts, Geoff (2011). "Leonard Hayflick and the limits of ageing". *The Lancet.* **377** (9783): 2075. doi:10.1016/S0140-6736(11)60908-2.

- Harley, Calvin B.; Futcher, A. Bruce; Greider, Carol W. (1990).  "Telomeres shorten during ageing of human fibroblasts". *Nature.* **345** (6274): 458–60. doi:10.1038/345458a0. PMID 2342578.

- Gavrilov LA, Gavrilova NS (1991).  *The Biology of Life Span: A Quantitative Approach.* New York: Harwood Academic Publisher. ISBN 3-7186-4983-7.

- Gavrilov LA, Gavrilova NS (1993). "How many cell divisions in 'old' cells?". *Int. J. Geriatric Psychiatry.* **8** (6): 528–528.

- Wang, Richard C.; Smogorzewska, Agata; De Lange, Titia (2004). "Homologous Recombination Generates T-Loop-Sized Deletions at Human Telomeres". *Cell.* **119** (3): 355–68.  doi:10.1016/j.cell.2004.10.011. PMID 15507207.

- Watson, J. M.; Shippen, D. E. (2006).  "Telomere Rapid Deletion Regulates Telomere Length in Arabidopsis thaliana". *Molecular and Cellular Biology.* **27** (5): 1706–15. doi:10.1128/MCB.02059-06. PMC 1820464⊙. PMID 17189431.

## 6.7  External links

- Cell immortality and cancer
- Historical review of studies on cell division limit

# Chapter 7

# Apoptosis

**Apoptosis** (from Ancient Greek ἀπόπτωσις "falling off") is a process of programmed cell death that occurs in multicellular organisms.[2] Biochemical events lead to characteristic cell changes (morphology) and death. These changes include blebbing, cell shrinkage, nuclear fragmentation, chromatin condensation, chromosomal DNA fragmentation, and global mRNA decay. Between 50 and 70 billion cells die each day due to apoptosis in the average human adult.[lower-alpha 1] For an average child between the ages of 8 and 14, approximately 20 billion to 30 billion cells die a day.[4]

In contrast to necrosis, which is a form of traumatic cell death that results from acute cellular injury, apoptosis is a highly regulated and controlled process that confers advantages during an organism's lifecycle. For example, the separation of fingers and toes in a developing human embryo occurs because cells between the digits undergo apoptosis. Unlike necrosis, apoptosis produces cell fragments called apoptotic bodies that phagocytic cells are able to engulf and quickly remove before the contents of the cell can spill out onto surrounding cells and cause damage to the neighboring cells.[5]

Because apoptosis cannot stop once it has begun, it is a highly regulated process. Apoptosis can be initiated through one of two pathways. In the *intrinsic pathway* the cell kills itself because it senses cell stress, while in the *extrinsic pathway* the cell kills itself because of signals from other cells. Both pathways induce cell death by activating caspases, which are proteases, or enzymes that degrade proteins. The two pathways both activate initiator caspases, which then activate executioner caspases, which then kill the cell by degrading proteins indiscriminately.

Research on apoptosis has increased substantially since the early 1990s. In addition to its importance as a biological phenomenon, defective apoptotic processes have been implicated in a wide variety of diseases. Excessive apoptosis causes atrophy, whereas an insufficient amount results in uncontrolled cell proliferation, such as cancer. Some factors like Fas receptors and caspases promote apoptosis, while some members of the Bcl-2 family of proteins inhibit apoptosis.

## 7.1 Discovery and etymology

Main article: History of apoptosis research

German scientist Karl Vogt was first to describe the principle of apoptosis in 1842. In 1885, anatomist Walther Flemming delivered a more precise description of the process of programmed cell death. However, it was not until 1965 that the topic was resurrected. While studying tissues using electron microscopy, John Foxton Ross Kerr at University of Queensland was able to distinguish apoptosis from traumatic cell death.[6] Following the publication of a paper describing the phenomenon, Kerr was invited to join Alastair R Currie, as well as Andrew Wyllie, who was Currie's graduate student,[7] at University of Aberdeen. In 1972, the trio published a seminal article in the British Journal of Cancer.[8] Kerr had initially used the term programmed cell necrosis, but in the article, the process of natural cell death was called *apoptosis*. Kerr, Wyllie and Currie credited James Cormack, a professor of Greek language at University of Aberdeen, with suggesting the term apoptosis. Kerr received the Paul Ehrlich and Ludwig Darmstaedter Prize on March 14, 2000, for his description of apoptosis. He shared the prize with Boston biologist H. Robert Horvitz.[9]

For many years, the terms "apoptosis" and "programmed cell death" were not highly cited. What transformed cell death from obscurity to a major field of research were two things: the identification of components of the cell death control and effector mechanisms, and the linkage of abnormalities in cell death to human disease, in particular cancer.

The 2002 Nobel Prize in Medicine was awarded to Sydney Brenner, Horvitz and John E. Sulston for their work identifying genes that control apoptosis. The genes were identified by studies in the nematode *C. elegans* and homologues of these genes function in humans to regulate apoptosis.

*John E. Sulston won the Nobel Prize in Medicine in 2002, for his pioneering research on apoptosis.*

In Greek, apoptosis translates to the "falling off" of leaves from a tree.[10] Cormack, professor of Greek language, reintroduced the term for medical use as it had a medical meaning for the Greeks over two thousand years before. Hippocrates used the term to mean "the falling off of the bones". Galen extended its meaning to "the dropping of the scabs". Cormack was no doubt aware of this usage when he suggested the name. Debate continues over the correct pronunciation, with opinion divided between a pronunciation with the second *p* silent (/æpəˈtoʊsɪs/ *ap-ə-TOH-sis*[11][12]) and the second *p* pronounced (/æpəpˈtoʊsɪs/),[11][13] as in the original Greek. In English, the *p* of the Greek *-pt-* consonant cluster is typically silent at the beginning of a word (e.g. pterodactyl, Ptolemy), but articulated when used in combining forms preceded by a vowel, as in helicopter or the orders of insects: diptera, lepidoptera, etc.

In the original Kerr, Wyllie & Currie paper,[8] there is a footnote regarding the pronunciation:

"We are most grateful to Professor James Cormack of the Department of Greek, University of Aberdeen, for suggesting this term. The word "apoptosis" (ἀπόπτωσις) is used in Greek to describe the "dropping off" or "falling off" of petals from flowers, or leaves from trees. To show the derivation clearly, we propose that the stress should be on the penultimate syllable, the second half of the word being pronounced like "ptosis" (with the "p" silent), which comes from the same root "to fall", and is already used to describe the

drooping of the upper eyelid."

## 7.2   Activation mechanisms

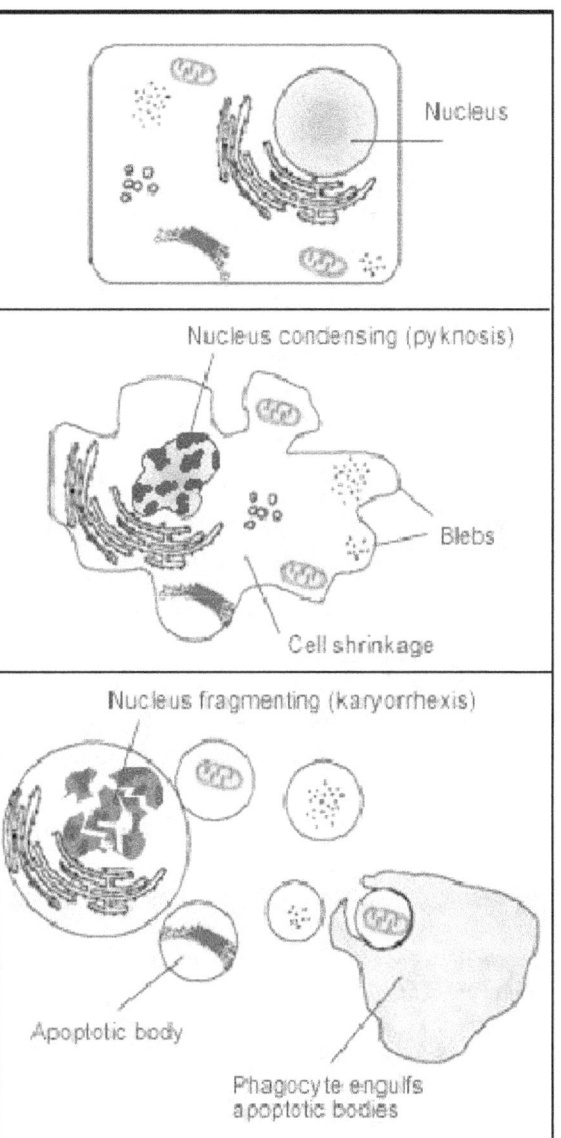

The initiation of apoptosis is tightly regulated by activation mechanisms, because once apoptosis has begun, it inevitably leads to the death of the cell.[14] [15] The two best-understood activation mechanisms are the intrinsic pathway (also called the mitochondrial pathway) and the extrinsic pathway.[16] The *intrinsic pathway* is activated by intracellular signals generated when cells are stressed and depends on the release of proteins from the intermembrane space of mitochondria.[17] The *extrinsic pathway* is activated by extracellular ligands binding to cell-surface death receptors,

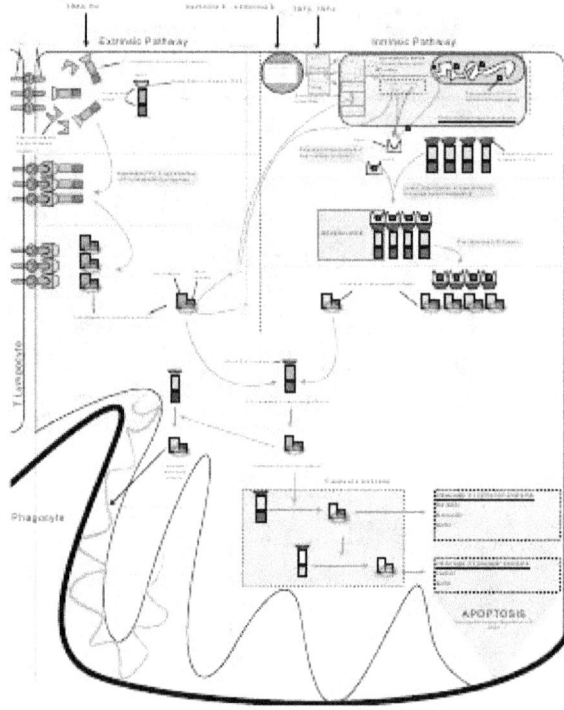

Control Of The Apoptotic Mechanisms

## 7.2.1  Intrinsic pathway

The mitochondria are essential to multicellular life. Without them, a cell ceases to respire aerobically and quickly dies. This fact forms the basis for some apoptotic pathways. Apoptotic proteins that target mitochondria affect them in different ways. They may cause mitochondrial swelling through the formation of membrane pores, or they may increase the permeability of the mitochondrial membrane and cause apoptotic effectors to leak out.[19][24] They are very closely related to intrinsic pathway, and tumors arise more frequently through intrinsic pathway than the extrinsic pathway because of sensitivity.[25] There is also a growing body of evidence indicating that nitric oxide is able to induce apoptosis by helping to dissipate the membrane potential of mitochondria and therefore make it more permeable.[26] Nitric oxide has been implicated in initiating and inhibiting apoptosis through its possible action as a signal molecule of subsequent pathways that activate apoptosis.[27]

Mitochondrial proteins known as SMACs (second mitochondria-derived activator of caspases) are released into the cell's cytosol following the increase in permeability of the mitochondia membranes. SMAC binds to *proteins that inhibit apoptosis* (IAPs) thereby deactivating them, and preventing the IAPs from arresting the process and therefore allowing apoptosis to proceed. IAP also normally suppresses the activity of a group of cysteine proteases called caspases,[28] which carry out the degradation of the cell. Therefore, the actual degradation enzymes can be seen to be indirectly regulated by mitochondrial permeability.

Cytochrome c is also released from mitochondria due to formation of a channel, the mitochondrial apoptosis-induced channel (MAC), in the outer mitochondrial membrane,[29] and serves a regulatory function as it precedes morphological change associated with apoptosis.[19] Once cytochrome c is released it binds with Apoptotic protease activating factor – 1 (*Apaf-1*) and ATP, which then bind to *pro-caspase-9* to create a protein complex known as an apoptosome. The apoptosome cleaves the pro-caspase to its active form of caspase-9, which in turn activates the effector *caspase-3*.

MAC (not to be confused with the Membrane Attack Complex formed by complement activation, also commonly denoted as MAC), also called "Mitochondrial Outer Membrane Permeabilization Pore" is regulated by various proteins, such as those encoded by the mammalian Bcl-2 family of anti-apoptopic genes, the homologs of the *ced-9* gene found in *C. elegans*.[30][31] *Bcl-2* proteins are able to promote or inhibit apoptosis by direct action on MAC/MOMPP. Bax and/or Bak form the pore, while Bcl-2, Bcl-xL or Mcl-1 inhibit its formation.

which leads to the formation of the death-inducing signaling complex (DISC).[18]

A cell initiates intracellular apoptotic signaling in response to a stress, which may bring about cell suicide. The binding of nuclear receptors by glucocorticoids,[19] heat,[19] radiation,[19] nutrient deprivation,[19] viral infection,[19] hypoxia[19] and increased intracellular calcium concentration,[20][21] for example, by damage to the membrane, can all trigger the release of intracellular apoptotic signals by a damaged cell. A number of cellular components, such as poly ADP ribose polymerase, may also help regulate apoptosis.[22]

Before the actual process of cell death is precipitated by enzymes, apoptotic signals must cause regulatory proteins to initiate the apoptosis pathway. This step allows those signals to cause cell death, or the process to be stopped, should the cell no longer need to die. Several proteins are involved, but two main methods of regulation have been identified: the targeting of mitochondria functionality,[23] or directly transducing the signal via adaptor proteins to the apoptotic mechanisms. An extrinsic pathway for initiation identified in several toxin studies is an increase in calcium concentration within a cell caused by drug activity, which also can cause apoptosis via a calcium binding protease calpain.

## 7.2.2  Extrinsic pathway

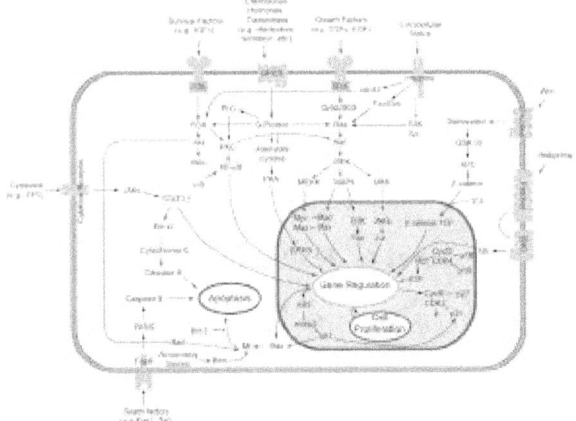

*Overview of signal transduction pathways.*

of TNF (left) and Fas (right) signalling in apoptosis, an example of direct signal transduction.

**Fas path**

The fas receptor (First apoptosis signal) – (also known as *Apo-1* or *CD95*) is a transmembrane protein of the TNF family which binds the Fas ligand (FasL).[32] The interaction between Fas and FasL results in the formation of the *death-inducing signaling complex* (DISC), which contains the FADD, caspase-8 and caspase-10. In some types of cells (type I), processed caspase-8 directly activates other members of the caspase family, and triggers the execution of apoptosis of the cell. In other types of cells (type II), the *Fas*-DISC starts a feedback loop that spirals into increasing release of proapoptotic factors from mitochondria and the amplified activation of caspase-8.[36]

**Common components**

Following *TNF-R1* and *Fas* activation in mammalian cells a balance between proapoptotic (BAX,[37] BID, BAK, or BAD) and anti-apoptotic (*Bcl-Xl* and *Bcl-2*) members of the *Bcl-2* family is established. This balance is the proportion of proapoptotic homodimers that form in the outer-membrane of the mitochondrion. The proapoptotic homodimers are required to make the mitochondrial membrane permeable

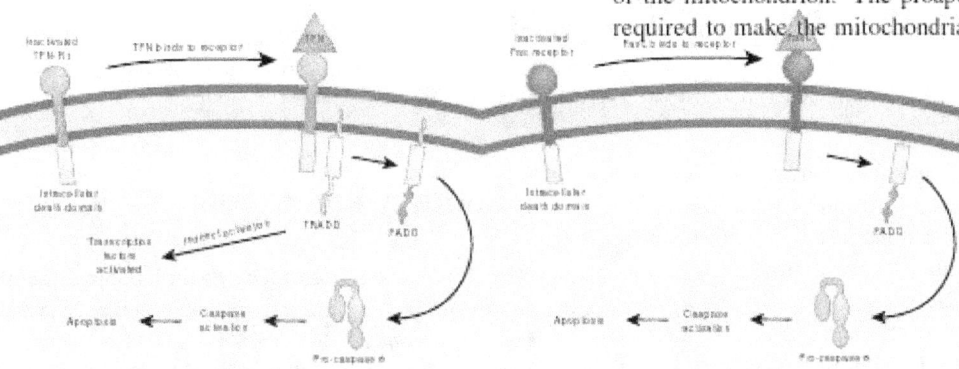

Two theories of the direct initiation of apoptotic mechanisms in mammals have been suggested: the *TNF-induced* (tumour necrosis factor) model and the *Fas-Fas ligand-mediated* model, both involving receptors of the *TNF receptor* (TNFR) family[32] coupled to extrinsic signals.

**TNF path**

TNF-alpha is a cytokine produced mainly by activated macrophages, and is the major extrinsic mediator of apoptosis. Most cells in the human body have two receptors for TNF-alpha: TNFR1 and TNFR2. The binding of TNF-alpha to TNFR1 has been shown to initiate the pathway that leads to caspase activation via the intermediate membrane proteins TNF receptor-associated death domain (TRADD) and Fas-associated death domain protein (FADD). cIAP1/2 can inhibit TNF-α signaling by binding to TRAF2. FLIP inhibits the activation of caspase-8.[33] Binding of this receptor can also indirectly lead to the activation of transcription factors involved in cell survival and inflammatory responses.[34] However, signalling through TNFR1 might also induce apoptosis in a caspase-independent manner.[35] The link between TNF-alpha and apoptosis shows why an abnormal production of TNF-alpha plays a fundamental role in several human diseases, especially in autoimmune diseases.

for the release of caspase activators such as cytochrome c and SMAC. Control of proapoptotic proteins under normal cell conditions of nonapoptotic cells is incompletely understood, but in general, Bax or Bak are activated by the activation of BH3-only proteins, part of the Bcl-2 family.

**Caspases** Caspases play the central role in the transduction of ER apoptotic signals. Caspases are proteins that are highly conserved, cysteine-dependent aspartate-specific proteases. There are two types of caspases: initiator caspases, caspase 2,8,9,10,11,12, and effector caspases, caspase 3,6,7. The activation of initiator caspases requires binding to specific oligomeric activator protein. Effector caspases are then activated by these active initiator caspases through proteolytic cleavage. The active effector caspases then proteolytically degrade a host of intracellular proteins to carry out the cell death program.

**Caspase-independent apoptotic pathway** There also exists a caspase-independent apoptotic pathway that is mediated by AIF (apoptosis-inducing factor).[38]

## 7.2.3  Apoptosis model in amphibians

Amphibian frog *Xenopus laevis* serves as an ideal model system for the study of the mechanisms of apoptosis. In fact, iodine and thyroxine also stimulate the spectacular apopto-

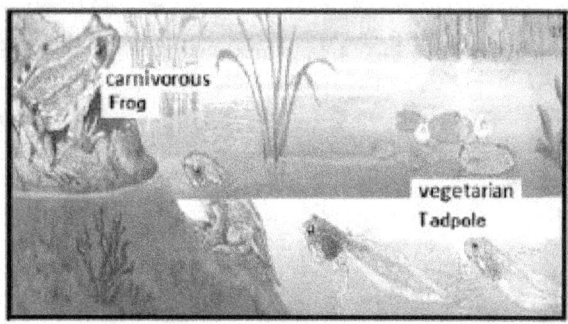

*Amphibian Metamorphosis*

sis of the cells of the larval gills, tail and fins in amphibians metamorphosis, and stimulate the evolution of their nervous system transforming the aquatic, vegetarian tadpole into the terrestrial, carnivorous frog.[39][40][41][42]

## 7.3 Proteolytic caspase cascade: Killing the cell

Many pathways and signals lead to apoptosis, but these converge on a single mechanism that actually causes the death of the cell. After a cell receives stimulus, it undergoes organized degradation of cellular organelles by activated proteolytic caspases. In addition to the destruction of cellular organelles, mRNA is rapidly and globally degraded by a mechanism that is not yet fully characterized.[43] mRNA decay is triggered very early in apoptosis. A cell undergoing apoptosis shows a characteristic morphology:

1. Cell shrinkage and rounding are shown because of the breakdown of the proteinaceous cytoskeleton by caspases.[44]

2. The cytoplasm appears dense, and the organelles appear tightly packed.

3. Chromatin undergoes condensation into compact patches against the nuclear envelope (also known as the perinuclear envelope) in a process known as pyknosis, a hallmark of apoptosis.[45][46]

4. The nuclear envelope becomes discontinuous and the DNA inside it is fragmented in a process referred to as karyorrhexis. The nucleus breaks into several discrete *chromatin bodies* or *nucleosomal units* due to the degradation of DNA.[47]

5. The cell membrane shows irregular buds known as blebs.

6. The cell breaks apart into multiple vesicles called *apoptotic bodies*, which are then phagocytosed.

Apoptosis progresses quickly and its products are quickly removed, making it difficult to detect or visualize. During karyorrhexis, endonuclease activation leaves short DNA fragments, regularly spaced in size. These give a characteristic "laddered" appearance on agar gel after electrophoresis. Tests for DNA laddering differentiate apoptosis from ischemic or toxic cell death.[48]

### 7.3.1 Removal of dead cells

The removal of dead cells by neighboring phagocytic cells has been termed efferocytosis.[49] Dying cells that undergo the final stages of apoptosis display phagocytotic molecules, such as phosphatidylserine, on their cell surface.[50] Phosphatidylserine is normally found on the inner leaflet surface of the plasma membrane, but is redistributed during apoptosis to the extracellular surface by a protein known as scramblase.[51] These molecules mark the cell for phagocytosis by cells possessing the appropriate receptors, such as macrophages.[52] The removal of dying cells by phagocytes occurs in an orderly manner without eliciting an inflammatory response.[53]

## 7.4 Pathway knock-outs

Many knock-outs have been made in the apoptosis pathways to test the function of each of the proteins. Several caspases, in addition to APAF-1 and FADD, have been mutated to determine the new phenotype. In order to create a tumor necrosis factor (TNF) knockout, an exon containing the nucleotides 3704-5364 was removed from the gene. This exon encodes a portion of the mature TNF domain, as well as the leader sequence, which is a highly conserved region necessary for proper intracellular processing. TNF-/- mice develop normally and have no gross structural or morphological abnormalities. However, upon immunization with SRBC (sheep red blood cells), these mice demonstrated a deficiency in the maturation of an antibody response; they were able to generate normal levels of IgM, but could not develop specific IgG levels. Apaf-1 is the protein that turns on caspase 9 by cleavage to begin the caspase cascade that leads to apoptosis. Since a -/- mutation in the APAF-1 gene is embryonic lethal, a gene trap strategy was used in order to generate an APAF-1 -/- mouse. This assay is used to disrupt gene function by creating an intragenic gene fusion. When an APAF-1 gene trap is introduced into cells, many morphological changes occur, such as spina bifida, the persistence of interdigital webs, and open brain. In addition, after embryonic day 12.5, the brain of the embryos showed several structural changes. APAF-1 cells are protected from apoptosis stimuli such as irradiation. A BAX-1 knock-out mouse exhibits normal forebrain formation and a decreased programmed cell death in some neuronal populations and in the spinal cord, leading to an increase in motor neurons.

The caspase proteins are integral parts of the apoptosis pathway, so it follows that knock-outs made have varying damaging results. A caspase 9 knock-out leads to a severe brain malformation. A caspase 8 knock-out leads to cardiac failure and thus embryonic lethality. However, with the use of cre-lox technology, a caspase 8 knock-out has been created that exhibits an increase in peripheral T cells, an impaired T cell response, and a defect in neural tube closure. These mice were found to be resistant to apoptosis mediated by CD95, TNFR, etc. but not resistant to apoptosis caused by UV irradiation, chemotherapeutic drugs, and other stimuli. Finally, a caspase 3 knock-out was characterized by ectopic cell masses in the brain and abnormal apoptotic features such as membrane blebbing or nuclear fragmentation. A remarkable feature of these KO mice is that they have a very restricted phenotype: Casp3, 9, APAF-1 KO mice have deformations of neural tissue and FADD and Casp 8 KO showed defective heart development, however in both types of KO other organs developed normally and some cell types were still sensitive to apoptotic stimuli suggesting that unknown proapoptotic pathways exist.

## 7.5 Methods for distinguishing apoptotic from necrotic (necroptotic) cells

In order to perform analysis of apoptotic versus necrotic (necroptotic) cells, one can do analysis of morphology by time-lapse microscopy, flow fluorocytometry, and transmission electron microscopy. There are also various biochemical techniques for analysis of cell surface markers (phosphatidylserine exposure versus cell permeability by flow fluorocytometry), cellular markers such as DNA fragmentation[54] (flow fluorocytometry), caspase activation, Bid cleavage, and cytochrome c release (Western blotting). It is important to know how primary and secondary necrotic cells can be distinguished by analysis of supernatant for caspases, HMGB1, and release of cytokeratin 18. However, no distinct surface or biochemical markers of necrotic cell death have been identified yet, and only negative markers are available. These include absence of apoptotic markers (caspase activation, cytochrome c release, and oligonucleosomal DNA fragmentation) and differential kinetics of cell death markers (phosphatidylserine exposure and cell membrane permeabilization). A selection of techniques that can be used to distinguish apoptosis from necroptotic cells could be found in these references.[55][56][57]

## 7.6 Implication in disease

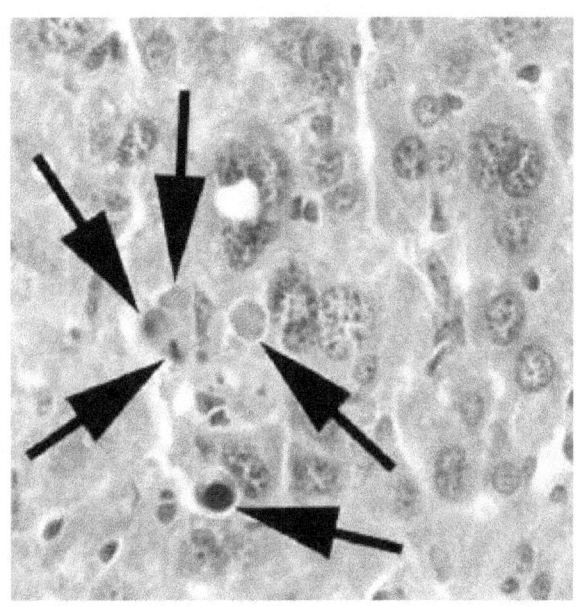

*A section of mouse liver showing several apoptotic cells, indicated by arrows*

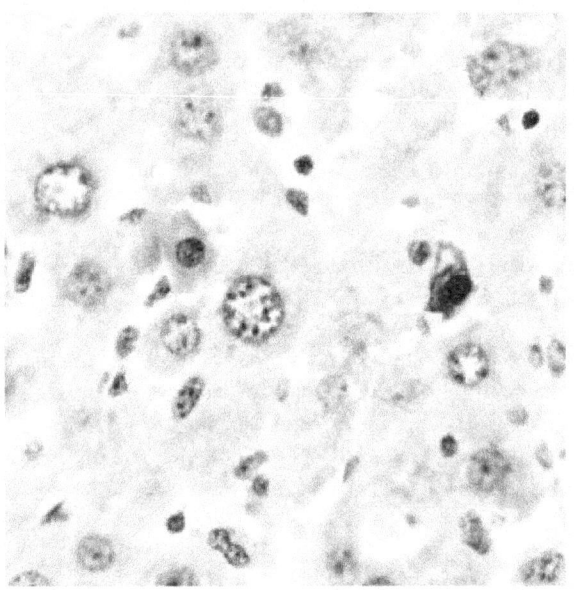

*A section of mouse liver stained to show cells undergoing apoptosis (orange)*

### 7.6.1 Defective pathways

The many different types of apoptotic pathways contain a multitude of different biochemical components, many of them not yet understood.[58] As a pathway is more or less

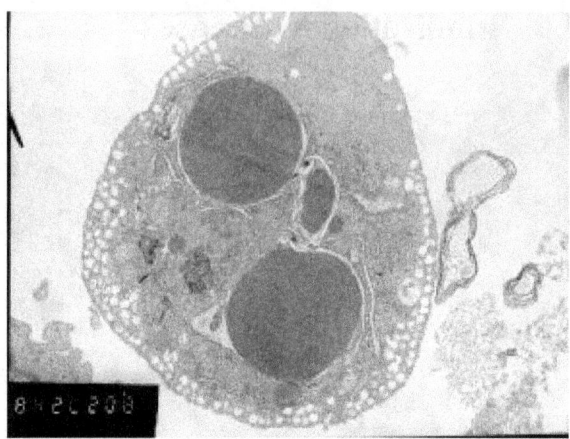

*Neonatal cardiomyocytes ultrastructure after anoxia-reoxygenation.*

sequential in nature, removing or modifying one component leads to an effect in another. In a living organism, this can have disastrous effects, often in the form of disease or disorder. A discussion of every disease caused by modification of the various apoptotic pathways would be impractical, but the concept overlying each one is the same: The normal functioning of the pathway has been disrupted in such a way as to impair the ability of the cell to undergo normal apoptosis. This results in a cell that lives past its "use-by-date" and is able to replicate and pass on any faulty machinery to its progeny, increasing the likelihood of the cell's becoming cancerous or diseased.

A recently described example of this concept in action can be seen in the development of a lung cancer called NCI-H460.[59] The *X-linked inhibitor of apoptosis protein* (XIAP) is overexpressed in cells of the H460 cell line. XIAPs bind to the processed form of caspase-9, and suppress the activity of apoptotic activator cytochrome c, therefore overexpression leads to a decrease in the amount of proapoptotic agonists. As a consequence, the balance of anti-apoptotic and proapoptotic effectors is upset in favour of the former, and the damaged cells continue to replicate despite being directed to die.

### Dysregulation of p53

The tumor-suppressor protein p53 accumulates when DNA is damaged due to a chain of biochemical factors. Part of this pathway includes alpha-interferon and beta-interferon, which induce transcription of the *p53* gene, resulting in the increase of p53 protein level and enhancement of cancer cell-apoptosis.[60] p53 prevents the cell from replicating by stopping the cell cycle at G1, or interphase, to give the cell time to repair, however it will induce apoptosis if damage is extensive and repair efforts fail. Any disruption to the reg-

ulation of the *p53* or interferon genes will result in impaired apoptosis and the possible formation of tumors.

### 7.6.2 Inhibition

Inhibition of apoptosis can result in a number of cancers, autoimmune diseases, inflammatory diseases, and viral infections. It was originally believed that the associated accumulation of cells was due to an increase in cellular proliferation, but it is now known that it is also due to a decrease in cell death. The most common of these diseases is cancer, the disease of excessive cellular proliferation, which is often characterized by an overexpression of IAP family members. As a result, the malignant cells experience an abnormal response to apoptosis induction: Cycle-regulating genes (such as p53, ras or c-myc) are mutated or inactivated in diseased cells, and further genes (such as bcl-2) also modify their expression in tumors.

### HeLa cell

Apoptosis in HeLa[lower-alpha 2] cells is inhibited by proteins produced by the cell; these inhibitory proteins target retinoblastoma tumor-suppressing proteins.[61] These tumor-suppressing proteins regulate the cell cycle, but are rendered inactive when bound to an inhibitory protein.[61] HPV E6 and E7 are inhibitory proteins expressed by the human papillomavirus, HPV being responsible for the formation of the cervical tumor from which HeLa cells are derived.[62] HPV E6 causes p53, which regulates the cell cycle, to become inactive.[63] HPV E7 binds to retinoblastoma tumor suppressing proteins and limits its ability to control cell division.[63] These two inhibitory proteins are partially responsible for HeLa cells' immortality by inhibiting apoptosis to occur.[64] CDV (Canine Distemper Virus) is able to induce apoptosis despite the presence of these inhibitory proteins. This is an important oncolytic property of CDV: this virus is capable of killing canine lymphoma cells. Oncoproteins E6 and E7 still leave p53 inactive, but they are not able to avoid the activation of caspases induced from the stress of viral infection. These oncolytic properties provided a promising link between CDV and lymphoma apoptosis, which can lead to development of alternative treatment methods for both canine lymphoma and human non-Hodgkin lymphoma. Defects in the cell cycle are thought to be responsible for the resistance to chemotherapy or radiation by certain tumor cells, so a virus that can induce apoptosis despite defects in the cell cycle is useful for cancer treatment.[64]

**Treatments**

The main method of treatment for death from signaling-related diseases involves either increasing or decreasing the susceptibility of apoptosis in diseased cells, depending on whether the disease is caused by either the inhibition of or excess apoptosis. For instance, treatments aim to restore apoptosis to treat diseases with deficient cell death, and to increase the apoptotic threshold to treat diseases involved with excessive cell death. To stimulate apoptosis, one can increase the number of death receptor ligands (such as TNF or TRAIL), antagonize the anti-apoptotic Bcl-2 pathway, or introduce Smac mimetics to inhibit the inhibitor (IAPs). The addition of agents such as Herceptin, Iressa, or Gleevec works to stop cells from cycling and causes apoptosis activation by blocking growth and survival signaling further upstream. Finally, adding p53-MDM2 complexes displaces p53 and activates the p53 pathway, leading to cell cycle arrest and apoptosis. Many different methods can be used either to stimulate or to inhibit apoptosis in various places along the death signaling pathway.[65]

Apoptosis is a multi-step, multi-pathway cell-death programme that is inherent in every cell of the body. In cancer, the apoptosis cell-division ratio is altered. Cancer treatment by chemotherapy and irradiation kills target cells primarily by inducing apoptosis.

### 7.6.3   Hyperactive apoptosis

On the other hand, loss of control of cell death (resulting in excess apoptosis) can lead to neurodegenerative diseases, hematologic diseases, and tissue damage. The progression of HIV is directly linked to excess, unregulated apoptosis. In a healthy individual, the number of CD4+ lymphocytes is in balance with the cells generated by the bone marrow; however, in HIV-positive patients, this balance is lost due to an inability of the bone marrow to regenerate CD4+ cells. In the case of HIV, CD4+ lymphocytes die at an accelerated rate through uncontrolled apoptosis, when stimulated.

**Treatments**

Treatments aiming to inhibit works to block specific caspases. Finally, the Akt protein kinase promotes cell survival through two pathways. Akt phosphorylates and inhibits Bas (a Bcl-2 family member), causing Bas to interact with the 14-3-3 scaffold, resulting in Bcl dissociation and thus cell survival. Akt also activates IKKα, which leads to NF-κB activation and cell survival. Active NF-κB induces the expression of anti-apoptotic genes such as Bcl-2, resulting in inhibition of apoptosis. NF-κB has been found to play both an antiapoptotic role and a proapoptotic role depending on the stimuli utilized and the cell type.[66]

### 7.6.4   HIV progression

The progression of the human immunodeficiency virus infection into AIDS is due primarily to the depletion of CD4+ T-helper lymphocytes in a manner that is too rapid for the body's bone marrow to replenish the cells, leading to a compromised immune system. One of the mechanisms by which T-helper cells are depleted is apoptosis, which results from a series of biochemical pathways:[67]

1. HIV enzymes deactivate anti-apoptotic *Bcl-2*. This does not directly cause cell death but primes the cell for apoptosis should the appropriate signal be received. In parallel, these enzymes activate proapoptotic *procaspase-8*, which does directly activate the mitochondrial events of apoptosis.

2. HIV may increase the level of cellular proteins that prompt Fas-mediated apoptosis.

3. HIV proteins decrease the amount of CD4 glycoprotein marker present on the cell membrane.

4. Released viral particles and proteins present in extracellular fluid are able to induce apoptosis in nearby "bystander" T helper cells.

5. HIV decreases the production of molecules involved in marking the cell for apoptosis, giving the virus time to replicate and continue releasing apoptotic agents and virions into the surrounding tissue.

6. The infected CD4+ cell may also receive the death signal from a cytotoxic T cell.

Cells may also die as direct consequences of viral infections. HIV-1 expression induces tubular cell G2/M arrest and apoptosis.[68] The progression from HIV to AIDS is not immediate or even necessarily rapid; HIV's cytotoxic activity toward CD4+ lymphocytes is classified as AIDS once a given patient's CD4+ cell count falls below 200.[69]

### 7.6.5   Viral infection

Viral induction of apoptosis occurs when one or several cells of a living organism are infected with a virus, leading to cell death. Cell death in organisms is necessary for the normal development of cells and the cell cycle maturation.[70] It is also important in maintaining the regular functions and activities of cells.

Viruses can trigger apoptosis of infected cells via a range of mechanisms including:

- Receptor binding

- Activation of protein kinase R (PKR)

- Interaction with p53

- Expression of viral proteins coupled to MHC proteins on the surface of the infected cell, allowing recognition by cells of the immune system (such as Natural Killer and cytotoxic T cells) that then induce the infected cell to undergo apoptosis.[71]

Canine distemper virus (CDV) is known to cause apoptosis in central nervous system and lymphoid tissue of infected dogs in vivo and in vitro.[72] Apoptosis caused by CDV is typically induced via the extrinsic pathway, which activates caspases that disrupt cellular function and eventually leads to the cells death.[61] In normal cells, CDV activates caspase-8 first, which works as the initiator protein followed by the executioner protein caspase-3.[61] However, apoptosis induced by CDV in HeLa cells does not involve the initiator protein caspase-8. HeLa cell apoptosis caused by CDV follows a different mechanism than that in vero cell lines.[61] This change in the caspase cascade suggests CDV induces apoptosis via the intrinsic pathway, excluding the need for the initiator caspase-8. The executioner protein is instead activated by the internal stimuli caused by viral infection not a caspase cascade.[61]

The Oropouche virus (OROV) is found in the family *Bunyaviridae*. The study of apoptosis brought on by *Bunyaviridae* was initiated in 1996, when it was observed that apoptosis was induced by the La Crosse virus into the kidney cells of baby hamsters and into the brains of baby mice.[73]

OROV is a disease that is transmitted between humans by the biting midge (*Culicoides paraensis*).[74] It is referred to as a zoonotic arbovirus and causes febrile illness, characterized by the onset of a sudden fever known as Oropouche fever.[75]

The Oropouche virus also causes disruption in cultured cells – cells that are cultivated in distinct and specific conditions. An example of this can be seen in HeLa cells, whereby the cells begin to degenerate shortly after they are infected.[73]

With the use of gel electrophoresis, it can be observed that OROV causes DNA fragmentation in HeLa cells. It can be interpreted by counting, measuring, and analyzing the cells of the Sub/G1 cell population.[73] When HeLA cells are infected with OROV, the cytochrome C is released from the membrane of the mitochondria, into the cytosol of the cells. This type of interaction shows that apoptosis is activated via an intrinsic pathway.[70]

In order for apoptosis to occur within OROV, viral uncoating, viral internalization, along with the replication of cells is necessary. Apoptosis in some viruses is activated by extracellular stimuli. However, studies have demonstrated that the OROV infection causes apoptosis to be activated through intracellular stimuli and involves the mitochondria.[73]

Many viruses encode proteins that can inhibit apoptosis.[76] Several viruses encode viral homologs of Bcl-2. These homologs can inhibit proapoptotic proteins such as BAX and BAK, which are essential for the activation of apoptosis. Examples of viral Bcl-2 proteins include the Epstein-Barr virus BHRF1 protein and the adenovirus E1B 19K protein.[77] Some viruses express caspase inhibitors that inhibit caspase activity and an example is the CrmA protein of cowpox viruses. Whilst a number of viruses can block the effects of TNF and Fas. For example, the M-T2 protein of myxoma viruses can bind TNF preventing it from binding the TNF receptor and inducing a response.[78] Furthermore, many viruses express p53 inhibitors that can bind p53 and inhibit its transcriptional transactivation activity. As a consequence, p53 cannot induce apoptosis, since it cannot induce the expression of proapoptotic proteins. The adenovirus E1B-55K protein and the hepatitis B virus HBx protein are examples of viral proteins that can perform such a function.[79]

Viruses can remain intact from apoptosis in particular in the latter stages of infection. They can be exported in the *apoptotic bodies* that pinch off from the surface of the dying cell, and the fact that they are engulfed by phagocytes prevents the initiation of a host response. This favours the spread of the virus.[78]

## 7.7 Plants

Programmed cell death in plants has a number of molecular similarities to that of animal apoptosis, but it also has differences, notable ones being the presence of a cell wall and the lack of an immune system that removes the pieces of the dead cell. Instead of an immune response, the dying cell synthesizes substances to break itself down and places them in a vacuole that ruptures as the cell dies. Whether this whole process resembles animal apoptosis closely enough to warrant using the name *apoptosis* (as opposed to the more general *programmed cell death*) is unclear.[80]

## 7.8 Caspase-independent apoptosis

The characterization of the caspases allowed the development of caspase inhibitors, which can be used to determine whether a cellular process involves active caspases. Using these inhibitors it was discovered that cells can die while displaying a morphology similar to apoptosis without caspase activation.[81] Later studies linked this phenomenon to the release of AIF (apoptosis-inducing factor) from the mitochondria and its translocation into the nucleus mediated by its NLS (nuclear localization signal). Inside the mitochondria, AIF is anchored to the inner membrane. In order to be released, the protein is cleaved by a calcium-dependent calpain protease.

## 7.9 Apoptosis protein subcellular location prediction

In 2003, a method was developed for predicting subcellular location of apoptosis proteins.[82] Subsequent to this, various modes of Chou's pseudo amino acid composition were developed for improving the quality of predicting subcellular localization of apoptosis proteins based on their sequence information alone.[83][84][85][86]

## 7.10 See also

- Anoikis
- Apaf-1
- Apo2.7
- Apoptotic DNA fragmentation
- Atromentin induces apoptosis in human leukemia U937 cells.[87]
- Autolysis
- Autophagy
- Cisplatin
- Cytotoxicity
- Entosis
- Immunology
- Necrobiosis
- Necrosis
- Necrotaxis
- p53
- Paraptosis
- Pseudoapoptosis
- PI3K/AKT/mTOR pathway

## 7.11 Footnotes

[1] Note that the average human adult has more than 13 trillion cells ($10^{13}$),[1] of which at most only 70 billion ($7.0 \times 10^{10}$) die per day. That is, about 5 out of every 1,000 cells (0.5%) die each day due to apoptosis.

[2] HeLa cells were cells taken from a cancer patent; it is an immortal cell line.

## 7.12 References

[1] "Quantitative phase contrast microscopy — label-free live cell imaging and quantification". Phase Holographic Imaging AB.

[2] Green, Douglas (2011). *Means to an End: Apoptosis and other Cell Death Mechanisms.* Cold Spring Harbor, NY: Cold Spring Harbor Laboratory Press. ISBN 978-0-87969-888-1.

[3] Alberts, p. 2.

[4] Karam, Jose A. (2009). *Apoptosis in Carcinogenesis and Chemotherapy.* Netherlands: Springer. ISBN 978-1-4020-9597-9.

[5] Alberts, Bruce; Johnson, Alexander; Lewis, Julian; Raff, Martin; Roberts, Keith; Walter, Peter (2008). "Chapter 18 Apoptosis: Programmed Cell Death Eliminates Unwanted Cells". *Molecular Biology of the Cell (textbook)* (5th ed.). Garland Science. p. 1115. ISBN 978-0-8153-4105-5.

[6] Kerr, JF. (1965). "A histochemical study of hypertrophy and ischaemic injury of rat liver with special reference to changes in lysosomes". *Journal of Pathology and Bacteriology.* **90** (2): 419–35. doi:10.1002/path.1700900210.

[7] Agency for Science, Technology and Research. "Prof Andrew H. Wyllie – Lecture Abstract". Archived from the original on 2007-11-13. Retrieved 2007-03-30.

[8] Kerr JF, Wyllie AH, Currie AR (August 1972). "Apoptosis: a basic biological phenomenon with wide-ranging implications in tissue kinetics". *Br. J. Cancer.* **26** (4): 239–57. doi:10.1038/bjc.1972.33. PMC 2008650. PMID 4561027.

[9] O'Rourke MG, Ellem KA (2000). "John Kerr and apoptosis". *Med. J. Aust.* **173** (11–12): 616–7. PMID 11379508.

[10] Alberts, p. 1021.

[11] American Heritage Dictionary Archived June 30, 2008, at the Wayback Machine.

[12] Apoptosis Interest Group (1999). "About apoptosis". Archived from the original on 28 December 2006. Retrieved 2006-12-15.

[13] Webster.com dictionary entry

[14] Alberts, p. 1029.

[15] Böhm I, Schild H (2003). "Apoptosis: the complex scenario for a silent cell death". *Mol Imaging Biol.* **5** (1): 2–14. doi:10.1016/S1536-1632(03)00024-6. PMID 14499155.

[16] Alberts, p. 1023.

[17] Alberts, p. 1032.

[18] Alberts, p. 1024.

[19] Cotran, Ramzi, S.; Kumar, Collins (1998). *Robbins Pathologic Basis of Disease*. Philadelphia: W.B Saunders Company. ISBN 0-7216-7335-X.

[20] Mattson MP, Chan SL (2003). "Calcium orchestrates apoptosis". *Nature Cell Biology*. **5** (12): 1041–1043. doi:10.1038/ncb1203-1041. PMID 14647298.

[21] Uguz AC, Naziroglu M, Espino J, Bejarano I, González D, Rodríguez AB, Pariente JA (November 2009). "Selenium Modulates Oxidative Stress-Induced Cell Apoptosis in Human Myeloid HL-60 Cells Through Regulationof Calcium Release and Caspase-3 and −9 Activities". J Membrane Biol **232**: 15-23. doi:[10.1007/s00232-009-9212-2].

[22] Chiarugi A, Moskowitz MA (2002). "PARP-1—a perpetrator of apoptotic cell death?". *Science*. **297** (5579): 259–63. doi:10.1126/science.1074592. PMID 12114611.

[23] Bejarano I, Espino J, González-Flores D, Casado JG, Redondo PC, Rosado JA, Barriga C, Pariente JA, Rodríguez AB (2009). "Role of Calcium Signals on Hydrogen Peroxide-Induced Apoptosis in Human Myeloid HL-60 Cells". *International Journal of Biomedical Science*. **5** (3): 246–256. PMC 3614781⊙. PMID 23675144.

[24] D. González, I. Bejarano, C. Barriga, A.B. Rodríguez, J.A. Pariente (2010). "Oxidative Stress-Induced Caspases are Regulated in Human Myeloid HL-60 Cells by Calcium Signal". Current Signal Transduction Therapy **5**: 181-186. doi:[10.2174/157436210791112172]

[25] Mohan S, Bustamam A, Abdelwahab SI, Al-Zubairi AS, Aspollah M, Abdullah R, Elhassan MM, Ibrahim MY, Syam S: Typhonium flagelliforme induces apoptosis in CEMss cells via activation of caspase-9, PARP cleavage and cytochrome c release: Its activation coupled with G0/G1 phase cell cycle arrest. Journal of Ethnopharmacology 2010

[26] Brüne B (August 2003). "Nitric oxide: NO apoptosis or turning it ON?". *Cell Death Differ.* **10** (8): 864–9. doi:10.1038/sj.cdd.4401261. PMID 12867993.

[27] Brune B, et al. (1999). "Nitric oxide (NO): an effector of apoptosis". *Cell Death and Differentiation*. **6** (10): 969–975. doi:10.1038/sj.cdd.4400582.

[28] Fesik SW, Shi Y (2001). "Controlling the caspases". *Science*. **294** (5546): 1477–8. doi:10.1126/science.1062236. PMID 11711663.

[29] Laurent M. Dejean; Sonia Martinez-Caballero; Kathleen W. Kinnally (2006). "Is MAC the knife that cuts cytochrome c from mitochondria during apoptosis?". *Cell Death and Differentiation*. **13** (8): 1387–5. doi:10.1038/sj.cdd.4401949. PMID 16676005.

[30] Dejean LM, Martinez-Caballero S, Manon S, Kinnally KW (February 2006). "Regulation of the mitochondrial apoptosis-induced channel, MAC, by BCL-2 family proteins". *Biochim. Biophys. Acta*. **1762** (2): 191–201. doi:10.1016/j.bbadis.2005.07.002. PMID 16055309.

[31] Lodish, Harvey; et al. (2004). *Molecular Cell Biology*. New York: W.H. Freedman and Company. ISBN 0-7167-4366-3.

[32] Wajant H (2002). "The Fas signaling pathway: more than a paradigm". *Science*. **296** (5573): 1635–6. doi:10.1126/science.1071553. PMID 12040174.

[33] Chen G, Goeddel DV (2002). "TNF-R1 signaling: a beautiful pathway". *Science*. **296** (5573): 1634–5. doi:10.1126/science.1071924. PMID 12040173.

[34] Goeddel, DV. "Connection Map for Tumor Necrosis Factor Pathway". *Science STKE*. **2007**: tw132. doi:10.1126/stke.3822007tw132.

[35] Chen W, Li N, Chen T, Han Y, Li C, Wang Y, et al. (2005). "The lysosome-associated apoptosis-inducing protein containing the pleckstrin homology (PH) and FYVE domains (LAPF), representative of a novel family of PH and FYVE domain-containing proteins, induces caspase-independent apoptosis via the lysosomal-mitochondrial pathway.". *Journal of Biological Chemistry*. **280** (49): 40985–40995. doi:10.1074/jbc.M502190200. PMID 16188880.

[36] Wajant, H. "Connection Map for Fas Signaling Pathway". *Science STKE*. **2007**: tr1. doi:10.1126/stke.3802007tr1.

[37] Murphy KM, Ranganathan V, Farnsworth ML, Kavallaris M, Lock RB (January 2000). "Bcl-2 inhibits Bax translocation from cytosol to mitochondria during drug-induced apoptosis of human tumor cells". *Cell Death Differ.* **7** (1): 102–11. doi:10.1038/sj.cdd.4400597. PMID 10713725.

[38] Susin SA, Lorenzo HK, Zamzami N (February 1999). "Molecular characterization of mitochondrial apoptosis-inducing factor". *Nature*. **397** (6718): 441–6. doi:10.1038/17135. PMID 9989411.

[39] Jewhurst K, Levin M, McLaughlin KA (2014). "Optogenetic Control of Apoptosis in Targeted Tissues of Xenopus laevis Embryos". *J Cell Death*. **7**: 25–31. doi:10.4137/JCD.S18368. PMC 4213186⊙. PMID 25374461.

[40] Venturi, Sebastiano (2011). "Evolutionary Significance of Iodine". *Current Chemical Biology-*. **5** (3): 155–162. doi:10.2174/187231311796765012. ISSN 1872-3136.

[41] Venturi, Sebastiano (2014). "Iodine, PUFAs and Iodolipids in Health and Disease: An Evolutionary Perspective". *Human Evolution-*. 29 (1-3): 185–205. ISSN 0393-9375.

[42] Tamura K, Takayama S, Ishii T, Mawaribuchi S, Takamatsu N, Ito M (2015). "Apoptosis and differentiation of Xenopus tail-derived myoblasts by thyroid hormone.". *J Mol Endocrinol*. **54** (3): 185–92. doi:10.1530/JME-14-0327. PMID 25791374.

[43] Thomas, M. P.; Liu, X; Whangbo, J; McCrossan, G; Sanborn, K. B.; Basar, E; Walch, M; Lieberman, J (2015). "Apoptosis Triggers Specific, Rapid, and Global mRNA Decay with 3' Uridylated Intermediates Degraded by DIS3L2". *Cell Reports*. **11**: 1079–89. doi:10.1016/j.celrep.2015.04.026. PMID 25959823.

[44] Böhm I (2003). "Disruption of the cytoskeleton after apoptosis induction by autoantibodies". *Autoimmunity*. **36**: 183–189. doi:10.1080/0891693031000105617.

[45] Susin, S; Daugas, E; Ravagnan, L; Samejima, K; Zamzami, N; Loeffler, M; Costantini, P; Ferri, KF; et al. (2000). "Two Distinct Pathways Leading to Nuclear Apoptosis". *Journal of Experimental Medicine*. **192** (4): 571–80. doi:10.1084/jem.192.4.571. PMC 2193229. PMID 10952727.

[46] Madeleine Kihlmark; Imreh, G; Hallberg, E (15 October 2001). "Sequential degradation of proteins from the nuclear envelope during apoptosis". *Journal of Cell Science*. **114** (20): 3643–53. PMID 11707516.

[47] Nagata S (April 2000). "Apoptotic DNA fragmentation". *Exp. Cell Res*. **256** (1): 12–8. doi:10.1006/excr.2000.4834. PMID 10739646.

[48] M Iwata; D Myerson; B Torok-Storb; RA Zager (1996). "An evaluation of renal tubular DNA laddering in response to oxygen deprivation and oxidant injury". Retrieved 2006-04-17.

[49] Vandivier RW, Henson PM, Douglas IS (June 2006). "Burying the dead: the impact of failed apoptotic cell removal (efferocytosis) on chronic inflammatory lung disease". *Chest*. **129** (6): 1673–82. doi:10.1378/chest.129.6.1673. PMID 16778289.

[50] Li MO; Sarkisian, MR; Mehal, WZ; Rakic, P; Flavell, RA (2003). "Phosphatidylserine receptor is required for clearance of apoptotic cells". *Science*. **302** (5650): 1560–3. doi:10.1126/science.1087621. PMID 14645847.

[51] Wang X; Wu, YC; Fadok, VA; Lee, MC; Gengyo-Ando, K; Cheng, LC; Ledwich, D; Hsu, PK; et al. (2003). "Cell corpse engulfment mediated by C. elegans phosphatidylserine receptor through CED-5 and CED-12". *Science*. **302** (5650): 1563–1566. doi:10.1126/science.1087641. PMID 14645848.

[52] Savill J, Gregory C, Haslett C. (2003). "Eat me or die". *Science*. **302** (5650): 1516–7. doi:10.1126/science.1092533. PMID 14645835.

[53] Krysko DV; Vandenabeele P. *Phagocytosis of dying cells: from molecular mechanisms to human diseases*. ISBN 978-1-4020-9292-3.

[54] Lozano GM, Bejarano I, Espino J, González D, Ortiz A, García JF, Rodríguez AB, Pariente JA (2009). "Density gradient capacitation is the most suitable method to improve fertilization and to reduce DNA fragmentation positive spermatozoa of infertile men". *Anatolian Journal of Obstetrics & Gynecology*. **3** (1): 1–7.

[55] Krysko DV, Vanden Berghe T, Parthoens E, D'Herde K, Vandenabeele P (2008). "Methods for distinguishing apoptotic from necrotic cells and measuring their clearance.". *Methods Enzymol*. **442**: 307–41. doi:10.1016/S0076-6879(08)01416-X. PMID 18662577.

[56] Krysko DV, Vanden Berghe T, D'Herde K, Vandenabeele P (2008). "Apoptosis and necrosis: detection, discrimination and phagocytosis.". *Methods*. **44** (3): 205–21. doi:10.1016/j.ymeth.2007.12.001. PMID 18314051.

[57] Vanden Berghe T, Grootjans S, Goossens V, Dondelinger Y, Krysko DV, Takahashi N, Vandenabeele P (2013). "Determination of apoptotic and necrotic cell death in vitro and in vivo.". *Methods*. **61** (2): 117–29. doi:10.1016/j.ymeth.2013.02.011. PMID 23473780.

[58] Thompson, CB (1995). "Apoptosis in the pathogenesis and treatment of disease". *Science*. **267** (5203): 1456–62. doi:10.1126/science.7878464. PMID 7878464.

[59] Yang L, Mashima T, Sato S (February 2003). "Predominant suppression of apoptosome by inhibitor of apoptosis protein in non-small cell lung cancer H460 cells: therapeutic effect of a novel polyarginine-conjugated Smac peptide". *Cancer Res*. **63** (4): 831–7. PMID 12591734.

[60] Takaoka A; Hayakawa, S; Yanai, H; Stoiber, D; Negishi, H; Kikuchi, H; Sasaki, S; Imai, K; et al. (2003). "Integration of interferon-alpha/beta signalling to p53 responses in tumour suppression and antiviral defence". *Nature*. **424** (6948): 516–23. doi:10.1038/nature01850. PMID 12872134.

[61] Del Puerto HL, Martins AS, Milsted A, et al. (2011). "Canine distemper virus induces apoptosis in cervical tumor derived cell lines". *Virol. J*. **8** (1): 334. doi:10.1186/1743-422X-8-334. PMC 3141686. PMID 21718481.

[62] Liu HC, Chen GG, Vlantis AC, Tse GM, Chan AT, van Hasselt CA (March 2008). "Inhibition of apoptosis in human laryngeal cancer cells by E6 and E7 oncoproteins of human papillomavirus 16". *J. Cell. Biochem*. **103** (4): 1125–43. doi:10.1002/jcb.21490. PMID 17668439.

[63] Niu XY, Peng ZL, Duan WQ, Wang H, Wang P (2006). "Inhibition of HPV 16 E6 oncogene expression by RNA interference in vitro and in vivo". *Int. J. Gynecol. Cancer*. **16** (2): 743–51. doi:10.1111/j.1525-1438.2006.00384.x. PMID 16681755.

[64] Liu, Y; McKalip, A; Herman, B (2000). "Human Papillomavirus Type 16 E6 and HPV-16 E6/E7 Sensitize Human Keratinocytes to Apoptosis Induced by Chemotherapeutic Agents: Roles of p53 and Caspase Activation". *Journal of Cellular Biochemistry*. **78**: 334–349. doi:10.1002/(sici)1097-4644(20000801)78:2<334::aid-jcb15>3.3.co;2-6.

[65] Boehm, I. (2006). "Apoptosis in physiological and pathological skin: Implications for therapy". *Current Molecular Medicine*. **6** (4): 375–394. doi:10.2174/156652406777435390. PMID 16900661.

[66] Farhana, Lulu. "Apoptosis Induction by a Novel Retinoid-Related Molecule Requires Nuclear Factor- κB Activation" (PDF). American Association of Cancer Research (AACR). Retrieved 2012-01-24.

[67] Judie B. Alimonti; T. Blake Ball; Keith R. Fowke (2003). "Mechanisms of CD4+ T lymphocyte cell death in human immunodeficiency virus infection and AIDS". *J. Gen. Virology*. **84** (7): 1649–61. doi:10.1099/vir.0.19110-0. PMID 12810858.

[68] Vashistha H, Husain M, Kumar D, Yadav A, Arora S, Singhal PC. (2008) Ren Fail. 2008;30(6):655-64.

[69] Indiana University Health. "AIDS Defining Criteria | Riley". IU Health. Retrieved 2013-01-20.

[70] Indran IR, Tufo G, Pervaiz S, Brenner C (June 2011). "Recent advances in apoptosis, mitochondria and drug resistance in cancer cells". *Biochim. Biophys. Acta*. **1807** (6): 735–45. doi:10.1016/j.bbabio.2011.03.010. PMID 21453675.

[71] Everett, H.; McFadden, G. (1999). "Apoptosis: an innate immune response to virus infection". *Trends Microbiol*. **7** (4): 160–5. doi:10.1016/S0966-842X(99)01487-0. PMID 10217831.

[72] Nishi T, Tsukiyama-Kohara K, Togashi K, Kohriyama N, Kai C (November 2004). "Involvement of apoptosis in syncytial cell death induced by canine distemper virus". *Comp. Immunol. Microbiol. Infect. Dis*. **27** (6): 445–55. doi:10.1016/j.cimid.2004.01.007. PMID 15325517.

[73] Acrani GO, Gomes R, Proença-Módena JL, et al. (April 2010). "Apoptosis induced by Oropouche virus infection in HeLa cells is dependent on virus protein expression". *Virus Res*. **149** (1): 56–63. doi:10.1016/j.virusres.2009.12.013. PMID 20080135.

[74] Azevedo RS, Nunes MR, Chiang JO, et al. (June 2007). "Reemergence of Oropouche fever, northern Brazil". *Emerging Infect. Dis*. **13** (6): 912–5. doi:10.3201/eid1306.061114. PMC 2792853. PMID 17553235.

[75] Santos RI, Rodrigues AH, Silva ML, et al. (December 2008). "Oropouche virus entry into HeLa cells involves clathrin and requires endosomal acidification". *Virus Res*. **138** (1–2): 139–43. doi:10.1016/j.virusres.2008.08.016. PMID 18840482.

[76] Teodoro, J.G. Branton, P.E. (1997). "Regulation of apoptosis by viral gene products". *J. Virol*. **71** (3): 1739–46. PMC 191242. PMID 9032302.

[77] Polster, B.M. Pevsner, J. and Hardwick, J.M. (2004). "Viral Bcl-2 homologs and their role in virus replication and associated diseases". *Biochim. Biophys. Acta*. **1644** (2–3): 211–27. doi:10.1016/j.bbamcr.2003.11.001. PMID 14996505.

[78] Hay, S.; Kannourakis, G. (2002). "A time to kill: viral manipulation of the cell death program". *J. Gen. Virol*. **83** (Pt 7): 1547–64. CiteSeerX 10.1.1.322.6923. doi:10.1099/0022-1317-83-7-1547. PMID 12075073.

[79] Wang, X.W. Gibson, M.K. Vermeulen, W. Yeh, H. Forrester, K. Sturzbecher, H.W. Hoeijmakers, J.H. and Harris, C.C. (1995). "Abrogation of p53-induced Apoptosis by the Hepatitis B Virus X Gene". *Cancer Res*. **55** (24): 6012–6. PMID 8521383.

[80] Cyrelys Collazo; Osmani Chacón; Orlando Borrás (2006). "Programmed cell death in plants resembles apoptosis of animals" (PDF). *Biotecnología Aplicada*. **23**: 1–10.

[81] Xiang J. et al., BAX-induced cell death may not require interleukin 1β-converting enzyme-likeproteases, 1996, cell biology

[82] Zhou, G. P. & Doctor, K. (2003). Subcellular location prediction of apoptosis proteins. PROTEINS: Structure, Function, and Genetics 50, 44-48.

[83] Chen Y. L.; Li Q. Z. (2007). "Prediction of apoptosis protein subcellular location using improved hybrid approach and pseudo amino acid composition". *Journal of Theoretical Biology*. **248** (2): 377–381. doi:10.1016/j.jtbi.2007.05.019. PMID 17572445.

[84] Ding, Y. S. & Zhang, T. L. (2008). Using Chou's pseudo amino acid composition to predict subcellular localization of apoptosis proteins: an approach with immune genetic algorithm-based ensemble classifier. Pattern Recognition Letters 29, 1887-1892.

[85] Jiang, X., Wei, R., Zhang, T. L. & Gu, Q. (2008). Using the concept of Chou's pseudo amino acid composition to predict apoptosis proteins subcellular location: an approach by approximate entropy. Protein & Peptide Letters 15, 392-396.

[86] Lin H, Wang H, Ding H, Chen YL, Li QZ (2009). "Prediction of Subcellular Localization of Apoptosis Protein Using Chou's Pseudo Amino Acid Composition". *Acta Biotheoretica*. **57**: 321–330. doi:10.1007/s10441-008-9067-4. PMID 19169652.

[87] "Atromentin-Induced Apoptosis in Human Leukemia U937 Cells". Kim Jin Hee and Choong Hwan Lee, *Journal of Microbiology and Biotechnology*, 2009, vol. 19, no. 9, pages 946–950, INIST:21945937

## 7.13   Bibliography

- Alberts, Bruce; Johnson, Alexander; Lewis, Julian; Morgan, David; Raff, Martin; Roberts, Keith; Walter, Peter (2015). *Molecular Biology of the Cell* (6th ed.). Garland Science. p. 2. ISBN 978-0815344322.

## 7.14   External links

- Apoptosis & cell surface

- Apoptosis & Caspase 3, The Proteolysis Map—animation

- Apoptosis & Caspase 8, The Proteolysis Map—animation

- Apoptosis & Caspase 7, The Proteolysis Map—animation

- Apoptosis MiniCOPE Dictionary—list of apoptosis terms and acronyms

- Apoptosis (Programmed Cell Death) – The Virtual Library of Biochemistry and Cell Biology

- Apoptosis Research Portal

- Apoptosis Info Apoptosis protocols, articles, news, and recent publications.

- Database of proteins involved in apoptosis

- Apoptosis Video

- Apoptosis Video (WEHI on YouTube )

- The Mechanisms of Apoptosis Kimball's Biology Pages. Simple explanation of the mechanisms of apoptosis triggered by internal signals (bcl-2), along the caspase-9, caspase-3 and caspase-7 pathway; and by external signals (FAS and TNF), along the caspase 8 pathway. Accessed 25 March 2007.

- WikiPathways – Apoptosis pathway

- Finding Cancer's Self-Destruct Button CR magazine (Spring 2007). Article on apoptosis and cancer.

- Xiaodong Wang's lecture: Introduction to Apoptosis

- Robert Horvitz's short clip: Discovering Programmed Cell Death

- The Bcl-2 Database

- DeathBase: a database of proteins involved in cell death, curated by experts

- European Cell Death Organization

# Chapter 8

# Programmed cell death

For the protein, see Programmed cell death protein 1.

**Programmed cell death** (or **PCD**) is the death of a cell in any form, mediated by an intracellular program.[1][2] PCD is carried out in a biological process, which usually confers advantage during an organism's life-cycle. For example, the differentiation of fingers and toes in a developing human embryo occurs because cells between the fingers apoptose; the result is that the digits are separate. PCD serves fundamental functions during both plant and animal tissue development. Apoptosis and autophagy are both forms of programmed cell death, but necrosis was long seen as a non-physiological process that occurs as a result of infection or injury.[3]

Necrosis is the death of a cell caused by external factors such as trauma or infection and occurs in several different forms. Recently a form of programmed necrosis, called necroptosis,[4] has been recognized as an alternate form of programmed cell death. It is hypothesized that necroptosis can serve as a cell-death backup to apoptosis when the apoptosis signaling is blocked by endogenous or exogenous factors such as viruses or mutations. Most recently, other types of regulated necrosis have been discovered as well, which share several signaling events with necroptosis and apoptosis.[5]

## 8.1  History

The concept of "programmed cell-death" was used by Lockshin & Williams[6] in 1964 in relation to insect tissue development, around eight years before "apoptosis" was coined. Since then, PCD has become the more general of these two terms.

The first insight into the mechanism came from studying BCL2, the product of a putative oncogene activated by chromosome translocations often found in follicular lymphoma. Unlike other cancer genes, which promote cancer by stimulating cell proliferation, BCL2 promoted

cancer by stopping lymphoma cells from being able to kill themselves.[7]

PCD has been the subject of increasing attention and research efforts. This trend has been highlighted with the award of the 2002 Nobel Prize in Physiology or Medicine to Sydney Brenner (United Kingdom), H. Robert Horvitz (US) and John E. Sulston (UK).[8]

## 8.2  Types

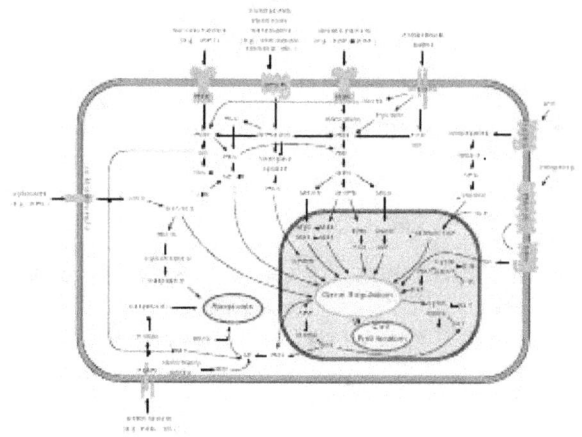

*Overview of signal transduction pathways involved in apoptosis.*

- Apoptosis or Type I cell-death.

- Autophagic or Type II cell-death. (*Cytoplasmic*: characterized by the formation of large vacuoles that eat away organelles in a specific sequence prior to the destruction of the nucleus.)[9]

### 8.2.1  Apoptosis

Apoptosis is the process of programmed cell death (PCD) that may occur in multicellular organisms.[10] Biochemical events lead to characteristic cell changes (morphology) and death. These changes include blebbing, cell shrinkage, nuclear fragmentation, chromatin condensation, and chromosomal DNA fragmentation. It is now thought that-in a developmental context- cells are induced to positively commit suicide whilst in a homeostatic context; the absence of certain survival factors may provide the impetus for suicide. There appears to be some variation in the morphology and indeed the biochemistry of these suicide pathways; some treading the path of "apoptosis", others following a more generalized pathway to deletion, but both usually being genetically and synthetically motivated. There is some evidence that certain symptoms of "apoptosis" such as endonuclease activation can be spuriously induced without engaging a genetic cascade, however, presumably true apoptosis and programmed cell death must be genetically mediated. It is also becoming clear that mitosis and apoptosis are toggled or linked in some way and that the balance achieved depends on signals received from appropriate growth or survival factors.[11]

## 8.2.2  Autophagy

Macroautophagy, often referred to as autophagy, is a catabolic process that results in the autophagosomic-lysosomal degradation of bulk cytoplasmic contents, abnormal protein aggregates, and excess or damaged organelles.

Autophagy is generally activated by conditions of nutrient deprivation but has also been associated with physiological as well as pathological processes such as development, differentiation, neurodegenerative diseases, stress, infection and cancer.

### Mechanism

A critical regulator of autophagy induction is the kinase mTOR, which when activated, suppresses autophagy and when not activated promotes it. Three related serine/threonine kinases, UNC-51-like kinase $-1, -2$, and $-3$ (ULK1, ULK2, UKL3), which play a similar role as the yeast Atg1, act downstream of the mTOR complex. ULK1 and ULK2 form a large complex with the mammalian homolog of an autophagy-related (Atg) gene product (mAtg13) and the scaffold protein FIP200. Class III PI3K complex, containing hVps34, Beclin-1, p150 and Atg14-like protein or ultraviolet irradiation resistance-associated gene (UVRAG), is required for the induction of autophagy.

The ATG genes control the autophagosome formation through ATG12-ATG5 and LC3-II (ATG8-II) complexes. ATG12 is conjugated to ATG5 in a ubiquitin-like reaction that requires ATG7 and ATG10. The Atg12–Atg5 conjugate then interacts non-covalently with ATG16 to form

a large complex. LC3/ATG8 is cleaved at its C terminus by ATG4 protease to generate the cytosolic LC3-I. LC3-I is conjugated to phosphatidylethanolamine (PE) also in a ubiquitin-like reaction that requires Atg7 and Atg3. The lipidated form of LC3, known as LC3-II, is attached to the autophagosome membrane.

Autophagy and apoptosis are connected both positively and negatively, and extensive crosstalk exists between the two. During nutrient deficiency, autophagy functions as a pro-survival mechanism, however, excessive autophagy may lead to cell death, a process morphologically distinct from apoptosis. Several pro-apoptotic signals, such as TNF, TRAIL, and FADD, also induce autophagy. Additionally, Bcl-2 inhibits Beclin-1-dependent autophagy, thereby functioning both as a pro-survival and as an anti-autophagic regulator.

## 8.2.3  Other types

Besides the above two types of PCD, other pathways have been discovered.[12] Called "non-apoptotic programmed cell-death" (or "caspase-independent programmed cell-death" or "necroptosis"), these alternative routes to death are as efficient as apoptosis and can function as either backup mechanisms or the main type of PCD.

Other forms of programmed cell death include anoikis, almost identical to apoptosis except in its induction; cornification, a form of cell death exclusive to the eyes; excitotoxicity; ferroptosis, an iron-dependent form of cell death[13] and Wallerian degeneration.

Necroptosis is a programmed form of necrosis, or inflammatory cell death. Conventionally, necrosis is associated with unprogrammed cell death resulting from cellular damage or infiltration by pathogens, in contrast to orderly, programmed cell death via apoptosis.

Eryptosis is a form of suicidal erythrocyte death.[14]

Aponecrosis is a hybrid of apoptosis and necrosis and refers to an incomplete apoptotic process that is completed by necrosis.[15]

NETosis is the process of cell-death generated by NETs.[16]

Plant cells undergo particular processes of PCD similar to autophagic cell death. However, some common features of PCD are highly conserved in both plants and metazoa.

## 8.3  Atrophic factors

An atrophic factor is a force that causes a cell to die. Only natural forces on the cell are considered to be atrophic factors, whereas, for example, agents of mechanical or chemical abuse or lysis of the cell are considered not to be atrophic factors. Common types of atrophic factors are:[17]

1. Decreased workload

2. Loss of innervation

3. Diminished blood supply

4. Inadequate nutrition

5. Loss of endocrine stimulation

6. Senility

7. Compression

## 8.4 Role in the development of the nervous system

The initial expansion of the developing nervous system is counterbalanced by the removal of neurons and their processes.[18] During the development of the nervous system almost 50% of developing neurons are naturally removed by programmed cell death (PCD).[19] PCD in the nervous system was first recognized in 1896 by John Beard.[20] Since then several theories were proposed to understand its biological significance during neural development.[21]

### 8.4.1 Role in neural development

PCD in the developing nervous system has been observed in proliferating as well as post-mitotic cells.[18] One theory suggests that PCD is an adaptive mechanism to regulate the number of progenitor cells. In humans, PCD in progenitor cells starts at gestational week 7 and remains until the first trimester.[22] This process of cell death has been identified in the germinal areas of the cerebral cortex, cerebellum, thalamus, brainstem, and spinal cord among other regions.[21] At gestational weeks 19-23, PCD is observed in post-mitotic cells.[23] The prevailing theory explaining this observation is the neurotrophic theory which states that PCD is required to optimize the connection between neurons and their afferent inputs and efferent targets.[21] Another theory proposes that developmental PCD in the nervous system occurs in order to correct for errors in neurons that have migrated ectopically, innervated incorrect targets, or have axons that have gone awry during path finding.[24] It is possible that PCD during the development of the nervous system serves different functions determined by the developmental stage, cell type, and even species.[21]

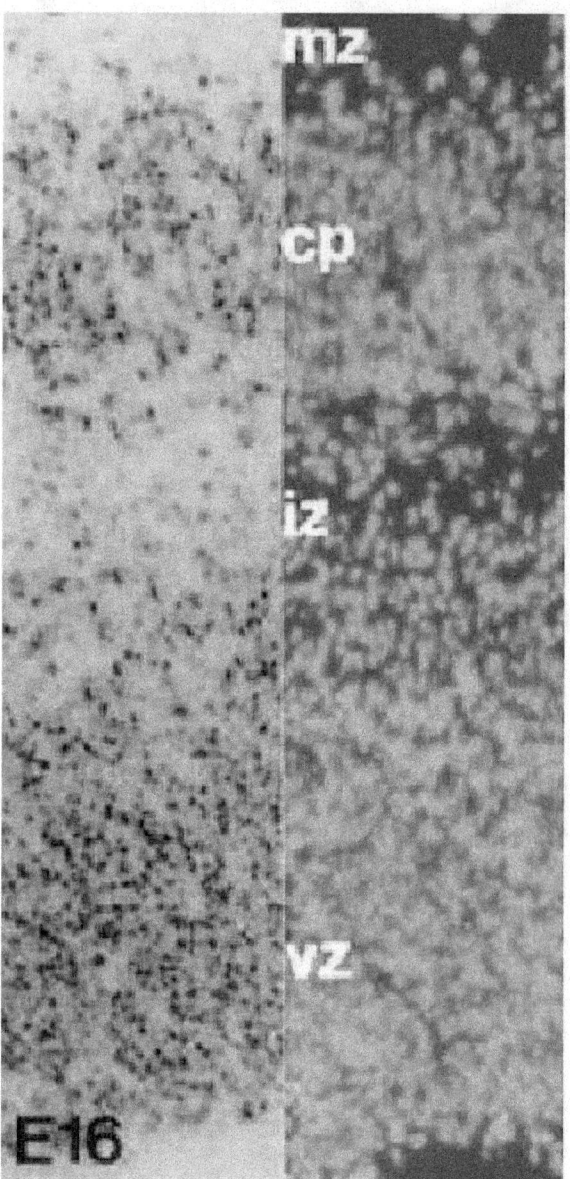

*Dying cells in the proliferate zone*

### 8.4.2 The neurotrophic theory

The neurotrophic theory is the leading hypothesis used to explain the role of programmed cell death in the developing nervous system. It postulates that in order to ensure optimal innervation of targets, a surplus of neurons is first produced which then compete for limited quantities of protective neurotrophic factors and only a fraction survive while others die by programmed cell death.[22] Furthermore, the theory states that predetermined factors regulate the amount of neurons that survive and the size of the innervating neuronal population directly correlates to the influence of their target field.[25]

The underlying idea that target cells secrete attractive or inducing factors and that their growth cones have a chemotactic sensitivity was first put forth by Santiago Ramon y Cajal in 1892.[26] Cajal presented the idea as an explanation for the "intelligent force" axons appear to take when finding their target but admitted that he had no empirical data.[26] The theory gained more attraction when experimental manipulation of axon targets yielded death of all innervating neurons. This developed the concept of target derived regulation which became the main tenet in the neurotrophic theory.[27][28] Experiments that further supported this theory led to the identification of the first neurotrophic factor, nerve growth factor (NGF).[29]

### 8.4.3   Peripheral versus central nervous system

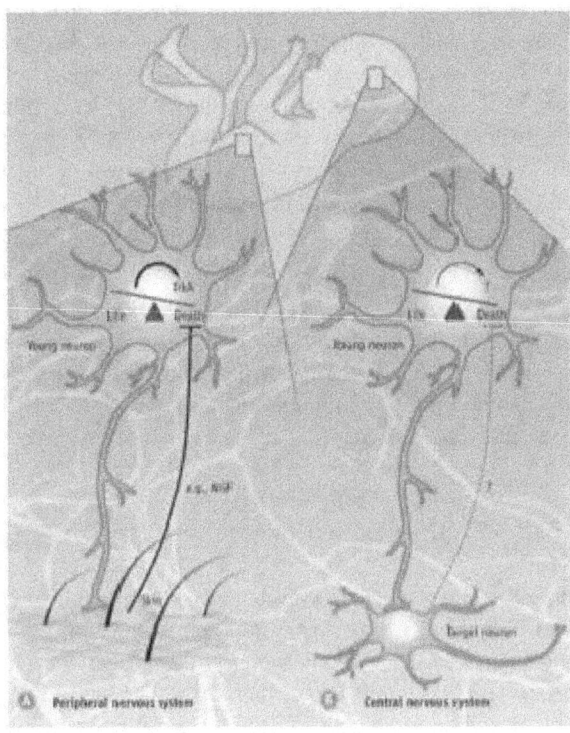

*Cell death in the peripheral vs central nervous system*

Different mechanisms regulate PCD in the peripheral nervous system (PNS) versus the central nervous system (CNS). In the PNS, innervation of the target is proportional to the amount of the target-released neurotrophic factors NGF and NT3.[30][31] Expression of neurotrophin receptors, TrkA and TrkC, is sufficient to induce apoptosis in the absence of their ligands.[19] Therefore, it is speculated that PCD in the PNS is dependent on the release of neurotrophic factors and thus follows the concept of the neurotrophic theory.

Programmed cell death in the CNS is not dependent on external growth factors but instead relies on intrinsically derived cues. In the neocortex, a 4:1 ratio of excitatory to inhibitory interneurons is maintained by apoptotic machinery that appears to be independent of the environment.[31] Supporting evidence came from an experiment where interneuron progenitors were either transplanted into the mouse neocortex or cultured in vitro.[32] Transplanted cells died at the age of two weeks, the same age at which endogenous interneurons undergo apoptosis. Regardless of the size of the transplant, the fraction of cells undergoing apoptosis remained constant. Furthermore, disruption of TrkB, a receptor for brain derived neurotrophic factor (Bdnf), did not affect cell death. It has also been shown that in mice null for the proapoptotic factor Bax (Bcl-2-associated X protein) a larger percentage of interneurons survived compared to wild type mice.[32] Together these findings indicate that programmed cell death in the CNS partly exploits Bax-mediated signaling and is independent of BDNF and the environment. Apoptotic mechanisms in the CNS are still not well understood, yet it is thought that apoptosis of interneurons is a self-autonomous process.

### 8.4.4   Nervous system development in its absence

Programmed cell death can be reduced or eliminated in the developing nervous system by the targeted deletion of pro-apoptotic genes or by the overexpression of anti-apoptotic genes. The absence or reduction of PCD can cause serious anatomical malformations but can also result in minimal consequences depending on the gene targeted, neuronal population, and stage of development.[21] Excess progenitor cell proliferation that leads to gross brain abnormalities is often lethal, as seen in caspase-3 or caspase-9 knockout mice which develop exencephaly in the forebrain.[33][34] The brainstem, spinal cord, and peripheral ganglia of these mice develop normally, however, suggesting that the involvement of caspases in PCD during development depends on the brain region and cell type.[35] Knockout or inhibition of apoptotic protease activating factor 1 (APAF1), also results in malformations and increased embryonic lethality.[36][37][38] Manipulation of apoptosis regulator proteins Bcl-2 and Bax (overexpression of Bcl-2 or deletion of Bax) produces an increase in the number of neurons in certain regions of the nervous system such as the retina, trigeminal nucleus, cerebellum, and spinal cord.[39][40][41][42][43][44][45] However, PCD of neurons due to Bax deletion or Bcl-2 overexpression does not result in prominent morphological or behavioral abnormalities in mice. For example, mice overexpressing Bcl-2 have generally normal motor skills and vision and only show impairment in complex behaviors such as learning and anxiety.[46][47][48] The normal behavioral phenotypes of these mice suggest that an adaptive mechanism may be involved to compensate for the excess neurons.[21]

### 8.4.5 Invertebrates and vertebrates

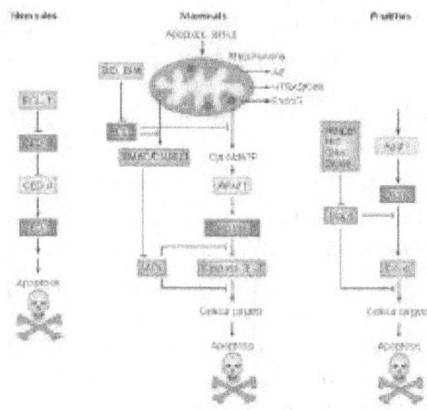

*A conserved apoptotic pathway in nematodes, mammals and fruit-flies*

Learning about PCD in various species is essential in understanding the evolutionary basis and reason for apoptosis in development of the nervous system. During the development of the invertebrate nervous system, PCD plays different roles in different species. The similarity of the asymmetric cell death mechanism in the nematode and the leech indicates that PCD may have an evolutionary significance in the development of the nervous system.[49] In the nematode, PCD occurs in the first hour of development leading to the elimination of 12% of non-gonadal cells including neuronal lineages.[50] Cell death in arthropods occurs first in the nervous system when ectoderm cells differentiate and one daughter cell becomes a neuroblast and the other undergoes apoptosis.[51] Furthermore, sex targeted cell death leads to different neuronal innervation of specific organs in males and females.[52] In Drosophila, PCD is essential in segmentation and specification during development.

In contrast to invertebrates, the mechanism of programmed cell death is found to be more conserved in vertebrates. Extensive studies performed on various vertebrates show that PCD of neurons and glia occurs in most parts of the nervous system during development. It has been observed before and during synaptogenesis in the central nervous system as well as the peripheral nervous system.[21] However, there are a few differences between vertebrate species. For example, mammals exhibit extensive arborization followed by PCD in the retina while birds do not.[53] Although synaptic refinement in vertebrate systems is largely dependent on PCD, other evolutionary mechanisms also play a role.[21]

## 8.5 In plant tissue

Programmed cell death in plants has a number of molecular similarities to animal apoptosis, but it also has differences, the most obvious being the presence of a cell wall and the lack of an immune system that removes the pieces of the dead cell. Instead of an immune response, the dying cell synthesizes substances to break itself down and places them in a vacuole that ruptures as the cell dies.[54]

In "APL regulates vascular tissue identity in Arabidopsis",[55] Martin Bonke and his colleagues had stated that one of the two long-distance transport systems in vascular plants, xylem, consists of several cell-types "the differentiation of which involves deposition of elaborate cell-wall thickenings and programmed cell-death." The authors emphasize that the products of plant PCD play an important structural role.

Basic morphological and biochemical features of PCD have been conserved in both plant and animal kingdoms.[56] It should be noted, however, that specific types of plant cells carry out unique cell-death programs. These have common features with animal apoptosis—for instance, nuclear DNA degradation—but they also have their own peculiarities, such as nuclear degradation triggered by the collapse of the vacuole in tracheary elements of the xylem.[57]

Janneke Balk and Christopher J. Leaver, of the Department of Plant Sciences, University of Oxford, carried out research on mutations in the mitochondrial genome of sun-flower cells. Results of this research suggest that mitochondria play the same key role in vascular plant PCD as in other eukaryotic cells.[58]

### 8.5.1 PCD in pollen prevents inbreeding

During pollination, plants enforce self-incompatibility (**SI**) as an important means to prevent self-fertilization. Research on the corn poppy (*Papaver rhoeas*) has revealed that proteins in the pistil on which the pollen lands, interact with pollen and trigger PCD in incompatible (i.e., *self*) pollen. The researchers, Steven G. Thomas and Veronica E. Franklin-Tong, also found that the response involves rapid inhibition of pollen-tube growth, followed by PCD.[59]

## 8.6 In slime molds

The social slime mold *Dictyostelium discoideum* has the peculiarity of either adopting a predatory amoeba-like behavior in its unicellular form or coalescing into a mobile slug-like form when dispersing the spores that will give birth to the next generation.[60]

The stalk is composed of dead cells that have undergone a type of PCD that shares many features of an autophagic cell-death: massive vacuoles forming inside cells, a degree of chromatin condensation, but no DNA fragmentation.[61] The structural role of the residues left by the dead cells is reminiscent of the products of PCD in plant tissue.

*D. discoideum* is a slime mold, part of a branch that might have emerged from eukaryotic ancestors about a billion years before the present. It seems that they emerged after the ancestors of green plants and the ancestors of fungi and animals had differentiated. But, in addition to their place in the evolutionary tree, the fact that PCD has been observed in the humble, simple, six-chromosome *D. discoideum* has additional significance: It permits the study of a developmental PCD path that does not depend on caspases characteristic of apoptosis.[62]

## 8.7   Evolutionary origin

The occurrence of programmed cell death in protists is possible,[63] but it remains controversial. Some categorize death in those organisms as unregulated, by necrosis or incidental death.[64]

Biologists had long suspected that mitochondria originated from bacteria that had been incorporated as endosymbionts ("living together inside") of larger eukaryotic cells. It was Lynn Margulis who from 1967 on championed this theory, which has since become widely accepted.[65] The most convincing evidence for this theory is the fact that mitochondria possess their own DNA and are equipped with genes and replication apparatus.

This evolutionary step would have been risky for the primitive eukaryotic cells, which began to engulf the energy-producing bacteria, as well as a perilous step for the ancestors of mitochondria, which began to invade their proto-eukaryotic hosts. This process is still evident today, between human white blood cells and bacteria. Most of the time, invading bacteria are destroyed by the white blood cells; however, it is not uncommon for the chemical warfare waged by prokaryotes to succeed, with the consequence known as infection by its resulting damage.

One of these rare evolutionary events, about two billion years before the present, made it possible for certain eukaryotes and energy-producing prokaryotes to coexist and mutually benefit from their symbiosis.[66]

Mitochondriate eukaryotic cells live poised between life and death, because mitochondria still retain their repertoire of molecules that can trigger cell suicide.[67] This process has now been evolved to happen only when programmed. Given certain signals to cells (such as feedback from neighbors,

stress or DNA damage), mitochondria release caspase activators that trigger the cell-death-inducing biochemical cascade. As such, the cell suicide mechanism is now crucial to all of our lives.

## 8.8   Programmed death of entire organisms

Main article: Phenoptosis

## 8.9   Clinical significance

### 8.9.1   ABL

The BCR-ABL oncogene has been found to be involved in the development of cancer in humans.[68]

### 8.9.2   c-Myc

c-Myc is involved in the regulation of apoptosis via its role in downregulating the Bcl-2 gene. Its role the disordered growth of tissue.[68]

### 8.9.3   Metastasis

A molecular characteristic of metastatic cells is their altered expression of several apoptotic genes.[68]

## 8.10   See also

- Anoikis

- Apoptosis-inducing factor

- Apoptosis versus Pseudoapoptosis

- Apoptosome

- Apoptotic DNA fragmentation

- Autolysis (biology)

- Autophagy

- Autoschizis

- Bcl-2

- BH3 interacting domain death agonist (BID)

- Calpains
- Caspases
- Cell damage
- Cornification
- Cytochrome c
- Cytotoxicity
- Diablo homolog
- Entosis
- Excitotoxicity
- Inflammasome
- Mitochondrial permeability transition pore
- Mitotic catastrophe
- Necrobiology
- Necroptosis
- Necrosis
- p53 upregulated modulator of apoptosis (PUMA)
- Paraptosis
- Parthanatos
- Pyroptosis
- RIP kinases
- Wallerian degeneration

## 8.11 Notes and references

- Srivastava, R. E. in Molecular Mechanisms (Humana Press, 2007).
- Kierszenbaum, A. L. & Tres, L. L. (ed Madelene Hyde) (ELSEVIER SAUNDERS, Philadelphia, 2012).

[1] Engelberg-Kulka H, Amitai S, Kolodkin-Gal I, Hazan R (2006). "Bacterial Programmed Cell Death and Multicellular Behavior in Bacteria". *PLoS Genetics*. **2** (10): e135. doi:10.1371/journal.pgen.0020135. PMC 1626106. PMID 17069462.

[2] Green, Douglas (2011). *Means To An End*. New York: Cold Spring Harbor Laboratory Press. ISBN 978-0-87969-887-4.

[3] Kierszenbaum, Abraham (2012). *Histology and Cell Biology - An Introduction to Pathology*. Philadelphia: ELSEVIER SAUNDERS.

[4] Degterev, Alexei; Huang, Zhihong; Boyce, Michael; Li, Yaqiao; Jagtap, Prakash; Mizushima, Noboru; Cuny, Gregory D.; Mitchison, Timothy J.; Moskowitz, Michael A. (2005-07-01). "Chemical inhibitor of nonapoptotic cell death with therapeutic potential for ischemic brain injury". *Nature Chemical Biology*. **1** (2): 112–119. doi:10.1038/nchembio711. ISSN 1552-4450. PMID 16408008.

[5] Vanden Berghe T, Linkermann A, Jouan-Lanhouet S, Walczak H, Vandenabeele P (2014). "Regulated necrosis: the expanding network of non-apoptotic cell death pathways". *Nat Rev Mol Cell Biol*. **15** (2): 135–147. doi:10.1038/nrm3737. PMID 24452471.

[6] Lockshin RA, Williams CM (1964). "Programmed cell death—II. Endocrine potentiation of the breakdown of the intersegmental muscles of silkmoths". *Journal of Insect Physiology*. **10** (4): 643–649. doi:10.1016/0022-1910(64)90034-4.

[7] Vaux DL, Cory S, Adams JM (September 1988). "Bcl-2 gene promotes haemopoietic cell survival and cooperates with c-myc to immortalize pre-B cells". *Nature*. **335** (6189): 440–2. doi:10.1038/335440a0. PMID 3262202.

[8] "The Nobel Prize in Physiology or Medicine 2002". The Nobel Foundation. 2002. Retrieved 2009-06-21.

[9] Schwartz LM, Smith SW, Jones ME, Osborne BA (1993). "Do all programmed cell deaths occur via apoptosis?". *PNAS*. **90** (3): 980–4. doi:10.1073/pnas.90.3.980. PMC 45794. PMID 8430112.; and, for a more recent view, see Bursch W, Ellinger A, Gerner C, Fröhwein U, Schulte-Hermann R (2000). "Programmed cell death (PCD). Apoptosis, autophagic PCD, or others?". *Annals of the New York Academy of Sciences*. **926**: 1–12. doi:10.1111/j.1749-6632.2000.tb05594.x. PMID 11193023.

[10] Green, Douglas (2011). *Means To An End*. New York: Cold Spring Harbor Laboratory Press. ISBN 978-0-87969-888-1.

[11] D. Bowen, Ivor (1993). *Cell Biology International 17*. Great Britain: Portland Press. pp. 365–380. ISSN 1095-8355.

[12] Kroemer G, Martin SJ (2005). "Caspase-independent cell death". *Nature Medicine*. **11** (7): 725–30. doi:10.1038/nm1263. PMID 16015365.

[13] Dixon Scott J.; Lemberg Kathryn M.; Lamprecht Michael R.; Skouta Rachid; Zaitsev Eleina M.; Gleason Caroline E.; Patel Darpan N.; Bauer Andras J.; Cantley Alexandra M.; et al. "Ferroptosis: An Iron-Dependent Form of Nonapoptotic Cell Death". *Cell*. **149** (5): 1060–1072. doi:10.1016/j.cell.2012.03.042.

[14] Lang, F; Lang, KS; Lang, PA; Huber, SM; Wieder, T. "Mechanisms and significance of eryptosis.". Antioxidants & Redox Signaling. 8 (7-8): 1183–92. doi:10.1089/ars.2006.8.1183. PMID 16910766.

[15] Formigli L et al aponecrosis: morphological and biochemical exploration of a syncretic process of cell death sharing apoptosis and necrosis. Journal Cellular Physiology 182(1): 41-49 (2000)

[16] Fadini, GP; Menegazzo, L; Scattolini, V; Gintoli, M; Albiero, M; Avogaro, A (25 November 2015). "A perspective on NETosis in diabetes and cardiometabolic disorders.". Nutrition, metabolism, and cardiovascular diseases : NMCD. 26: 1–8. doi:10.1016/j.numecd.2015.11.008. PMID 26719220.

[17] Chapter 10: All the Players on One Stage from PsychEducation.org

[18] Tau, GZ (2009). "Normal development of brain circuits". Neuropsychopharmacology. 35 (1): 147–168. doi:10.1038/npp.2009.115. PMC 3055433. PMID 19794405.

[19] Dekkers, MP (2013). "Death of developing neurons: new insights and implications for connectivity". Journal of Cell Biology. 203 (3): 385–393. doi:10.1083/jcb.201306136. PMC 3824005. PMID 24217616.

[20] Oppenheim, RW (1981). Neuronal cell death and some related regressive phenomena during neurogenesis: a selective historical review and progress report. In Studies in Developmental Neurobiology: Essays in Honor of Viktor Hamburger: Oxford University Press. pp. 74–133.

[21] Buss, RR (2006). "Adaptive roles of programmed cell death during nervous system development". Annual Review of Neuroscience. 29: 1–35. doi:10.1146/annurev.neuro.29.051605.112800.

[22] De la Rosa, EJ; De Pablo, F (October 23, 2000). "Cell death in early neural development: beyond the neurotrophic theory". Trends in Neuroscience. 23 (10): 454–458. doi:10.1016/s0166-2236(00)01628-3.

[23] Lossi, L; Merighi, A (April 2003). "In vivo cellular and molecular mechanisms of neuronal apoptosis in the mammalian CNS". Progress in Neurobiology. 69 (5): 287–312. doi:10.1016/s0301-0082(03)00051-0.

[24] Finlay, BL (1989). "Control of cell number in the developing mammalian visual system". Progress in Neurobiology. 32: 207–234. doi:10.1016/0301-0082(89)90017-8.

[25] Rubenstein, John; Pasko Rakic (2013). "Regulation of Neuronal Survival by Neurotrophins in the Developing Peripheral Nervous System". Patterning and Cell Type Specification in the Developing CNS and PNS: Comprehensive Developmental Neuroscience. Academic Press. ISBN 978-0-12-397348-1.

[26] Constantino, Sotelo (2002). "The chemotactic hypothesis of Cajal: a century behind". Progress in Brain Research. 136: 11–20. doi:10.1016/s0079-6123(02)36004-7.

[27] Oppenheim, Ronald (1989). "The neurotrophic theory and naturally occurring motorneuron death". Trends in Neuroscience. 12 (7): 252–255. doi:10.1016/0166-2236(89)90021-0.

[28] Dekkers, MP; Nikoletopoulou, V; Barde, YA (November 11, 2013). "Cell biology in neuroscience: Death of developing neurons: new insights and implications for connectivity". J Cell Biol. 203 (3): 385–393. doi:10.1083/jcb.201306136. PMC 3824005. PMID 24217616.

[29] Cowan, WN (2001). "Viktor Hamburger and Rita Levi-Montalcini: the path to the discovery of nerve growth factor". Annual Review of Neuroscience. 24: 551–600. doi:10.1146/annurev.neuro.24.1.551. PMID 11283321.

[30] Weltman, JK (February 8, 1987). "The 1986 Nobel Prize for Physiology or Medicine awarded for discovery of growth factors: Rita Levi-Montalcini, M.D., and Stanley Cohen, Ph.D.". New England regional allergy proceedings. 8 (1): 47–8. doi:10.2500/108854187779045385. PMID 3302667.

[31] Dekkers, M (April 5, 2013). "Programmed Cell Death in Neuronal Development". Science. 340 (6128): 39–41. doi:10.1126/science.1236152.

[32] Southwell, D.G. (November 2012). "Intrinsically determined cell death of developing cortical interneurons". Nature. 491 (7422): 109–115. doi:10.1038/nature11523.

[33] Kuida, K (1998). "Reduced apoptosis and cytochrome c-mediated caspase activation in mice lacking caspase 9". Cell. 94: 325–337. doi:10.1016/s0092-8674(00)81476-2. PMID 9708735.

[34] Kuida, K (1996). "Decreased apoptosis in the brain and premature lethality in CPP32-deficient mice". Nature. 384 (6607): 368–372. doi:10.1038/384368a0.

[35] Oppenheim, RW (2001). "Programmed cell death of developing mammalian neurons after genetic deletion of caspases". Journal of Neuroscience. 21: 4752–4760.

[36] Cecconi, F (1998). "Apaf1 (CED-4 homolog) regulates programmed cell death in mammalian development". Cell. 94: 727–737. doi:10.1016/s0092-8674(00)81732-8. PMID 9753320.

[37] Hao, Z (2005). "Specific ablation of the apoptotic functions of cytochrome c reveals a differential requirement for cytochrome c and Apaf-1 in apoptosis". Cell. 121: 579–591. doi:10.1016/j.cell.2005.03.016.

[38] Yoshida, H (1998). "Apaf1 is required for mitochondrial pathways of apoptosis and brain development". Cell. 94: 739–750. doi:10.1016/s0092-8674(00)81733-x.

[39] Bonfanti, L (1996). "Protection of retinal ganglion cells from natural and axotomy-induced cell death in neonatal transgenic mice overexpressing bcl-2". *Journal of Neuroscience*. **16**: 4186–4194.

[40] Martinou, JC (1994). "Overexpression of BCL-2 in transgenic mice protects neurons from naturally occurring cell death and experimental ischemia". *Neuron*. **13**: 1017–1030. doi:10.1016/0896-6273(94)90266-6.

[41] Zanjani, HS (1996). "Increased cerebellar Purkinje cell numbers in mice overexpressing a human bcl-2 transgene". *Journal of Computational Neurology*. **374**: 332–341. doi:10.1002/(sici)1096-9861(19961021)374:3<332::aid-cne2>3.0.co;2-2.

[42] Zup, SL (2003). "Overexpression of bcl-2 reduces sex differences in neuron number in the brain and spinal cord". *Journal of Neuroscience*. **23**: 2357–2362.

[43] Fan, H (2001). "Elimination of Bax expression in mice increases cerebellar Purkinje cell numbers but not the number of granule cells". *Journal of Computational Neurology*. **436**: 82–91. doi:10.1002/cne.1055.abs.

[44] Mosinger, Ogilvie (1998). "Suppression of developmental retinal cell death but not of photoreceptor degeneration in Bax-deficient mice". *Investigative Ophthalmology & Visual Science*. **39**: 1713–1720.

[45] White, FA (1998). "Widespread elimination of naturally occurring neuronal death in Bax-deficient mice". *Journal of Neuroscience*. **18**: 1428–1439.

[46] Gianfranceschi, L (1999). "Behavioral visual acuity of wild type and bcl2 transgenic mouse". *Vision Research Journal*. **39**: 569–574. doi:10.1016/s0042-6989(98)00169-2.

[47] Rondi-Reig, L (2002). "To die or not to die, does it change the function? Behavior of transgenic mice reveals a role for developmental cell death". *Brain Research Bulletin*. **57**: 85–91. doi:10.1016/s0361-9230(01)00639-6.

[48] Rondi-Reig, L (2001). "Transgenic mice with neuronal overexpression of bcl-2 gene present navigation disabilities in a water task". *Neuroscience*. **104**: 207–215. doi:10.1016/s0306-4522(01)00050-1.

[49] Sulston, JE (1980). "The Caenorhabditis elegans male: postembryonic development of nongonadal structures". *Developmental Biology*. **78**: 542–576. doi:10.1016/0012-1606(80)90352-8.

[50] Sulston2, JE (1983). "The embryonic cell lineage of the nematode Caenorhabditis elegans". *Developmental Biology*. **100** (1): 64–119. doi:10.1016/0012-1606(83)90201-4. PMID 6684600.

[51] Doe, Cq (1985). "Development and segmental differences in the pattern of neuronal precursor cells". *Journal of Developmental Biology*. **111**: 193–205.

[52] Giebultowicz, JM (1984). "Sexual differentiation in the terminal ganglion of the moth Manduca sexta: role of sex-specific neuronal death". *Journal of Comparative Neurology*. **226**: 87–95. doi:10.1002/cne.902260107.

[53] Cook, B (1998). "Developmental neuronal death is not a universal phenomenon among cell types in the chick embryo retina". *Journal of Comparative Neurology*. **396**: 12–19. doi:10.1002/(sici)1096-9861(19980622)396:1<12::aid-cne2>3.0.co;2-1.

[54] Collazo C, Chacón O, Borrás O (2006). "Programmed cell death in plants resembles apoptosis of animals" (PDF). *Biotecnología Aplicada*. **23**: 1–10.

[55] Bonke M, Thitamadee S, Mähönen AP, Hauser MT, Helariutta Y (2003). "APL regulates vascular tissue identity in Arabidopsis". *Nature*. **426** (6963): 181–6. doi:10.1038/nature02100. PMID 14614507.

[56] Solomon M, Belenghi B, Delledonne M, Menachem E, Levine A (1999). "The involvement of cysteine proteases and protease inhibitor genes in the regulation of programmed cell death in plants". *The Plant Cell*. **11** (3): 431–44. doi:10.2307/3870871. JSTOR 3870871. PMC 144188. PMID 10072402. See also related articles in *The Plant Cell Online*

[57] Ito J, Fukuda H (2002). "ZEN1 Is a Key Enzyme in the Degradation of Nuclear DNA during Programmed Cell Death of Tracheary Elements". *The Plant Cell*. **14** (12): 3201–11. doi:10.1105/tpc.006411. PMC 151212. PMID 12468737.

[58] Balk J, Leaver CJ (2001). "The PET1-CMS Mitochondrial Mutation in Sunflower Is Associated with Premature Programmed Cell Death and Cytochrome c Release". *The Plant Cell*. **13** (8): 1803–18. doi:10.1105/tpc.13.8.1803. PMC 139137. PMID 11487694.

[59] Thomas SG, Franklin-Tong VE (2004). "Self-incompatibility triggers programmed cell death in Papaver pollen". *Nature*. **429** (6989): 305–9. doi:10.1038/nature02540. PMID 15152254.

[60] Crespi B, Springer S (2003). "Ecology. Social slime molds meet their match". *Science*. **299** (5603): 56–7. doi:10.1126/science.1080776. PMID 12511635.

[61] Levraud JP, Adam M, Luciani MF, de Chastellier C, Blanton RL, Golstein P (2003). "Dictyostelium cell death: early emergence and demise of highly polarized paddle cells". *Journal of Cell Biology*. **160** (7): 1105–14. doi:10.1083/jcb.200212104. PMC 2172757. PMID 12654899.

[62] Roisin-Bouffay C, Luciani MF, Klein G, Levraud JP, Adam M, Golstein P (2004). "Developmental cell death in dictyostelium does not require paracaspase". *Journal of Biological Chemistry*. **279** (12): 11489–94. doi:10.1074/jbc.M312741200. PMID 14681218.

[63] Deponte, M (2008). "Programmed cell death in protists". *Biochimica et Biophysica Acta (BBA) - Molecular Cell Research*. **1783** (7): 1396–1405. doi:10.1016/j.bbamcr.2008.01.018.

[64] Proto, W. R.; Coombs, G. H.; Mottram, J. C. (2012). "Cell death in parasitic protozoa: regulated or incidental?" (PDF). *Nature Reviews Microbiology*. **11** (1): 58–66. doi:10.1038/nrmicro2929.

[65] de Duve C (1996). "The birth of complex cells". *Scientific American*. **274** (4): 50–7. doi:10.1038/scientificamerican0496-50. PMID 8907651.

[66] Dyall SD, Brown MT, Johnson PJ (2004). "Ancient invasions: from endosymbionts to organelles". *Science*. **304** (5668): 253–7. doi:10.1126/science.1094884. PMID 15073369.

[67] Chiarugi A, Moskowitz MA (2002). "Cell biology. PARP-1--a perpetrator of apoptotic cell death?". *Science*. **297** (5579): 200–1. doi:10.1126/science.1074592. PMID 12114611.

[68] Srivastava, Rakesh (2007). *Apoptosis, Cell Signaling, and Human Diseases*. Humana Press.

## 8.12   External links

- Apoptosis and Cell Death Labs
- International Cell Death Society
- The Bcl-2 Family Database

# Chapter 9

# Cell division

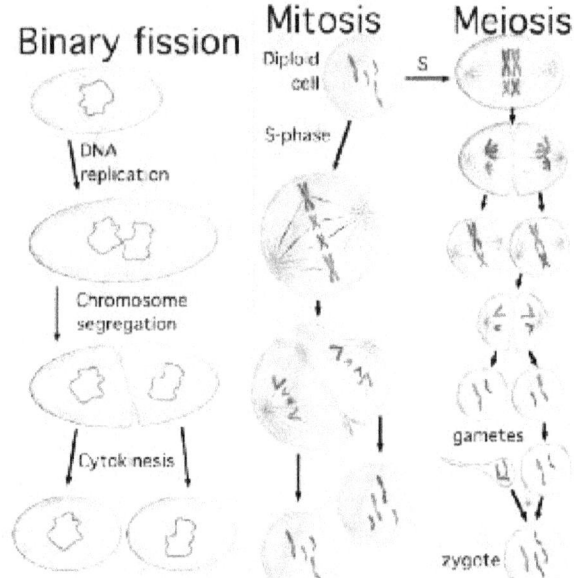

*Three types of cell division*

**Cell division** is the process by which a parent cell divides into two or more daughter cells.[1] Cell division usually occurs as part of a larger cell cycle. In eukaryotes, there are two distinct types of cell division: a vegetative division, whereby each daughter cell is genetically identical to the parent cell (mitosis),[2] and a reproductive cell division, whereby the number of chromosomes in the daughter cells is reduced by half to produce haploid gametes (meiosis). Meiosis results in four haploid daughter cells by undergoing one round of DNA replication followed by two divisions. Homologous chromosomes are separated in the first division, and sister chromatids are separated in the second division. Both of these cell division cycles are used in the process of sexual reproduction at some point in their life cycle. Both are believed to be present in the last eukaryotic common ancestor. Prokaryotes undergo a vegetative cell division known as binary fission, where their genetic material is segregated equally into two daughter cells. All cell divisions, regardless of organism, are preceded by a single round of DNA replication.

For simple unicellular organisms[Note 1] such as the amoeba, one cell division is equivalent to reproduction – an entire new organism is created. On a larger scale, mitotic cell division can create progeny from multicellular organisms, such as plants that grow from cuttings. Mitotic cell division enables sexually reproducing organisms to develop from the one-celled zygote, which itself was produced by meiotic cell division from gametes. After growth, cell division by mitosis allows for continual construction and repair of the organism.[3] The human body experiences about 10 quadrillion cell divisions in a lifetime.[4]

The primary concern of cell division is the maintenance of the original cell's genome. Before division can occur, the genomic information that is stored in chromosomes must be replicated, and the duplicated genome must be separated cleanly between cells. A great deal of cellular infrastructure is involved in keeping genomic information consistent between generations.

## 9.1 Phases of cell division

### 9.1.1 Interphase

Interphase is the process a cell must go through before mitosis, meiosis, and cytokinesis.[5] Interphase consists of four main stages: G1, S, G2, and M. G1 is a time of growth for the cell. If the cell does not progress through G1, the cell then enters a stage called G0. In G0, cells are still living but they are put on hold. The cells may later be called back into interphase if needed at a later time. There are checkpoints during interphase that allow the cell to be either progressed or denied further development. In S phase, the chromosomes are replicated in order for the genetic content to be maintained. During G2, the cell undergoes the final stages of growth before it enters the M phase. The M phase, can be either mitosis or meiosis depending on the type of cell. Germ cells undergo meiosis, while somatic cells will undergo mitosis. After the cell proceeds through successfully through the M phase, it may then undergo cell division through cytokinesis. The control of each check-point is controlled by cyclin and cyclin dependent kinases. The progression of interphase is the result of the increased amount of cyclin. As the amount of cyclin increases, more and more cyclin dependent kinases attach to cyclin signaling the cell further into interphase. The peak of the cyclin attached to the cyclin dependent kinases this system pushes the cell out of interphase and into the M phase, where mitosis, meiosis, and cytokinesis occur.

### 9.1.2 Prophase

Prophase is the first stage of division. The nuclear envelope is broken down, long strands of chromatin condense to form shorter more visible strands called chromosomes, the nucleolus disappears, and microtubules attach to the chromosomes at the kinetochores present in the centromere.[6] Microtubules associated with the alignment and separation of chromosomes are referred to as the spindle and spindle fibers. Chromosomes will also be visible under a microscope and will be connected at the centromere. During this condensation and alignment period, homologous chromosomes may swap portions of their DNA in a process known as crossing over.

### 9.1.3 Metaphase

Metaphase is the stage in cell division when the chromosomes line up in the middle of the cell. The chromosomes are still condensing and are currently at one step away from being the most coiled and condensed they will be.[7] Spindle and spindle fibers have already connected to the kinetochores. At this point, the chromosomes are ready to split into opposite poles of the cell towards the spindle to which they are connected.

### 9.1.4 Anaphase

Anaphase is a very short stage of the cell cycle and occurs after the chromosomes align at the mitotic plate. After the chromosomes line up in the middle of the cell, the spindle fibers will pull them apart. The chromosomes are split apart as the sister chromatids move to opposite sides of the cell.[8]

### 9.1.5 Telophase

Telophase is the last stage of the cell cycle. Two cells form around the chromatin at the two poles of the cell. Two nuclear membranes begin to reform and the chromatin begin to unwind.[9]

## 9.2   Variants

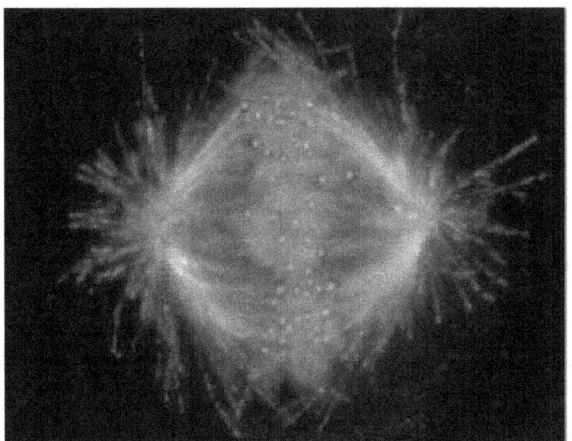

*Image of the mitotic spindle in a human cell showing microtubules in green, chromosomes (DNA) in blue, and kinetochores in red.*

**Cells** are broadly classified into two main categories: simple, non-nucleated prokaryotic cells, and complex, nucleated eukaryotic cells. Owing to their structural differences, eukaryotic and prokaryotic cells do not divide in the same way. Also, the pattern of cell division that transforms eukaryotic stem cells into gametes (sperm cells in males or egg cells in females), termed meiosis, is different from that of the division of somatic cells in the body.

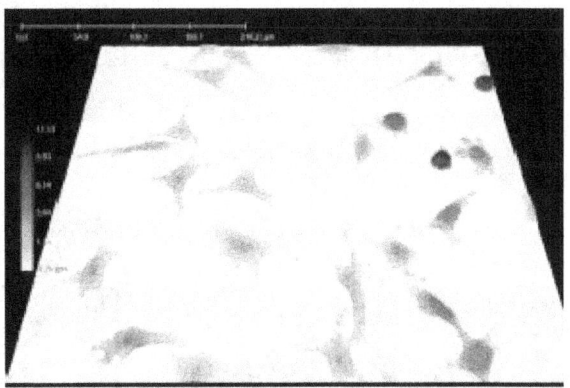

*Cell division over 42. The cells were directly imaged in the cell culture vessel, using non-invasive quantitative phase contrast time-lapse microscopy.*[10]

## 9.3   Degradation

Multicellular organisms replace worn-out cells through cell division. In some animals, however, cell division eventually halts. In humans this occurs, on average, after 52 divisions, known as the Hayflick limit. The cell is then referred to as senescent. Cells stop dividing because the telomeres, protective bits of DNA on the end of a chromosome required for replication, shorten with each copy, eventually being consumed. Cancer cells, on the other hand, are not thought to degrade in this way, if at all. An enzyme called telomerase, present in large quantities in cancerous cells, rebuilds the telomeres, allowing division to continue indefinitely.

## 9.4 See also

- Mitosis
- Meiosis
- Binary fission
- Cell growth
- Labile cells, cells that constantly divide
- Klerokinesis

## 9.5 Notes

[1] Single-celled organisms. See introduction of the article on microorganisms.

## 9.6 References

[1] Robert.S Hine. ed. (2008). *Oxford Dictionary Biology* (6th ed.). New York: Oxford University Press. p. 113. ISBN 978-0-19-920462-5.

[2] Griffiths, Anthony J.F.; Wessler, Susan R.; Carroll, Sean B.; Doebley, John (2012). *Introduction to Genetic Analysis* (10 ed.). New York: W.H. Freeman and Company. p. 35. ISBN 978-1-4292-2943-2.

[3] Maton, Anthea (1997). *Cells: Building Blocks of Life*. New Jersey: Prentice Hall. pp. 70–74. ISBN 0-13-423476-6.

[4] Quammen, David (April 2008). "Contagious cancer: The evolution of a killer". *Harper's*. **316** (1895): 42. Retrieved 24 September 2012.

[5] Marieb, Elaine (2000). *Essentials of human anatomy and physiology*. San Francisco: Benjamin Cummings. ISBN 0-8053-4940-5.

[6] Schermelleh, Lothar; Carlton, Peter M.; Haase, Sebastian; Shao, Lin; Winoto, Lukman; Kner, Peter; Burke, Brian; Cardoso, M. Cristin; Agard, David A. (2008-06-06). "Subdiffraction Multicolor Imaging of the Nuclear Periphery with 3D Structured Illumination Microscopy". *Science*.

**320** (5881): 1332–1336. doi:10.1126/science.1156947. ISSN 0036-8075. PMC 2916659. PMID 18535242.

[7] "Researchers Shed Light On Shrinking Of Chromosomes". *ScienceDaily*. June 12, 2007. Retrieved 2017-02-02.

[8] "The Cell Cycle". *www.biology-pages.info*. Retrieved 2017-02-02.

[9] Hetzer, Martin W. (2017-02-02). "The Nuclear Envelope". *Cold Spring Harbor Perspectives in Biology*. **2** (3). doi:10.1101/cshperspect.a000539. ISSN 1943-0264. PMC 2829960. PMID 20300205.

[10] Phase Holographic Imaging. *Cell Division*

## 9.7 Further reading

- Morgan HI. (2007). "The Cell Cycle: Principles of Control" London: New Science Press.

- J.M.Turner *Fetus into Man* (1978, 1989). Harvard University Press. ISBN 0-674-30692-9

- Cell division: binary fission and mitosis

## 9.8 External links

- How Cells Divide: Mitosis vs. Meiosis

- The Mitosis and Cell Cycle Control Section from the *Landmark Papers in Cell Biology* (Gall JG, McIntosh JR, eds.) contains commentaries on and links to seminal research papers on mitosis and cell division. Published online in the Image & Video Library of The American Society for Cell Biology

- The Image & Video Library of The American Society for Cell Biology contains many videos showing the cell division.

- Videos of the first cell divisions in Xenopus laevis embryos (side view and top view), acquired by MRI (DOI of paper)

- Images : *Calanthe discolor* Lindl. - Flavon's Secret Flower Garden

- Tyson's model of cell division and a Description on BioModels Database

- WormWeb.org: Interactive Visualization of the *C. elegans* Cell Lineage - Visualize the entire set of cell divisions of the nematode *C. elegans*

# Chapter 10

# Telomere

For other uses, see Telomere (disambiguation).
A **telomere** is a region of repetitive nucleotide sequences

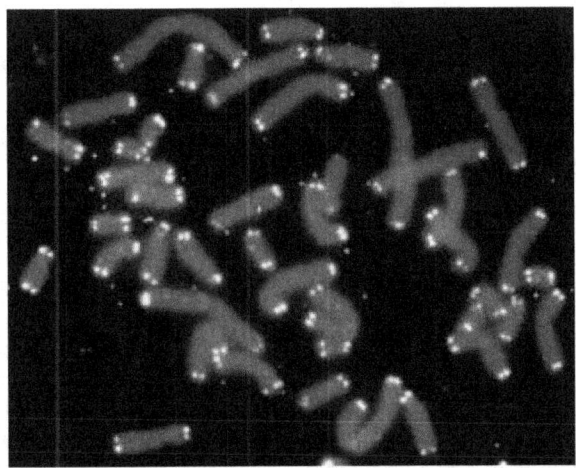

*Human chromosomes (grey) capped by telomeres (white)*

at each end of a chromosome, which protects the end of the chromosome from deterioration or from fusion with neighboring chromosomes. Its name is derived from the Greek nouns telos (τέλος) "end" and meros (μέρος, root: μερ-) "part". For vertebrates, the sequence of nucleotides in telomeres is TTAGGG, with the complementary DNA strand being AATCCC, with a single-stranded TTAGGG overhang. This sequence of TTAGGG is repeated approximately 2,500 times in humans.[1] In humans, average telomere length declines from about 11 kilobases at birth [2] to less than 4 kilobases in old age,[3] with average rate of decline being greater in men than in women.[4]

During chromosome replication, the enzymes that duplicate DNA cannot continue their duplication all the way to the end of a chromosome, so in each duplication the end of the chromosome is shortened[5] (this is because the synthesis of Okazaki fragments requires RNA primers attaching ahead on the lagging strand). The telomeres are disposable buffers at the ends of chromosomes which are truncated during cell division; their presence protects the genes before them on the chromosome from being truncated instead. The telom-eres themselves are protected by a complex of shelterin proteins, as well as by the RNA that telomeric DNA encodes (TERRA).

Over time, due to each cell division, the telomere ends become shorter.[6] They are replenished by an enzyme, telomerase reverse transcriptase.

## 10.1 Discovery

In the early 1970s, Russian theorist Alexei Olovnikov first recognized that chromosomes could not completely replicate their ends. Building on this, and to accommodate Leonard Hayflick's idea of limited somatic cell division, Olovnikov suggested that DNA sequences are lost every time a cell/DNA replicates until the loss reaches a critical level, at which point cell division ends.[7][8] However, Olovnikov's prediction was not widely known except by a handful of researchers studying cellular aging and immortalization.[9]

In 1975–1977, Elizabeth Blackburn, working as a postdoctoral fellow at Yale University with Joseph Gall, discovered the unusual nature of telomeres, with their simple repeated DNA sequences composing chromosome ends.[10] Blackburn, Carol Greider, and Jack Szostak were awarded the 2009 Nobel Prize in Physiology or Medicine for the discovery of how chromosomes are protected by telomeres and the enzyme telomerase.[11]

Nevertheless, in the 1970s there was no recognition that the telomere-shortening mechanism normally limits cells to a fixed number of divisions, nor was there any animal study suggesting that this could be responsible for aging on the cellular level. There was also no recognition that the mechanism set a limit on lifespans.[12][13]

It remained for a privately funded collaboration from biotechnology company Geron to isolate the genes for the RNA and protein component of human telomerase in order to establish the role of telomere shortening in cellular aging and telomerase reactivation in cell immortalization.[14]

## 10.2 Nature and function

### 10.2.1 Structure, function and evolutionary biology

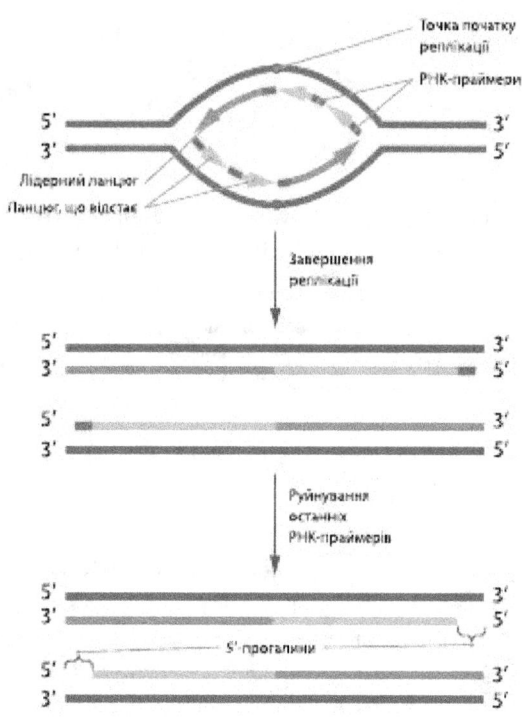

loss of genetic material can be caused be telomere shortening.

Telomeres are repetitive nucleotide sequences located at the termini of linear chromosomes of most eukaryotic organisms. For vertebrates, the sequence of nucleotides in telomeres is TTAGGG. Most prokaryotes, lacking this linear arrangement, do not have telomeres. Telomeres compensate for incomplete semi-conservative DNA replication at chromosomal ends. A protein complex known as shelterin serves to protect the ends of telomeres from being recognised as double-strand breaks by inhibiting homologous recombination (HR) and non-homologous end joining (NHEJ).[15][16]

In most prokaryotes, chromosomes are circular and, thus, do not have ends to suffer premature replication termination. A small fraction of bacterial chromosomes (such as those in *Streptomyces*, *Agrobacterium*, and *Borrelia*) are linear and possess telomeres, which are very different from those of the eukaryotic chromosomes in structure and functions. The known structures of bacterial telomeres take the form of proteins bound to the ends of linear chromosomes, or hairpin loops of single-stranded DNA at the ends of the linear chromosomes.[17]

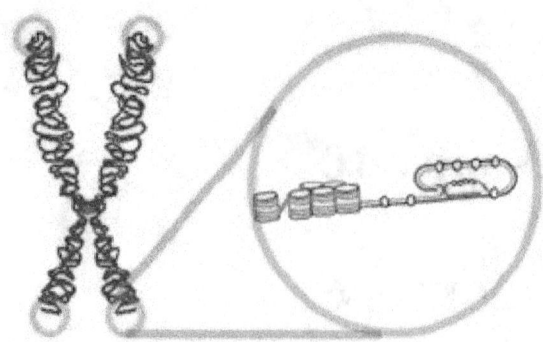

Telomeres are found at the termini of chromosomes. The end of a telomere inserts back into the main body of the telomere to form a T-loop

While replicating DNA, the eukaryotic DNA replication enzymes (the DNA polymerase protein complex) cannot replicate the sequences present at the ends of the chromosomes (or more precisely the chromatid fibres). Hence, these sequences and the information they carry may get lost. This is the reason telomeres are so important in context of successful cell division: They "cap" the end-sequences and themselves get lost in the process of DNA replication. But the cell has an enzyme called telomerase, which carries out the task of adding repetitive nucleotide sequences to the ends of the DNA. Telomerase, thus, "replenishes" the telomere "cap" of the DNA. In most multicellular eukaryotic organisms, telomerase is active only in germ cells, some types of stem cells such as embryonic stem cells, and certain white blood cells. Telomerase can be re activated and telomeres reset back to an embryonic state by somatic cell nuclear transfer.[18] There are theories that claim that the steady shortening of telomeres with each replication in somatic (body) cells may have a role in senescence and in the prevention of cancer. This is because the telomeres act as a sort of time-delay "fuse", eventually running out after a certain number of cell divisions and resulting in the eventual loss of vital genetic information from the cell's chromosome with future divisions.

Telomere length varies greatly between species, from approximately 300 base pairs in yeast[19] to many kilobases in humans, and usually is composed of arrays of guanine-rich, six- to eight-base-pair-long repeats. Eukaryotic telomeres normally terminate with 3′ single-stranded-DNA overhang, which is essential for telomere maintenance and capping. Multiple proteins binding single- and double-stranded telomere DNA have been identified.[20] These function in both telomere maintenance and capping. Telomeres form large loop structures called telomere loops, or T-loops. Here, the single-stranded DNA curls around in a long circle, stabilized by telomere-binding proteins.[21] At the very end of the T-loop, the single-stranded telomere DNA is held onto a region of double-stranded DNA by the telomere strand disrupting the double-helical DNA, and base pairing to one of the two strands. This triple-stranded structure is called a displacement loop or D-loop.[22]

Telomere shortening in humans can induce replicative senescence, which blocks cell division. This mechanism appears to prevent genomic instability and development of cancer in human aged cells by limiting the number of cell divisions. However, shortened telomeres impair immune function that might also increase cancer susceptibility.[23] If telomeres become too short, they have the potential to unfold from their presumed closed structure. The cell may detect this uncapping as DNA damage and then either stop growing, enter cellular old age (senescence), or begin programmed cell self-destruction (apoptosis) depending on the cell's genetic background (p53 status). Uncapped telomeres also result in chromosomal fusions. Since this damage cannot be repaired in normal somatic cells, the cell may even go into apoptosis. Many aging-related diseases are linked to shortened telomeres. Organs deteriorate as more and more of their cells die off or enter cellular senescence.

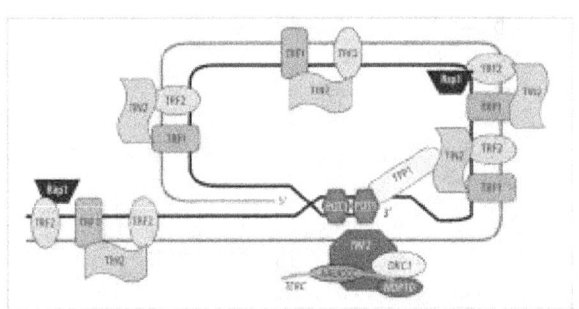

*Shelterin co-ordinates the T-loop formation of telomeres*

### 10.2.2   Shelterin

Main article: Shelterin

At the very distal end of the telomere is a 300 bp single-stranded portion, which forms the T-Loop. This loop is analogous to a *knot*, which stabilizes the telomere, preventing the telomere ends from being recognized as break points by the DNA repair machinery. Should non-homologous end joining occur at the telomeric ends, chromosomal fusion will result. The T-loop is held together by several proteins, the most notable ones being TRF1, TRF2, POT1, TIN1, and TIN2, collectively referred to as the shelterin complex. In humans, the shelterin complex consists of six proteins identified as TRF1, TRF2, TIN2, POT1, TPP1, and RAP1.[15]

## 10.3   Shortening

Telomeres shorten in part because of the *end replication problem* that is exhibited during DNA replication in eukaryotes only. Because DNA replication does not begin at either end of the DNA strand, but starts in the center, and considering that all known DNA polymerases move in the 5' to 3' direction, one finds a leading and a lagging strand on the DNA molecule being replicated.

On the leading strand, DNA polymerase can make a complementary DNA strand without any difficulty because it goes from 5' to 3'. However, there is a problem going in the other direction on the lagging strand. To counter this, short sequences of RNA acting as primers attach to the lagging strand a short distance ahead of where the initiation site was. The DNA polymerase can start replication at that point and go to the end of the initiation site. This causes the formation of Okazaki fragments. More RNA primers attach further on the DNA strand and DNA polymerase comes along and continues to make a new DNA strand.

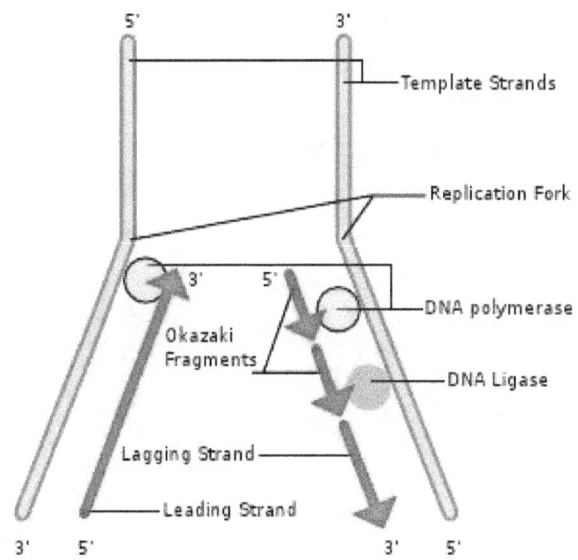

*Lagging strand during DNA replication*

Eventually, the last RNA primer attaches, and DNA polymerase, RNA nuclease, and DNA ligase come along to convert the RNA (of the primers) to DNA and to seal the gaps in between the Okazaki fragments. But, in order to change RNA to DNA, there must be another DNA strand in front of the RNA primer. This happens at all the sites of the lagging strand, but it does not happen at the end where the last RNA primer is attached. Ultimately, that RNA is destroyed by enzymes that degrade any RNA left on the DNA. Thus, a section of the telomere is lost during each cycle of replication at the 5' end of the lagging strand's daughter.

However, test-tube studies have shown that telomeres are

highly susceptible to oxidative stress. There is evidence that oxidative stress-mediated DNA damage is an important determinant of telomere shortening.[24] Telomere shortening due to free radicals explains the difference between the estimated loss per division because of the end-replication problem (c. 20 bp) and actual telomere shortening rates (50–100 bp), and has a greater absolute impact on telomere length than shortening caused by the end-replication problem. Population-based studies have also indicated an interaction between anti-oxidant intake and telomere length. In the Long Island Breast Cancer Study Project (LIBCSP), authors found a moderate increase in breast cancer risk among women with the shortest telomeres and lower dietary intake of beta carotene, vitamin C or E.[25] These results suggest that cancer risk due to telomere shortening may interact with other mechanisms of DNA damage, specifically oxidative stress.

Telomere shortening is associated with aging, mortality and aging-related diseases. In 2003, Richard Cawthon discovered that those with longer telomeres lead longer lives than those with short telomeres.[26] However, it is not known whether short telomeres are just a sign of cellular age or actually contribute to the aging process.

## 10.4   Lengthening

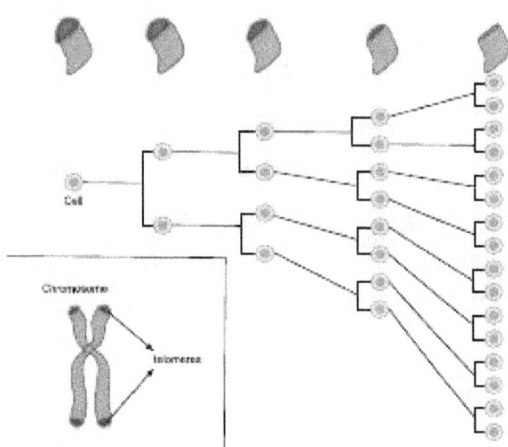

*The average cell will divide between 50 and 70 times before cell death. As the cell divides the telomeres on the end of the chromosome get smaller. The Hayflick limit is the theoretical limit to the number of times a cell may divide until the telomere becomes so short that division is inhibited and the cell enters senescence.*

The phenomenon of limited cellular division was first observed by Leonard Hayflick, and is now referred to as the Hayflick limit.[27][28] Significant discoveries were subsequently made by a group of scientists organized at Geron Corporation by Geron's founder Michael D. West that tied telomere shortening with the Hayflick limit.[29] The cloning

of the catalytic component of telomerase enabled experiments to test whether the expression of telomerase at levels sufficient to prevent telomere shortening was capable of immortalizing human cells. Telomerase was demonstrated in a 1998 publication in *Science* to be capable of extending cell lifespan, and now is well-recognized as capable of immortalizing human somatic cells.[30]

It is becoming apparent that reversing shortening of telomeres through temporary activation of telomerase may be a potent means to slow aging. The reason that this would extend human life is because it would extend the Hayflick limit. Three routes have been proposed to reverse telomere shortening: drugs, gene therapy, or metabolic suppression, so-called, torpor/hibernation. So far these ideas have not been proven in humans, but it has been demonstrated that telomere shortening is reversed in hibernation and aging is slowed (Turbill, et al. 2012 & 2013) and that hibernation prolongs life-span (Lyman et al. 1981). It has also been demonstrated that telomere extension has successfully reversed some signs of aging in laboratory mice [31][32] and the nematode worm species *Caenorhabditis elegans*.[33] It has been hypothesized that longer telomeres and especially telomerase activation might cause increased cancer (e.g. Weinstein and Ciszek, 2002). However, longer telomeres might also protect against cancer, because short telomeres are associated with cancer. It has also been suggested that longer telomeres might cause increased energy consumption.[23]

Techniques to extend telomeres could be useful for tissue engineering, because they might permit healthy, noncancerous mammalian cells to be cultured in amounts large enough to be engineering materials for biomedical repairs.

Two recent studies on long-lived seabirds demonstrate that the role of telomeres is far from being understood. In 2003, scientists observed that the telomeres of Leach's storm-petrel (*Oceanodroma leucorhoa*) seem to lengthen with chronological age, the first observed instance of such behaviour of telomeres.[34] In 2006, Juola *et al.*[35] reported that in another unrelated, long-lived seabird species, the great frigatebird (*Fregata minor*), telomere length did decrease until at least c. 40 years of age (i.e. probably over the entire lifespan), but the speed of decrease slowed down massively with increasing ages, and that rates of telomere length decrease varied strongly between individual birds. They concluded that in this species (and probably in frigatebirds and their relatives in general), telomere length could not be used to determine a bird's age sufficiently well. Thus, it seems that there is much more variation in the behavior of telomere length than initially believed.

Furthermore, Gomes et al. found, in a study of the comparative biology of mammalian telomeres, that telomere length of different mammalian species correlates inversely, rather than directly, with lifespan, and they concluded that the contribution of telomere length to lifespan remains controversial.[36] Harris et al. found little evidence that, in humans, telomere length is a significant biomarker of normal aging with respect to important cognitive and physical abilities.[37] Gilley and Blackburn tested whether cellular senescence in paramecium is caused by telomere shortening, and found that telomeres were not shortened during senescence.[38]

### 10.4.1    Exercise-induced lengthening

A 2013 pilot study from UCSF took 35 men with localized early-stage prostate cancer and had 10 of them begin "lifestyle changes that included: a plant-based diet (high in fruits, vegetables and unrefined grains, and low in fat and refined carbohydrates); moderate exercise (walking 30 minutes a day, six days a week); stress reduction (gentle yoga-based stretching, breathing, meditation)" and also "weekly group support". When compared to the other 25 study participants, "The group that made the lifestyle changes experienced a 'significant' increase in telomere length of approximately 10 percent. Further, the more people changed their behavior by adhering to the recommended lifestyle program, the more dramatic their improvements in telomere length."[39] A 2014 study entitled "Stand up for health – avoiding sedentary behaviour might lengthen your telomeres: secondary outcomes from a physical activity RCT in older people" indicated somewhat contradictory results, stating, "In the intervention group, there was a negative correlation between changes in time spent exercising and changes in telomere length (rho=−0.39, p=0.07). On the other hand, in the intervention group, telomere lengthening was significantly associated with reduced sitting time (rho=−0.68, p=0.02).[49]

## 10.5    Sequences

Known, up-to-date telomere nucleotide sequences are listed in Telomerase Database website.

## 10.6    Cancer

Telomeres are critical for maintaining genomic integrity and studies show that telomere dysfunction or shortening is commonly acquired during the process of tumor development.[43] Short telomeres can lead to genomic instability, chromosome loss and the formation of non-reciprocal translocations; and telomeres in tumor cells and their precursor lesions are significantly shorter than surrounding normal tissue.[44][45]

Observational studies have found shortened telomeres in many cancers: including pancreatic, bone, prostate, bladder, lung, kidney, and head and neck. In addition, people with many types of cancer have been found to possess shorter leukocyte telomeres than healthy controls.[46] Recent meta-analyses suggest 1.4 to 3.0 fold increased risk of cancer for those with the shortest vs. longest telomeres.[47][48] However the increase in risk varies by age, sex, tumor type and differences in lifestyle factors.

Some of the same lifestyle factors which increase risk of developing cancer have also been associated with shortened telomeres: including stress, smoking, physical inactivity and diet high in refined sugars [48] Diet and physical activity influence inflammation and oxidative stress. These factors are thought to influence telomere maintenance.[49] Psychologic stress has also been linked to accelerated cell aging, as reflected by decreased telomerase activity and short telomeres.[50] It has been suggested that a combination of lifestyle modifications, including healthy diet, exercise and stress reduction, have the potential to increase telomere length, reverse cellular aging, and reduce the risk for aging-related diseases. In a recent clinical trial for early-stage prostate cancer patients, comprehensive lifestyle changes resulted in a short-term increase in telomerase activity and long-term modification in telomere length.[51][52] Lifestyle modifications have the potential to naturally regulate telomere maintenance without promoting tumorigenesis, as traditional mechanisms of telomere lengthening involve the use of telomerase activating agents.

Cancer cells require a mechanism to maintain their telomeric DNA in order to continue dividing indefinitely (immortalization). A mechanism for telomere elongation or maintenance is one of the key steps in cellular immortalization and can be used as a diagnostic marker in the clinic. Telomerase, the enzyme complex responsible for elongating telomeres through the addition of telomere repeats to the ends of chromosomes, is activated in approximately 80% of tumors.[53] However, a sizeable fraction of cancerous cells employ alternative lengthening of telomeres (ALT),[54] a non-conservative telomere lengthening pathway involving the transfer of telomere tandem repeats between sister-chromatids.[55]

### 10.6.1    Telomerase and cancer

Telomerase is the natural enzyme that promotes telomere lengthening. It is active in stem cells, germ cells, hair follicles, and 90 percent of cancer cells, but its expression is low or absent in somatic cells. Telomerase functions by adding bases to the ends of the telomeres. Cells with sufficient telomerase activity are considered immortal in the sense that they can divide past the Hayflick limit without entering senescence or apoptosis. For this reason, telomerase is viewed as a potential target for anti-cancer drugs (such as Geron's Imetelstat currently in human clinical trials and telomestatin).[56]

Studies using knockout mice have demonstrated that the role of telomeres in cancer can both be limiting to tumor growth, as well as promote tumorigenesis, depending on the cell type and genomic context.[57][58]

Telomerase is a "ribonucleoprotein complex" composed of a protein component and an RNA primer sequence that acts to protect the terminal ends of chromosomes from being broken down by enzymes. The telomeres (and the actions of telomerase) are necessary because, during replication, DNA polymerase can synthesize DNA in only a 5' to 3' direction (each DNA strand having a polarity that is determined by the precise manner in which sugar molecules of the strand's "backbone" are linked together) and can do so only by adding nucleotides to RNA primers (that have already been placed at various points along the length of the DNA). The RNA strands are replaced with newly synthesized DNA, but DNA polymerase can only "backfill" deoxyribonucleotides if there is already DNA "upstream" from (i.e., located 5' to) the RNA primer. At the chromosome terminal, however, there is no nucleotide sequence in the 5' direction (and therefore no upstream RNA primer or DNA), so DNA polymerase cannot function and genetic sequence might be lost through chromosomal fraying. Chromosomal ends might also be processed as breaks in double-strand DNA with chromosome-to-chromosome telomere fusions resulting.

Telomeres at the end of DNA prevent the chromosome from growing shorter during replications (with loss of genetic information) by employing "telomerases" to synthesize DNA at the chromosome terminal. These include a protein subgroup of specialized reverse transcriptase enzymes known as TERT (telomerase reverse transcriptases) and are involved in synthesis of telomeres in humans and many other, but not all, organisms. Because DNA replication mechanisms are affected by oxidative stress and because TERT expression is very low in most types of human cell, telomeres shorten every time a cell divides. Among cell types characterized by extensive cell division (such as stem cells and certain white blood cells), however, TERT is expressed at higher levels and telomere shortening is partially or fully prevented.

In addition to its TERT protein component, telomerase also contains a piece of template RNA known as the TERC (telomerase RNA component) or TR (telomerase RNA). In humans, this TERC telomere sequence is a repeating string of TTAGGG, between 3 and 20 kilobases in length. There are an additional 100-300 kilobases of telomere-associated repeats between the telomere and the rest of the chromosome. Telomere sequences vary from species to species, but, in general, one strand is rich in G with fewer Cs. These G-rich sequences can form four-stranded structures (G-quadruplexes), with sets of four bases held in plane and then stacked on top of each other, with either a sodium or a potassium ion between the planar quadruplexes.

Mammalian (and other) somatic cells without telomerase gradually lose telomeric sequences as a result of incomplete replication (Counter et al., 1992). As mammalian telomeres shorten, eventually cells reach their replicative limit and progress into senescence or old age. Senescence involves p53 and pRb pathways and leads to the halting of cell proliferation (Campisi, 2005). Senescence may play an important role in suppression of cancer emergence, although inheriting shorter telomeres probably does not protect against cancer.[23] With critically shortened telomeres, further cell proliferation can be achieved by inactivation of p53 and pRb pathways. Cells entering proliferation after inactivation of p53 and pRb pathways undergo crisis. Crisis is characterized by gross chromosomal rearrangements and genome instability, and almost all cells die.

## ALT (Alternative Lengthening of Telomeres) and cancer

About 5–10% of human cancers activate the alternative lengthening of telomeres (ALT) pathway, which relies on recombination-mediated elongation.[59] Rarely, cells emerge from crisis immortalized through telomere lengthening by either activated telomerase or ALT (Colgina and Reddel, 1999; Reddel and Bryan, 2003). The first description of an ALT cell line demonstrated that their telomeres are highly heterogeneous in length and predicted a mechanism involving recombination (Murnane et al., 1994). Subsequent studies have confirmed a role for recombination in telomere maintenance by ALT (Dunham et al., 2000), however the exact mechanism of this pathway is yet to be determined. ALT cells produce abundant T-circles, possible products of intratelomeric recombination and T-loop resolution (Tomaska et al., 2000; 2009; Cesare and Griffith, 2004; Wang et al., 2004).

## Evolutionary aspects

Since shorter telomeres are thought by some to be a cause of aging, this raises the question of why longer telomeres are not selected for to ameliorate these effects. A prominent explanation suggests that inheriting longer telomeres would cause increased cancer rates (e.g. Weinstein and Ciszek, 2002). However, a recent literature review and analysis [23] suggests this is unlikely, because shorter telomeres and telomerase inactivation is more often associated with increased cancer rates, and the mortality from cancer occurs late in life when the force of natural selection is very low. An alternative explanation to the hypothesis that long telomeres are selected against due to their cancer promoting effects is the "thrifty telomere" hypothesis, which suggests that the cellular proliferation effects of longer telomeres causes increased energy expenditures.[23] In environments of energetic limitation, shorter telomeres might be an energy sparing mechanism.

### Relation to breast cancer

In a healthy female breast, a proportion of cells called luminal progenitors that line the milk ducts have proliferative and differentiation potential and most of them contain critically short telomeres with DNA damage foci. These cells are believed to be the possible common cellular loci where cancers of the breast involving telomere dysregulation may arise.[60] The telomere shortening in these progenitors is not age dependent but is speculated to be basal to luminal epithelial differentiation program-dependent. Also, the telomerase activity is unusually high in these cells when isolated from younger women, but declines with age.[61]

## 10.7   Measurement

Several techniques are currently employed to assess average telomere length in eukaryotic cells. One method is the Terminal Restriction Fragment (TRF) southern blot,[62] which involves hybridization of a radioactive 32P-(TTAGGG)n oligonucleotide probe to Hinf / Rsa I digested genomic DNA embedded on a nylon membrane and subsequently exposed to autoradiographic film or phosphoimager screen. Another histochemical method, termed Q-FISH, involves fluorescent in situ hybridization (FISH).[63] Q-FISH, however, requires significant amounts of genomic DNA (2-20 micrograms) and labor that renders its use limited in large epidemiological studies. Some of these impediments have been overcome with a Real-Time PCR assay for telomere length and Flow-FISH. Real-time PCR assay involves determining the Telomere-to-Single Copy Gene (T/S)ratio,[64] which is demonstrated to be proportional to the average telomere length in a cell.

Another technique, referred to as single telomere elongation length analysis (STELA), was developed in 2003 by Duncan Baird. This technique allows investigations that can target specific telomere ends, which is not possible with TRF analysis. However, due to this technique's being PCR-based, telomeres larger than 25Kb cannot be amplified and

there is a bias towards shorter telomeres.

While multiple companies offer telomere length measurement services,[65][66][67] the utility of these measurements for widespread clinical or personal use has been questioned by prominent scientists without financial interests in these companies.[68][69] Nobel Prize winner Elizabeth Blackburn, who was the co-founder of one of these companies and has prominently promoted the clinical utility of telomere length measures,[70] resigned from the company in June 2013 "owing to an impending change in the control of Telome Health".[71]

## 10.8   In popular culture

The opening track of the 2016 album, Curve of the Earth, by the UK, indie rock band, Mystery Jets, is named Telomere and contains the following stanza:

In the telomere that lives inside us
And the people walking down below
Crawling home alone like spiders
As the cancer slowly starts to grow.[1]

1. ^ "Telomere Lyrics".

## 10.9   See also

- Biological clock
- Epigenetic clock
- Centromere
- DNA damage theory of aging
- Immortality
- Maximum life span
- Rejuvenation (aging)
- Senescence, biological aging

## 10.10   References

[1]  Sadava, D., Hillis, D., Heller, C., & Berenbaum, M. (2011). *Life: The science of biology* (9th ed.), Sunderland, MA: Sinauer Associates Inc.

[2]  Okuda K, Bardeguez A, Gardner JP, Rodriguez P, Ganesh V, Kimura M, Skurnick J, Awad G, Aviv A (2002). "Telomere length in the newborn" (PDF). *Pediatric Research*. **52** (3): 377–81. doi:10.1203/00006450-200209000-00012. PMID 12193671.

[3] Arai Y, Martin-Ruiz CM, Takayama M, Abe Y, Takebayashi T, Koyasu S, Suematsu M, Hirose N, von Zglinicki T (2015). "Inflammation, But Not Telomere Length, Predicts Successful Ageing at Extreme Old Age: A Longitudinal Study of Semi-supercentenarians". *EBioMedicine*. **2** (10): 1549–48. doi:10.1016/j.ebiom.2015.07.029. PMC 4634197Ⓒ. PMID 26629551.

[4] Dalgård C, Benetos A, Verhulst S, Labat C, Kark JD, Christensen K, Kimura M, Kyvik KO, Aviv A (2015). "Leukocyte telomere length dynamics in women and men: menopause vs age effects". *International Journal of Epidemiology*. **44** (5): 1688–95. doi:10.1093/ije/dyv165. PMC 4681111Ⓒ. PMID 26385867.

[5] Talks at Google (20 August 2008). "Dr. Elizabeth Blackburn" – via YouTube.

[6] Passarge, Eberhard. *Color atlas of genetics*, 2007.

[7] Olovnikov, Alexei M. (1971). Принцип маргинотомии в матричном синтезе полинуклеотидов [Principle of marginotomy in template synthesis of polynucleotides]. *Doklady Akademii Nauk SSSR* (in Russian). **201** (6): 1496–99. PMID 5158754.

[8] Olovnikov AM (September 1973). "A theory of marginotomy. The incomplete copying of template margin in enzymic synthesis of polynucleotides and biological significance of the phenomenon". *J. Theor. Biol.* **41** (1): 181–90. doi:10.1016/0022-5193(73)90198-7. PMID 4754905.

[9] "No Nobel physiology and medicine award for Russian gerontologist Aleksey Olovnikov". *Telegraph*. October 21, 2009.

[10] Blackburn AM; Gall, Joseph G. (March 1978). "A tandemly repeated sequence at the termini of the extrachromosomal ribosomal RNA genes in Tetrahymena". *J. Mol. Biol.* **120** (1): 33–53. doi:10.1016/0022-2836(78)90294-2. PMID 642006.

[11] "The 2009 Nobel Prize in Physiology or Medicine – Press Release". Nobelprize.org. 2009-10-05. Retrieved 2012-06-12.

[12] Harrison's Principles of Internal Medicine. Ch. 69, Cancer cell biology and angiogenesis. Robert G. Fenton and Dan L. Longo, p. 454.

[13] "Portfolio".

[14] "Unravelling the secret of ageing". *COSMOS: The Science of Everything*. October 5, 2009. Archived from the original on January 14, 2015.

[15] Blasco, Maria; Paula Martinez (21 Jun 2010). "Role of shelterin in cancer and aging". *Aging Cell*. **9** (5): 653–66. doi:10.1111/j.1474-9726.2010.00596.x. PMID 20569239.

[16] Lundblad, 2000; Ferreira *et al.*, 2004

[17] Maloy, Stanley (July 12, 2002). "Bacterial Chromosome Structure". Retrieved 2008-06-22.

[18] Robert P. Lanza, Jose B. Cibelli, Catherine Blackwell, Vincent J. Cristofalo, Mary Kay Francis, Gabriela M. Baerlocher, Jennifer Mak, Michael Schertzer, Elizabeth A. Chavez, Nancy Sawyer, Peter M. Lansdorp, Michael D. West1 (28 April 2000). "Extension of Cell Life-Span and Telomere Length in Animals Cloned from Senescent Somatic Cells" (PDF). Science.

[19] Shampay , Szostak J.W., Blackburn E.H.; Szostak; Blackburn (1984). "DNA sequences of telomeres maintained in yeast". *Nature*. **310** (5973): 154–57. doi:10.1038/310154a0. PMID 6330571.

[20] Williams, TL; Levy, DL; Maki-Yonekura, S; Yonekura, K; Blackburn, EH (2010). "Characterization of the yeast telomere nucleoprotein core: Rap1 binds independently to each recognition site". *J. Biol. Chem.* **285**: 35814–24. doi:10.1074/jbc.M110.170167. PMC 2975205Ⓒ. PMID 20826803.

[21] Griffith J, Comeau L, Rosenfield S, Stansel R, Bianchi A, Moss H, de Lange T; Comeau; Rosenfield; Stansel; Bianchi; Moss; De Lange (1999). "Mammalian telomeres end in a large duplex loop". *Cell*. **97** (4): 503–14. doi:10.1016/S0092-8674(00)80760-6. PMID 10338214.

[22] Burge S, Parkinson G, Hazel P, Todd A, Neidle S; Parkinson; Hazel; Todd; Neidle (2006). "Quadruplex DNA: sequence, topology and structure". *Nucleic Acids Res.* **34** (19): 5402–15. doi:10.1093/nar/gkl655. PMC 1636468Ⓒ. PMID 17012276.

[23] Eisenberg DTA (2011). "An evolutionary review of human telomere biology: The thrifty telomere hypothesis and notes on potential adaptive paternal effects". *American Journal of Human Biology*. **23** (2): 149–67. doi:10.1002/ajhb.21127. PMID 21319244.

[24] Richter, T; von Zglinicki, T (2007). "A continuous correlation between oxidative stress and telomere shortening in fibroblasts". *Exp Gerontol.* **42** (11): 1039–42. doi:10.1016/j.exger.2007.08.005. PMID 17869047.

[25] Shen, J; Gammon, MD; Terry, MB; Wang, Q; Bradshaw, P; Teitelbaum, SL; Neugut, AI; Santella, RM (Apr 2009). "Telomere length, oxidative damage, antioxidants and breast cancer risk". *Int J Cancer*. **124** (7): 1637–43. doi:10.1002/ijc.24105.

[26] Cawthon, RM; Smith, KR; O'Brien, E; Sivatchenko, A; Kerber, RA (2003). "Association between telomere length in blood and mortality in people aged 60 years or older". *Lancet*. **361** (9355): 393–95. doi:10.1016/s0140-6736(03)12384-7.

[27] Hayflick L, Moorhead PS; Moorhead (1961). "The serial cultivation of human diploid cell strains". *Exp Cell Res.* **25** (3): 585–621. doi:10.1016/0014-4827(61)90192-6. PMID 13905658.

[28] Hayflick L. (1965). "The limited in vitro lifetime of human diploid cell strains". *Exp. Cell Res.* **37** (3): 614–36. doi:10.1016/0014-4827(65)90211-9. PMID 14315085.

[29] Feng J, Funk WD, Wang SS, Weinrich SL, Avilion AA, Chiu CP, Adams RR, Chang E, Allsopp RC, Yu J; Funk; Wang; Weinrich; Avilion; Chiu; Adams; Chang; Allsopp; Yu (September 1995). "The RNA component of human telomerase". *Science.* **269** (5228): 1236–41. doi:10.1126/science.7544491. PMID 7544491.

[30] Bodnar, A.G.; Ouellette, M.; Frolkis, M.; Holt, S.E.; Chiu, C.P.; Morin, G.B.; Harley, C.B.; Shay, J.W.; Lichtsteiner, S.; Wright, W.E. (1998). "Extension of life-span by introduction of telomerase into normal human cells". *Science.* **279** (5349): 349–52. doi:10.1126/science.279.5349.349.

[31] Sample, Ian (November 28, 2010). "Harvard scientists reverse the ageing process in mice – now for humans". *The Guardian.* London.

[32] Jaskelioff, Mariela; Muller, Florian L.; Paik, Ji-Hye; Thomas, Emily; Jiang, Shan; Adams, Andrew C.; Sahin, Ergun; Kost-Alimova, Maria; Protopopov, Alexei; Cadiñanos, Juan; Horner, James W.; Maratos-Flier, Eleftheria; DePinho, Ronald A. (6 January 2011). "Telomerase reactivation reverses tissue degeneration in aged telomerase-deficient mice". *Nature.* **469** (7328): 102–06. doi:10.1038/nature09603. PMC 3057569. PMID 21113150 – via www.nature.com.

[33] Joeng KS, Song EJ, Lee KJ, Lee J; Song; Lee; Lee (2004). "Long lifespan in worms with long telomeric DNA". *Nature Genetics.* **36** (6): 607–11. doi:10.1038/ng1356. PMID 15122256.

[34] Nakagawa S, Gemmell NJ, Burke T; Gemmell; Burke (September 2004). "Measuring vertebrate telomeres: applications and limitations". *Mol. Ecol.* **13** (9): 2523–33. doi:10.1111/j.1365-294X.2004.02291.x. PMID 15315667.

[35] Juola, Frans A; Haussmann, Mark F; Dearborn, Donald C; Vleck, Carol M (2006). "Telomere shortening in a long-lived marine bird: Cross-sectional analysis and test of an aging tool". *The Auk.* **123** (3): 775. doi:10.1642/0004-8038(2006)123[775:TSIALM]2.0.CO;2. ISSN 0004-8038.

[36] Gomes, NM; Ryder, OA; Houck, ML; Charter, SJ; Walker, W; Forsyth, NR; Austad, SN; Venditti, C; Pagel, M; Shay, JW; Wright, WE (2011). "Comparative biology of mammalian telomeres: hypotheses on ancestral states and the roles of telomeres in longevity determination". *Aging Cell.* **10** (5): 761–68. doi:10.1111/j.1474-9726.2011.00718.x. PMC 3387546. PMID 21518243.

[37] Harris, SE; Martin-Ruiz, C; von Zglinicki, T; Starr, JM; Deary, IJ (2010). "Telomere length and aging biomarkers in 70-year-olds: the Lothian Birth Cohort 1936". *Neurobiol Aging.* **33** (7): 1486.e3–1486.e8. doi:10.1016/j.neurobiolaging.2010.11.013. PMID 21194798.

[38] Gilley, D; Blackburn, EH (1994). "Lack of telomere shortening during senescence in Paramecium". *Proc Natl Acad Sci U S A.* **91** (5): 1955–58. doi:10.1073/pnas.91.5.1955. PMC 43283. PMID 8127914.

[39] Fernandez, Elizabeth (2013-09-16). "Lifestyle Changes May Lengthen Telomeres, A Measure of Cell Aging". *http://www.ucsf.edu/.* University of California, San Francisco. Retrieved 2015-03-16. External link in |website= (help)

[40] Sjögren, P; Fisher, R; Kallings, L; Svenson, U; Roos, G; Hellénius, M (2014-09-03). "Stand up for health – avoiding sedentary behaviour might lengthen your telomeres: secondary outcomes from a physical activity RCT in older people.". *Br J Sports Med.* **48**: 1407–09. doi:10.1136/bjsports-2013-093342. PMID 25185586.

[41] Peška, Vratislav; Fajkus, Petr; Fojtová, Miloslava; Dvořáčková, Martina; Hapala, Jan; Dvořáček, Vojtěch; Polanská, Pavla; Leitch, Andrew R.; Sýkorová, Eva; Fajkus, Jiří (May 2015). "Characterisation of an unusual telomere motif (TTTTTTAGGG) in the plant (Solanaceae), a species with a large genome". *The Plant Journal.* **82** (4): 644–54. doi:10.1111/tpj.12839.

[42] Fajkus, Petr; Peška, Vratislav; Sitová, Zdeňka; Fulnečková, Jana; Dvořáčková, Martina; Gogela, Roman; Sýkorová, Eva; Hapala, Jan; Fajkus, Jiří (2016). "Allium telomeres unmasked: the unusual telomeric sequence (CTCGGTTATGGG)n is synthesized by telomerase". *The Plant Journal.* **85** (3): 337–47. doi:10.1111/tpj.13115.

[43] Raynaud, CM; Sabatier, L; Philipot, O; Olaussen, KA; Soria, JC (2008). "Telomere length, telomeric proteins and genomic instability during the multistep carcinogenic process". *Crit Rev Oncol Hematol.* **66**: 99–117. doi:10.1016/j.critrevonc.2007.11.006.

[44] Blasco, MA; Lee, HW; Hande, MP; Samper, E; Lansdorp, PM; et al. (1997). "Telomere shortening and tumor formation by mouse cells lacking telomerase RNA". *Cell.* **91** (1): 25–34. doi:10.1016/s0092-8674(01)80006-4. PMID 9335332.

[45] Artandi, SE; Chang, S; Lee, SL; Alson, S; Gottlieb, GJ; et al. (2000). "Telomere dysfunction promotes non-reciprocal translocations and epithelial cancers in mice". *Nature.* **406**: 641–45. doi:10.1038/35020592.

[46] Willeit Peter, Willeit Johann, Mayr Anita, Weger Siegfried, Oberhollenzer Friedrich, Brandstätter Anita, Kronenberg Florian, Kiechl Stefan; Willeit; Mayr; Weger; Oberhollenzer; Brandstätter; Kronenberg; Kiechl (2010). "Telomere length and risk of incident cancer and cancer mortality". *JAMA.* **304** (1): 69–75. doi:10.1001/jama.2010.897. PMID 20606151.

[47] Ma, H; Zhou, Z; Wei, S; et al. (2011). "Shortened telomere length is associated with increased risk of cancer: a meta-analysis". *PLOS ONE.* **6** (6): e20466. doi:10.1371/journal.pone.0020466.

[48] Wentzensen, IM; Mirabello, L; Pfeiffer, RM; Savage, SA (2011). "The association of telomere length and cancer: a meta-analysis". *Cancer Epidemiol Biomarkers Prev*. **20** (6): 1238–50. doi:10.1158/1055-9965.epi-11-0005.

[49] Paul, L (Oct 2011). "Diet, nutrition and telomere length". *J Nutr Biochem*. **22** (10): 895–901. doi:10.1016/j.jnutbio.2010.12.001.

[50] Epel, ES; Lin, J; Wilhelm, FH; Wolkowitz, OM; Cawthon, R; Adler, NE; Dolbier, C; Mendes, WB; Blackburn, EH (April 2006). "Cell aging in relation to stress arousal and cardiovascular disease risk factors". *Psychoneuroendocrinology*. **31** (3): 277–87. doi:10.1016/j.psyneuen.2005.08.011.

[51] Ornish, D; Lin, J; Chan, JM; Epel, E; Kemp, C; Weidner, G; Marlin, R; Frenda, SJ; Magbanua, MJ; Daubenmier, J; Estay, I; Hills, NK; Chainani-Wu, N; Carroll, PR; Blackburn, EH (Oct 2013). "Effect of comprehensive lifestyle changes on telomerase activity and telomerelength in men with biopsy-proven low-risk prostate cancer: 5-year follow-up of a descriptive pilot study". *Lancet Oncol*. **14** (11): 1112–20. doi:10.1016/S1470-2045(13)70366-8.

[52] Ornish, D; Lin, J; Daubenmier, J; Weidner, G; Epel, E; Kemp, C; Magbanua, MJ; Marlin, R; Yglecias, L; Carroll, PR; Blackburn, EH (Nov 2008). "Increased telomerase activity and comprehensive lifestyle changes: a pilot study". *Lancet Oncol*. **9** (11): 1048–57. doi:10.1016/S1470-2045(08)70234-1.

[53] Aschacher; Wolf; Enzmann; Kienzl (2015). "ALINE-1 induces hTERT and ensures telomere maintenance in tumour cell lines". *Oncogene*. **35**: 94–104. doi:10.1038/onc.2015.65. PMID 25798839.

[54] Henson JD, Neumann AA, Yeager TR, Reddel RR; Neumann; Yeager; Reddel (2002). "Alternative lengthening of telomeres in mammalian cells". *Oncogene*. **21** (4): 598–610. doi:10.1038/sj.onc.1205058. PMID 11850785.

[55] Chris Molenaar; Karien Wiesmeijer; Nico P. Verwoerd; Shadi Khazen; Roland Eils; Hans J. Tanke & Roeland W. Dirks (2003-12-15). "Visualizing telomere dynamics in living mammalian cells using PNA probes". *The EMBO Journal*. The European Molecular Biology Organization. **22** (24): 6631–41. doi:10.1093/emboj/cdg633. PMC 291828. PMID 14657034.

[56] Philippi C, Loretz B, Schaefer UF, Lehr CM.; Loretz; Schaefer; Lehr (April 2010). "Telomerase as an emerging target to fight cancer – Opportunities and challenges for nanomedicine". *Journal of Controlled Release*. **146** (2): 228–40. doi:10.1016/j.jconrel.2010.03.025. PMID 20381558.

[57] Chin L, Artandi SE, Shen Q, et al. (May 1999). "p53 deficiency rescues the adverse effects of telomere loss and cooperates with telomere dysfunction to accelerate carcinogenesis". *Cell*. **97** (4): 527–38. doi:10.1016/S0092-8674(00)80762-X. PMID 10338216.

[58] Greenberg RA, Chin L, Femino A, et al. (May 1999). "Short dysfunctional telomeres impair tumorigenesis in the INK4a(delta2/3) cancer-prone mouse". *Cell*. **97** (4): 515–25. doi:10.1016/S0092-8674(00)80761-8. PMID 10338215.

[59] Henson, JD; Neumann, AA; Yeager, TR; Reddel, RR (2002). "Alternative lengthening of telomeres in mammalian cells". *Oncogene*. **21** (4): 598–610. doi:10.1038/sj.onc.1205058. PMID 11850785.

[60] BBC, World/Mundo. "Resuelven misterio sobre el origen del cáncer de mama".

[61] Kannan, Nagarajan; Nazmul Huda, LiRen Tu, Radina Droumeva, Geraldine Aubert, Elizabeth Chavez, Ryan R. Brinkman, Peter Lansdorp, Joanne Emerman, Satoshi Abe, Connie Eaves, David Gilley (4 June 2013). "The Luminal Progenitor Compartment of the Normal Human Mammary Gland Constitutes a Unique Site of Telomere Dysfunction". *Stem Cell Reports*. **1** (1): 28–31. doi:10.1016/j.stemcr.2013.04.003. PMID 24052939.

[62] Allshire RC; et al. (1989). "Human telomeres contain at least three types of G-rich repeat distributed nonrandomly". *Nucleic Acids Res*. **17** (12): 4611–27. doi:10.1093/nar/17.12.4611. PMC 318019. PMID 2664709.

[63] Rufer N; et al. (1998). "Telomere length dynamics in human lymphocyte subpopulations measured by flow cytometry". *Nat Biotechnol*. **16** (8): 743–47. doi:10.1038/nbt0898-743. PMID 9702772.

[64] Cawthon, RM (2002). "Telomere measurement by quantitative PCR". *Nucleic Acids Research*. **30** (10): e47. doi:10.1093/nar/30.10.e47. PMC 115301. PMID 12000852.

[65] "Titanovo, Inc". Titanovo.com. Retrieved 2015-04-15.

[66] "Telome Health, Inc". Telomehealth.com. Retrieved 2013-07-13.

[67] "TeloMe Home". Telome.com. Retrieved 2013-07-13.

[68] "A Blood Test Offers Clues to Longevity".

[69] Zglinicki, T. v. (13 March 2012). "Will your telomeres tell your future?" (PDF). *BMJ*. **344** (mar13 1): e1727. doi:10.1136/bmj.e1727.

[70] Jo Marchant. "Spit test offers guide to health : Nature News". Nature.com. Retrieved 2013-07-13.

[71] "Elizabeth Blackburn calls time on 'fountain of youth' firm Telome Health".

## 10.11    Further reading

- Aubert G.; Lansdorp P.M. (April 2008). "Telomeres and Aging". *Physiological Reviews*. **88** (2): 557–79. doi:10.1152/physrev.00026.2007. PMID 18391173.

- Cong YS, Wright WE, Shay JW (September 2002). "Human telomerase and its regulation". *Microbiol. Mol. Biol. Rev.* **66** (3): 407–25, table of contents.   doi:10.1128/MMBR.66.3.407-425.2002. PMC 120798. PMID 12208997.

- Eisenberg DTA (2011).   "An evolutionary review of human telomere biology: The thrifty telomere hypothesis and notes on potential adaptive paternal effects". *American Journal of Human Biology*. **23** (2): 149–67. doi:10.1002/ajhb.21127. PMID 21319244.

- Tomaska L.; Nosek J.; Kramara J.; Griffith J.D. (2009). "Telomeric circles: universal players in telomere maintenance". *Nature Structural & Molecular Biology*. **16** (10): 1010–15.   doi:10.1038/nsmb.1660. PMID 19809492.

- Weinstein BS, Ciszek D; Lansdorp (May 2002). "The reserve-capacity hypothesis: evolutionary origins and modern implications of the trade-off between tumor-suppression and tissue-repair". *Exp. Gerontol.* **37** (5): 615–27.   doi:10.1016/S0531-5565(02)00012-8. PMID 11909679. – A paper detailing the evolutionary origins and medical implications of the vertebrate telomere system, including the pervasive trade-off between cancer prevention and damage repair. Also addresses the probable danger posed by the elongation of telomeres in lab mice.

## 10.12    External links

- Elizabeth Blackburn's seminars:   "Telomeres and Telomerase"

- Telomeres and Telomerase: The Means to the End Nobel Lecture by Elizabeth Blackburn, which includes a reference to the impact of stress, and pessimism on telomere length

- Telomerase and the Consequences of Telomere Dysfunction Nobel Lecture by Carol Greider

- DNA Ends: Just the Beginning Nobel Lecture by Jack Szostak

# Chapter 11

# Telomerase

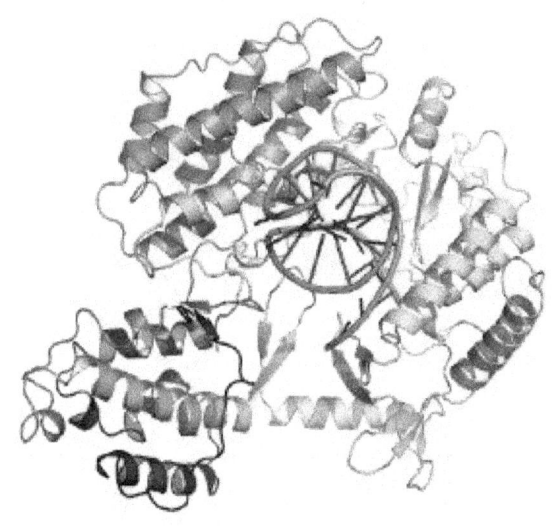

Tribolium castaneum *telomerase catalytic subunit, TERT, bound to putative RNA template and telomeric DNA (PDB 3KYL)*

**Telomerase**, also called **terminal transferase**,[1] is a ribonucleoprotein that adds a species-dependent telomere repeat sequence to the 3' end of telomeres. A telomere is a region of repetitive sequences at each end of a eukaryotic chromosomes in most eukaryotes. Telomeres protect the end of the chromosome from DNA damage or from fusion with neighbouring chromosomes. The fruit fly Drosophila melanogaster lacks telomerase, but instead uses retrotransposons to maintain telomeres.[2]

Telomerase is a reverse transcriptase enzyme that carries its own RNA molecule (e.g., with the sequence "CCCAAUCCC" in vertebrates) which is used as a template when it elongates telomeres. Telomerase, active in normal stem cells and most cancer cells, is normally absent from, or at very low levels in, most somatic cells.

## 11.1 History

The existence of a compensatory mechanism for telomere shortening was first found by Soviet biologist Alexey Olovnikov in 1973,[3] who also suggested the telomere hypothesis of aging and the telomere's connections to cancer.

Telomerase in the ciliate *Tetrahymena* was discovered by Carol W. Greider and Elizabeth Blackburn in 1984.[4] Together with Jack W. Szostak, Greider and Blackburn were awarded the 2009 Nobel Prize in Physiology or Medicine for their discovery.[5]

The role of telomeres and telomerase in cell aging and cancer was established by scientists at biotechnology company Geron with the cloning of the RNA and catalytic components of human telomerase[6] and the development of a polymerase chain reaction (PCR) based assay for telomerase activity called the TRAP assay, which surveys telomerase activity in multiple types of cancer.[7]

## 11.2 Human telomerase structure

The molecular composition of the human telomerase enzyme complex was determined by Dr Scott Cohen and his team at the Children's Medical Research Institute (Sydney Australia) and consists of two molecules each of human telomerase reverse transcriptase (TERT), telomerase RNA (TR or TERC), and dyskerin (DKC1).[8] The genes of telomerase subunits, which include TERT,[9] TERC,[10] DKC1[11] and TEP1,[12] are located on different chromosomes. The human TERT gene (hTERT) is translated into a protein of 1132 amino acids.[13] TERT polypeptide folds with (and carries) TERC, a non-coding RNA (451 nucleotides long). TERT has a 'mitten' structure that allows it to wrap around the chromosome to add single-stranded telomere repeats.

TERT is a reverse transcriptase, which is a class of enzyme that creates single-stranded DNA using single-stranded RNA as a template.

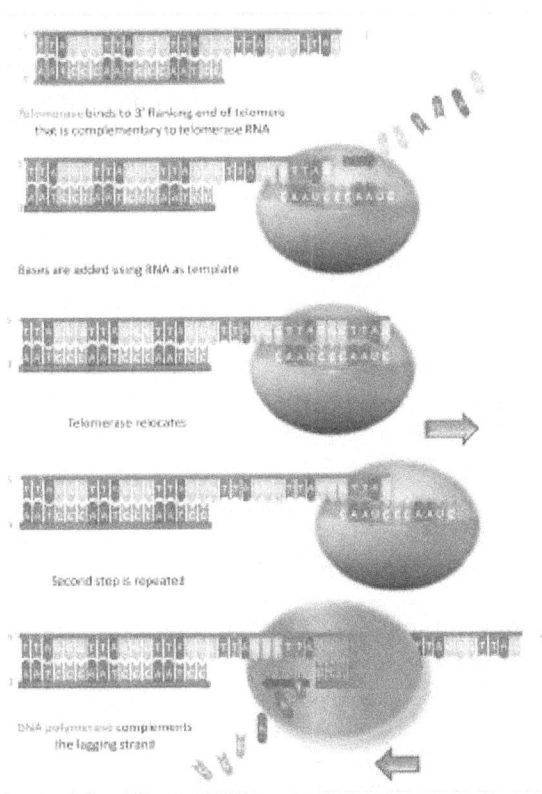

*An image illustrating how telomerase elongates telomere ends progressively.*

The protein consists of four conserved domains (RNA-Binding Domain (TRBD), fingers, palm and thumb), organized into a ring configuration that shares common features with retroviral reverse transcriptases, viral RNA polymerases and bacteriophage B-family DNA polymerases.[14][15]

TERT proteins from many eukaryotes have been sequenced.[16]

## 11.3   Function

By using TERC, TERT can add a six-nucleotide repeating sequence, 5'-TTAGGG (in vertebrates, the sequence differs in other organisms) to the 3' strand of chromosomes. These TTAGGG repeats (with their various protein binding partners) are called telomeres. The template region of TERC is 3'-CAAUCCCAAUC-5'.[17]

Telomerase can bind the first few nucleotides of the template to the last telomere sequence on the chromosome, add a new telomere repeat (5'-GGTTAG-3') sequence, let go, realign the new 3'-end of telomere to the template, and repeat the process. Telomerase reverses telomere shortening.

## 11.4   Clinical implications

### 11.4.1   Aging

Telomerase replaces short bits of DNA known as telomeres, which are otherwise shortened when a cell divides via mitosis.

In normal circumstances, absent telomerase, if a cell divides recursively, at some point the progeny reach their Hayflick limit,[18] which is believed to be between 50–70 cell divisions. At the limit the cells become senescent and cell division stops.[19] Telomerase allows each offspring to replace the lost bit of DNA allowing the cell line to divide without ever reaching the limit. This same unbounded growth is a feature of cancerous growth.[20]

Embryonic stem cells express telomerase, which allows them to divide repeatedly and form the individual. In adults, telomerase is highly expressed only in cells that need to divide regularly especially in male sperm cells but also in epidermal cells,[21] in activated T cell[22] and B cell[23] lymphocytes, as well as in certain adult stem cells, but in the great majority of cases somatic cells do not express telomerase.[24]

A comparative biology study of mammalian telomeres indicated that telomere length of some mammalian species correlates inversely, rather than directly, with lifespan, and concluded that the contribution of telomere length to lifespan is unresolved.[25] Telomere shortening does not occur with age in some postmitotic tissues, such as in the rat brain.[26] In humans, skeletal muscle telomere lengths remain stable from ages 23 –74.[27] In baboon skeletal muscle, which consists of fully differentiated post-mitotic cells, less than 3% of myonuclei contain damaged telomeres and this percentage does not increase with age.[28] Thus, telomere shortening does not appear to be a major factor in the aging of the differentiated cells of brain or skeletal muscle. In human liver, cholangiocytes and hepatocytes show no age-related telomere shortening.[29] Another study found little evidence that, in humans, telomere length is a significant biomarker of normal aging with respect to important cognitive and physical abilities.[30]

Some experiments have raised questions on whether telomerase can be used as an anti-aging therapy, namely, the fact that mice with elevated levels of telomerase have higher cancer incidence and hence do not live longer. Telomerase also favors tumorogenesis, which leads to questions about its potential as an anti-aging therapy.[31] On the other hand, one study showed that activating telomerase in cancer-resistant mice by overexpressing its catalytic subunit extended lifespan.[32]

Exposure of T lymphocytes from HIV-infected human

donors to a small molecule telomerase activator (TAT2) retards telomere shortening, increases proliferative potential and enhances cytokine/chemokine production and antiviral activity.[33]

A study that focused on Ashkenazi Jews found that long-lived subjects inherited a hyperactive version of telomerase.[34]

Mice engineered to block the gene that produces telomerase, unless they are given a certain drug, aged at a much faster rate, and died at about six months, instead of reaching the average mouse lifespan, about three years. Administering the drug at 6 months turned on telomerase production and caused their organs to be "rejuvenated," restored fertility, and normalized their ability to detect or process odors.[35][36]

A 2012 study reported that introducing the TERT gene into healthy one-year-old mice using an engineered adeno-associated virus led to a 24% increase in lifespan, without any increase in cancer.[37]

**Premature aging**

Premature aging syndromes including Werner syndrome, Ataxia telangiectasia, Ataxia-telangiectasia like disorder, Bloom syndrome, Fanconi anemia and Nijmegen breakage syndrome are associated with short telomeres.[38] However, the genes that have mutated in these diseases all have roles in the repair of DNA damage and the increased DNA damage may, itself, be a factor in the premature aging (see DNA damage theory of aging). An additional role in maintaining telomere length is an active area of investigation.

## 11.4.2  Cancer

*In vitro*, when cells approach the Hayflick limit, the time to senescence can be extended by inactivating the tumor suppressor proteins - p53 and Retinoblastoma protein (pRb). Cells that have been so-altered eventually undergo an event termed a "crisis" when the majority of the cells in the culture die. Sometimes, a cell does not stop dividing once it reaches crisis. In a typical situation, the telomeres are shortened[39] and chromosomal integrity declines with every subsequent cell division. Exposed chromosome ends are interpreted as double-stranded breaks (DSB) in DNA; such damage is usually repaired by reattaching (religating) the broken ends together. When the cell does this due to telomere-shortening, the ends of different chromosomes can be attached to each other. This solves the problem of lacking telomeres, but during cell division anaphase, the fused chromosomes are randomly ripped apart, causing many mutations and chromosomal abnormalities. As

this process continues, the cell's genome becomes unstable. Eventually, either fatal damage is done to the cell's chromosomes (killing it via apoptosis), or an additional mutation that activates telomerase occurs.

With telomerase activation some types of cells and their offspring become immortal (bypass the Hayflick limit), thus avoiding cell death as long as the conditions for their duplication are met. Many cancer cells are considered 'immortal' because telomerase activity allows them to live much longer than any other somatic cell, which, combined with uncontrollable cell proliferation[40] is why they can form tumors. A good example of immortal cancer cells is HeLa cells, which have been used in laboratories as a model cell line since 1951.

While this method of modeling human cancer in cell culture is effective and has been used for many years by scientists, it is also very imprecise. The exact changes that allow for the formation of the tumorigenic clones in the above-described experiment are not clear. Scientists addressed this question by the serial introduction of multiple mutations present in a variety of human cancers. This has led to the identification of mutation combinations that form tumorigenic cells in a variety of cell types. While the combination varies by cell type, the following alterations are required in all cases: TERT activation, loss of p53 pathway function, loss of pRb pathway function, activation of the Ras or myc proto-oncogenes, and aberration of the PP2A protein phosphatase. That is to say, the cell has an activated telomerase, eliminating the process of death by chromosome instability or loss, absence of apoptosis-induction pathways, and continued mitosis activation.

This model of cancer in cell culture accurately describes the role of telomerase in actual human tumors. Telomerase activation has been observed in ~90% of all human tumors,[41] suggesting that the immortality conferred by telomerase plays a key role in cancer development. Of the tumors without TERT activation,[42] most employ a separate pathway to maintain telomere length termed Alternative Lengthening of Telomeres (ALT ).[43] The exact mechanism behind telomere maintenance in the ALT pathway is unclear, but likely involves multiple recombination events at the telomere.

Elizabeth Blackburn *et al.*, identified the upregulation of 70 genes known or suspected in cancer growth and spread through the body, and the activation of glycolysis, which enables cancer cells to rapidly use sugar to facilitate their programmed growth rate (roughly the growth rate of a fetus).[44]

Approaches to controlling telomerase and telomeres for cancer therapy include gene therapy, immunotherapy, small-molecule and signal pathway inhibitors.[45]

## Drugs

The ability to maintain functional telomeres may be one mechanism that allows cancer cells to grow *in vitro* for decades.[46] Telomerase activity is necessary to preserve many cancer types and is inactive in somatic cells, creating the possibility that telomerase inhibition could selectively repress cancer cell growth with minimal side effects.[47] If a drug can inhibit telomerase in cancer cells, the telomeres of successive generations will progressively shorten, limiting tumor growth.[48]

Telomerase is a good biomarker for cancer detection because most human cancers cells express high levels of it. Telomerase activity can be identified by its catalytic protein domain (hTERT). This is the rate-limiting step in telomerase activity. It is associated with many cancer types. Various cancer cells and fibroblasts transformed with hTERT cDNA have high telomerase activity, while somatic cells do not. Cells testing positive for hTERT have positive nuclear signals. Epithelial stem cell tissue and its early daughter cells are the only noncancerous cells in which hTERT can be detected. Since hTERT expression is dependent only on the number of tumor cells within a sample, the amount of hTERT indicates the severity of a cancer.[49]

The expression of hTERT can also be used to distinguish benign tumors from malignant tumors. Malignant tumors have higher hTERT expression than benign tumors. Real-time reverse transcription polymerase chain reaction (RT-PCR) quantifying hTERT expression in various tumor samples verified this varying expression.[50]

The lack of telomerase does not affect cell growth, until the telomeres are short enough to cause cells to "die or undergo growth arrest". However, inhibiting telomerase alone is not enough to destroy large tumors. It must be combined with surgery, radiation, chemotherapy or immunotherapy.[49]

Cells may reduce their telomere length by only 50-252 base pairs per cell division, which can lead to a long lag phase.[51][52]

**Immunotherapy** Immunotherapy successfully treats some kinds of cancer, such as melanoma. This treatment involves manipulating a human's immune system to destroy cancerous cells. Humans have two major antigen identifying lymphocytes: CD8+ cytotoxic T-lymphocytes (CTL) and CD4+ helper T-lymphocytes that can destroy cells. Antigen receptors on CTL can bind to a 9-10 amino acid chain that is presented by the major histocompatibility complex (MHC) as in Figure 4. HTERT is a potential target antigen. Immunotargeting should result in relatively few side effects since hTERT expression is associated only with telomerase and is not essential in almost all somatic cells.[53] GV1001 uses this pathway.[45] Experimental drug

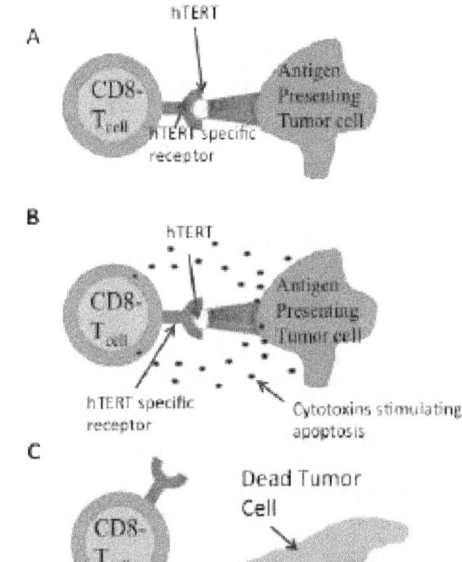

*Figure 4:A) Tumor cells expressing hTERT will actively degrade some of the protein and process for presenting. The major histocompatibility complex 1(MHC1), can then present the hTERT epitote. CD8- T cells that have antibodies against hTERT will then bind to the presented epitote. B) As a result of the antigenic binding, the T cells will release cytotoxins, which can be absorbed by the affected cell. C) These cytotoxins induce multiple proteases and results in apoptosis (or cell death).*

and vaccine therapies targeting active telomerase have been tested in mouse models, and clinical trials have begun.

In 2014 Geron Corporation received permission to resume a trial of its drug imetelstat for myelofibrosis after addressing FDA concerns over liver toxicity.[54][55] Geron licensee Merck had approval of an IND for one vaccine type. Imetelstat (GRN163L) binds directly to the telomerase's RNA template. One 2015 study reported that Imetelstat caused partial or complete remission in seven of 33 patients, while a second reported that it decreased blood platelet levels in all 18 study patients with essential thrombocythemia, a disorder in which the body overproduces blood platelets, increasing the risk of blood clots.[56]

Most of the harmful cancer-related effects of telomerase are dependent on an intact RNA template. Cancer stem cells that use an alternative method of telomere maintenance are still killed when telomerase's RNA template is blocked or damaged.

**Telomerase Vaccines**

Two telomerase vaccines have been developed: GRNVAC1 and GV1001. GRNVAC1 isolates dendritic cells and the RNA that codes for the telomerase protein and puts them back into the patient to make cytotoxic T cells that kill the telomerase-active cells. GV1001 is a peptide from the active site of hTERT and is recognized by the immune system that reacts by killing the telomerase-active cells.[45]

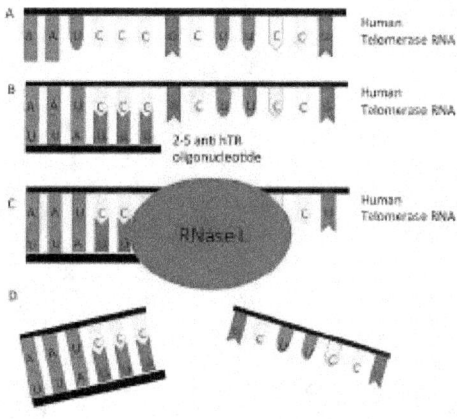

*Figure 5: A) Human telomerase RNA (hTR) is present in the cell and can be targeted. B) 2-5 anti-hTR oligonucleotides is a specialized antisense oligo that can bind to the telomerase RNA. C) Once bound, the 2-5 anti-hTR oligonucleotide recruits RNase L to the sequence. Once recruited, the RNase L creates a single cleavage in the RNA (D) and causes dissociation of the RNA sequence.*

**Targeted apoptosis** Another independent approach is to use oligoadenylated anti-telomerase antisense oligonucleotides and ribozymes to target telomerase RNA, inducing dissociation and apoptosis (Figure 5). The fast induction of apoptosis through antisense binding may be a good alternative to the slower telomere shortening.[51]

### 11.4.3 Heart disease, diabetes and quality of life

Blackburn also discovered that mothers caring for very sick children have shorter telomeres when they report that their emotional stress is at a maximum and that telomerase was active at the site of blockages in coronary artery tissue, possibly accelerating heart attacks.

In 2009, it was shown that the amount of telomerase activity significantly increased following psychological stress. Across the sample of patients telomerase activity in increased peripheral blood mononuclear cells by 18% one hour after the end of the stress.[57]

E. V. Gostjeva *et al.* found no differences between colon cancer stem cells and fetal colon stem cells.[58]

A study in 2010 found that there was "significantly greater" telomerase activity in participants than controls after a three-month meditation retreat.[59]

Telomerase deficiency has been linked to diabetes mellitus and impaired insulin secretion in mice, due to loss of pancreatic insulin-producing cells.[60]

### 11.4.4 Rare human diseases

Mutations in TERT have been implicated in predisposing patients to aplastic anemia, a disorder in which the bone marrow fails to produce blood cells, in 2005.[61]

Cri du chat syndrome (CdCS) is a complex disorder involving the loss of the distal portion of the short arm of chromosome 5. TERT is located in the deleted region, and loss of one copy of TERT has been suggested as a cause or contributing factor of this disease.[62]

Dyskeratosis congenita (DC) is a disease of the bone marrow that can be caused by some mutations in the telomerase subunits.[63] In the DC cases, about 35% cases are X-linked-recessive on the DKC1 locus[64] and 5% cases are autosomal dominant on the TERT[65] and TERC[66] loci.

Patients with DC have severe bone marrow failure manifesting as abnormal skin pigmentation, leucoplakia (a white thickening of the oral mucosa) and nail dystrophy, as well as a variety of other symptoms. Individuals with either TERC or DKC1 mutations have shorter telomeres and defective telomerase activity *in vitro* versus other individuals of the same age.[67]

In one family autosomal dominant DC was linked to a heterozygous TERT mutation.[68] These patients also exhibited an increased rate of telomere-shortening, and genetic anticipation (i.e., the DC phenotype worsened with each generation).

## 11.5   See also

- DNA repair
- Imetelstat
- TA-65
- Telomere

## 11.6   References

[1]  What are telomeres and telomerase?

[2] Pardue ML, DeBaryshe PG (2011). "Retrotransposons that maintain chromosome ends" (PDF). *PNAS.* **108** (51): 20317–24. doi:10.1073/pnas.1100278108. PMC 3251079. PMID 21821789.

[3] Olovnikov AM (September 1973). "A theory of marginotomy. The incomplete copying of template margin in enzymic synthesis of polynucleotides and biological significance of the phenomenon". *J. Theor. Biol.* **41** (1): 181–90. doi:10.1016/0022-5193(73)90198-7. PMID 4754905.

[4] Greider CW, Blackburn EH (December 1985). "Identification of a specific telomere terminal transferase activity in Tetrahymena extracts". *Cell.* **43** (2 Pt 1): 405–13. doi:10.1016/0092-8674(85)90170-9. PMID 3907856.

[5] "The Nobel Prize in Physiology or Medicine 2009". The Nobel Foundation. 2009-10-05. Retrieved 2010-10-23.

[6] Feng J, Funk WD, Wang SS, Weinrich SL, Avilion AA, Chiu CP, Adams RR, Chang E, Allsopp RC, Yu J (September 1995). "The RNA component of human telomerase". *Science.* **269** (5228): 1236–41. doi:10.1126/science.7544491. PMID 7544491.

[7] Kim, N.; Piatyszek, M.; Prowse, K.; Harley, C.; West, M.; Ho, P.; Coviello, G.; Wright, W.; Weinrich, S. (1994). "Specific association of human telomerase activity with immortal cells and cancer". *Science.* **266** (5193): 2011–15. doi:10.1126/science.7605428. PMID 7605428.

[8] Cohen S, Graham M, Lovrecz G, Bache N, Robinson P, Reddel R (2007). "Protein composition of catalytically active human telomerase from immortal cells". *Science.* **315** (5820): 1850–3. doi:10.1126/science.1138596. PMID 17395830.

[9] "HGNC database of human gene names - HUGO Gene Nomenclature Committee". *genenames.org.*

[10] HGNC - TERC

[11] HGNC - DKC1

[12] HGNC - TEP1

[13] NCBI - telomerase reverse transcriptase isoform 1

[14] Gillis AJ, Schuller AP, Skordalakes E. Structure of the Tribolium castaneum telomerase catalytic subunit TERT. Nature. 2008 Oct 2;455(7213):633-7

[15] Mitchell M, Gillis A, Futahashi M, Fujiwara H, Skordalakes E. Structural basis for telomerase catalytic subunit TERT binding to RNA template and telomeric DNA. Nat Struct Mol Biol. 2010 Apr;17(4):513-8

[16] NCBI - telomerase reverse transcriptase

[17] Gavory G, Farrow M, Balasubramanian S (October 2002). "Minimum length requirement of the alignment domain of human telomerase RNA to sustain catalytic activity in vitro". *Nucleic Acids Res.* **30** (20): 4470–80. doi:10.1093/nar/gkf575. PMC 137139. PMID 12384594.

[18] Hayflick L, Moorhead PS (1961). "The serial cultivation of human diploid cell strains". *Exp Cell Res.* **25** (3): 585–621. doi:10.1016/0014-4827(61)90192-6. PMID 13905658.

[19] Siegel, L. (2013). Are Telomeres the Key to Aging and Cancer? The University of Utah. Retrieved 30 September 2013

[20] Hanahan D, Weinberg RA (March 2011). "Hallmarks of cancer: the next generation". *Cell.* **144** (5): 646–74. doi:10.1016/j.cell.2011.02.013. PMID 21376230.

[21] Härle-Bachor C, Boukamp P (1996). "Telomerase activity in the regenerative basal layer of the epidermis in human skin and in immortal and carcinoma-derived skin keratinocytes". *PNAS.* **93** (13): 6476–6481. doi:10.1073/pnas.93.13.6476. PMC 39048. PMID 8692840.

[22] Barsov EV (2011). "Telomerase and primary T cells: biology and immortalization for adoptive immunotherapy". *Immunotherapy (journal).* **3** (3): 407–421. doi:10.2217/imt.10.107. PMC 3120014. PMID 21395382.

[23] Bougel S, Renaud S, Braunschweig R, Loukinov D, Morse HC 3rd, Bosman FT, Lobanenkov V, Benhattar J (2010). "PAX5 activates the transcription of the human telomerase reverse transcriptase gene in B cells". *The Journal of Pathology.* **220** (1): 87–96. doi:10.1002/path.2620. PMC 3422366. PMID 19806612.

[24] Cong YS (2002). "Human Telomerase and Its Regulation". *Microbiology and Molecular Biology Reviews.* **66** (3): 407–425. doi:10.1128/MMBR.66.3.407-425.2002. PMC 120798. PMID 12208997.

[25] Gomes, NM; Ryder, OA; Houck, ML; Charter, SJ; Walker, W.; Forsyth, NR; Austad, SN; Venditti, C; Pagel, M; Shay, JW; Wright, WE (2011). "Comparative biology of mammalian telomeres: hypotheses on ancestral states and the roles of telomeres in longevity determination.". *Aging Cell.* **10** (5): 761–768. doi:10.1111/j.1474-9726.2011.00718.x. PMC 3387546. PMID 21518243.

[26] Cherif H, Tarry JL, Ozanne SE, Hales CN (2003). "Ageing and telomeres: a study into organ- and gender-specific telomere shortening". *Nucleic Acids Res.* **31** (5): 1576–1583. doi:10.1093/nar/gkg208. PMC 149817. PMID 12595567.

[27] Renault, V; Thornell, LE; Eriksson, PO; Butler-Browne, G; Mouly, V (Dec 2002). "Regenerative potential of human skeletal muscle during aging.". *Aging Cell.* **1** (2): 132–9. doi:10.1046/j.1474-9728.2002.00017.x. PMID 12882343.

[28] Jeyapalan JC, Ferreira M, Sedivy JM, Herbig U (2007). "Accumulation of senescent cells in mitotic tissue of aging primates". *Mech Ageing Dev.* **128** (1): 36–44. doi:10.1016/j.mad.2006.11.008. PMC 3654105. PMID 17116315.

[29] Verma S, Tachtatzis P, Penrhyn-Lowe S, Scarpini C, Jurk D, Von Zglinicki T, Coleman N, Alexander GJ (2012). "Sustained telomere length in hepatocytes and cholangiocytes with increasing age in normal liver". *Hepatology*. **56** (4): 1510–1520. doi:10.1002/hep.25787. PMID 22504828.

[30] Harris, SE; Martin-Ruiz, C; von Zglinicki, T; Starr, JM; Deary, IJ (Jul 2012). "Telomere length and aging biomarkers in 70-year-olds: the Lothian Birth Cohort 1936.". *Neurobiol Aging*. **33** (7): 1486.e3–8. doi:10.1016/j.neurobiolaging.2010.11.013. PMID 21194798.

[31] de Magalhães JP, Toussaint O (2004). "Telomeres and telomerase: a modern fountain of youth?". *Rejuvenation Res*. **7** (2): 126–33. doi:10.1089/1549168041553044. PMID 15312299.

[32] Tomás-Loba A, Flores I, Fernández-Marcos PJ, Cayuela ML, Maraver A, Tejera A, Borrás C, Matheu A, Klatt P, Flores JM, Viña J, Serrano M, Blasco MA (November 2008). "Telomerase reverse transcriptase delays aging in cancer-resistant mice". *Cell*. **135** (4): 609–22. doi:10.1016/j.cell.2008.09.034. PMID 19013273.

[33] Fauce SR, Jamieson BD, Chin AC, Mitsuyasu RT, Parish ST, Ng HL, Kitchen CM, Yang OO, Harley CB, Effros RB (November 2008). "Telomerase-Based Pharmacologic Enhancement of Antiviral Function> of Human CD8+ T Lymphocytes". *J. Immunol*. **181** (10): 7400–6. doi:10.4049/jimmunol.181.10.7400. PMC 2682219⊚. PMID 18981163.

[34] Atzmon G, Cho M, Cawthon RM, Budagov T, Katz M, Yang X, Siegel G, Bergman A, Huffman DM, Schechter CB, Wright WE, Shay JW, Barzilai N, Govindaraju DR, Suh Y (January 2010). "Genetic variation in human telomerase is associated with telomere length in Ashkenazi centenarians". *Proc. Natl. Acad. Sci. U.S.A.* 107 Suppl 1 (suppl_1): 1710–7. doi:10.1073/pnas.0906191106. PMC 2868292⊚. PMID 19915151. Lay summary – *LiveScience*.

[35] Jaskelioff, M; Muller, FL; Paik, JH; Thomas, E; Jiang, S; Adams, AC; Sahin, E; Kost-Alimova, M; Protopopov, A; Cadiñanos, J; Horner, JW; Maratos-Flier, E; Depinho, RA (November 2010). "Telomerase reactivation reverses tissue degeneration in aged telomerase deficient mice". *Nature*. **469** (7328): 102–6. doi:10.1038/nature09603. PMC 3057569⊚. PMID 21113150. Lay summary – *news.discovery.com*.

[36] "Stop, rewind: the scientists slowing the ageing process". *BBC News*. 2011-01-26.

[37] Bernardes de Jesus, B; Vera, E; Schneeberger, K; Tejera, AM; Ayuso, E; Bosch, F; Blasco, MA (August 2012). "Telomerase gene therapy in adult and old mice delays aging and increases longevity without increasing cancer". *EMBO Molecular Medicine*. **4** (8): 691–704. doi:10.1002/emmm.201200245. PMC 3494070⊚. PMID 22585399.

[38] Blasco, MA (August 2005). "Telomeres and human disease: ageing, cancer and beyond". *Nature Reviews Genetics*. **6** (8): 611–22. doi:10.1038/nrg1656. PMID 16136653.

[39] Skloot, Rebecca (2010). *The Immortal Life of Henrietta Lacks*. New York: Broadway Paperbacks. pp. 216, 217. ISBN 978-1-4000-5218-9.

[40] Dr. Todd Hennessey, 2016 University at Buffalo

[41] Shay, JW; Bacchetti, S (April 1997). "A survey of telomerase activity in human cancer". *Eur. J. Cancer*. **33** (5): 787–91. doi:10.1016/S0959-8049(97)00062-2. PMID 9282118.

[42] Bryan, TM; Englezou, A; Gupta, J; Bacchetti, S; Reddel, RR (September 1995). "Telomere elongation in immortal human cells without detectable telomerase activity". *EMBO J*. **14** (17): 4240–8. PMC 394507⊚. PMID 7556065.

[43] Henson, JD; Neumann, AA; Yeager, TR; Reddel, RR (January 2002). "Alternative lengthening of telomeres in mammalian cells". *Oncogene*. **21** (4): 598–610. doi:10.1038/sj.onc.1205058. PMID 11850785.

[44] Blackburn, EH (February 2005). "Telomeres and telomerase: their mechanisms of action and the effects of altering their functions". *FEBS Lett*. **579** (4): 859–62. doi:10.1016/j.febslet.2004.11.036. PMID 15680963.

[45] Tian, X; Chen, B; Liu, X (March 2013). "Telomere and Telomerase as Targets for Cancer Therapy". *Applied Biochemistry and Biotechnology*. **160** (5): 1460–1472. doi:10.1007/s12010-009-8633-9.

[46] Griffiths, Anthony J. F.; Wessler, Susan R.; Carroll, Sean B.; Doebley, John (2008). *Introduction to Genetic Analysis*. W. H. Freeman. ISBN 978-0-7167-6887-6.

[47] Williams, SC (January 2013). "No end in sight for telomerase-targeted cancer drugs.". *Nat Med*. **19** (1): 6. doi:10.1038/nm0113-6. PMID 23295993.

[48] Blasco, MA (2001). "Telomeres in Cancer Therapy". *J Biomed Biotechnol*. **1** (1): 3–4. doi:10.1155/S1110724301000109. PMC 79678⊚. PMID 12488618.

[49] Shay, Jerry W.; Ying, Zou; Hiyama, Eiso; Wright, Woodring E. (2001). "Telomerase and Cancer". *Human Molecular Genetics*. **10** (7): 677–685. doi:10.1093/hmg/10.7.677. Retrieved June 2015. Check date values in: |access-date= (help)

[50] Gul, Ilhami; Dundar, Ozgur; Bodur, Serkan; Tunca, Yusuf; Tutuncu, Levent (2013). "The Status of Telomerase Enzyme Activity in Benign and Malignant Gynecologic Pathologies". *Balkan Medical Journal*. **30**: 287–292. doi:10.5152/balkanmedj.2013.7328. PMC 4115914⊚. PMID 25207121.

[51] Saretzki, Gabriele (2003). "Telomerase inhibition as Cancer Therapy". *Cancer Letter.* **194** (2): 209–219. doi:10.1016/s0304-3835(02)00708-5. PMID 12757979.

[52] Stoyanov, V (2009). "T-loop deletion factor showing speeding aging of Homo telomere diversity and evolution". *Rejuvenation Research.* **12** (1): 52.

[53] Patel, Kunal P.; Robert H., Vonderheide (2004). "Telomerase as a tumor-associated antigen for cancer immunotherapy". *Cytotechnology.* **45** (1–2): 91–99. doi:10.1007/s10616-004-5132-2. PMC 3449959⊙. PMID 19003246.

[54] "Geron's cancer drug shakes off one FDA hold but remains on pause". *FierceBiotech.* Retrieved 2015-06-28.

[55] "J&J bets up to $935M that Geron's drug can shake a checkered past". *FierceBiotech.* Retrieved 2015-06-28.

[56] Johnson, Steven Ross (September 2, 2015). "Experimental blood disorder therapy shows promise in new studies". *Modern Healthcare.* Retrieved September 2015. Check date values in: |access-date= (help)

[57] Epel, ES; Lin, J; Dhabhar, FS; Wolkowitz, OM; Puterman, E; Karan, L; Blackburn, EH (2010). "Dynamics of telomerase activity in response to acute psychological stress". *Brain Behav Immun.* **24** (4): 531–9. doi:10.1016/j.bbi.2009.11.018. PMC 2856774⊙. PMID 20018236.

[58] Gostjeva, EV; Thilly, WG (2005). "Stem cell stages and the origins of colon cancer: a multidisciplinary perspective.". *Stem Cell Rev.* **1** (3): 243–51. doi:10.1385/SCR:1:3:243. PMID 17142861.

[59] Jacobs TL, et al. (June 2011). "Intensive meditation training, immune cell telomerase activity, and psychological mediators.". *nih.gov.* **36**: 664–81. doi:10.1016/j.psyneuen.2010.09.010. PMID 21035949.

[60] Ristow, Michael (2010). "Telomerase deficiency impairs glucose metabolism and insulin secretion" (PDF). *Aging.* **2** (10): 650–658. PMC 2993795⊙. PMID 20876939.

[61] Yamaguchi, H; Calado, RT; Ly, H; Kajigaya, S; Baerlocher, GM; Chanock, SJ; Lansdorp, PM; Young, NS (April 2005). "Mutations in TERT, the gene for telomerase reverse transcriptase, in aplastic anemia". *N. Engl. J. Med.* **352** (14): 1413–24. doi:10.1056/NEJMoa042980. PMID 15814878.

[62] Zhang, A; Zheng, C; Hou, M; Lindvall, C; Li, KJ; Erlandsson, F; Björkholm, M; Gruber, A; Blennow, E; Xu, D (April 2003). "Deletion of the Telomerase Reverse Transcriptase Gene and Haploinsufficiency of Telomere Maintenance in Cri du Chat Syndrome". *Am. J. Hum. Genet.* **72** (4): 940–8. doi:10.1086/374565. PMC 1180356⊙. PMID 12629597.

[63] Yamaguchi, H (June 2007). "Mutations of telomerase complex genes linked to bone marrow failures". *J Nippon Med Sch.* **74** (3): 202–9. doi:10.1272/jnms.74.202. PMID 17625368.

[64] Heiss, NS; Knight, SW; Vulliamy, TJ; Klauck, SM; Wiemann, S; Mason, PJ; Poustka, A; Dokal, I (May 1998). "X-linked dyskeratosis congenita is caused by mutations in a highly conserved gene with putative nucleolar functions". *Nat. Genet.* **19** (1): 32–8. doi:10.1038/ng0598-32. PMID 9590285.

[65] Vulliamy, TJ; Walne, A; Baskaradas, A; Mason, PJ; Marrone, A; Dokal, I (2005). "Mutations in the reverse transcriptase component of telomerase (TERT) in patients with bone marrow failure". *Blood Cells Mol. Dis.* **34** (3): 257–63. doi:10.1016/j.bcmd.2004.12.008. PMID 15885610.

[66] Vulliamy, T; Marrone, A; Goldman, F; Dearlove, A; Bessler, M; Mason, PJ; Dokal, I (September 2001). "The RNA component of telomerase is mutated in autosomal dominant dyskeratosis congenita". *Nature.* **413** (6854): 432–5. doi:10.1038/35096585. PMID 11574891.

[67] Marrone, A; Walne, A; Dokal, I (June 2005). "Dyskeratosis congenita: telomerase, telomeres and anticipation". *Current Opinion in Genetics & Development.* **15** (3): 249–57. doi:10.1016/j.gde.2005.04.004. PMID 15917199.

[68] Armanios, M; Chen, JL; Chang, YP; Brodsky, RA; Hawkins, A; Griffin, CA; Eshleman, JR; Cohen, AR; Chakravarti, A; Hamosh, A; Greider, CW (November 2005). "Haploinsufficiency of telomerase reverse transcriptase leads to anticipation in autosomal dominant dyskeratosis congenita". *Proc. Natl. Acad. Sci. U.S.A.* **102** (44): 15960–4. doi:10.1073/pnas.0508124102. PMC 1276104⊙. PMID 16247010.

## 11.7  Further reading

- *The Immortal Cell*, by Michael D. West, Doubleday (2003) ISBN 978-0-385-50928-2

## 11.8  External links

- The Telomerase Database - A Web-based tool for telomerase research.

- Three-dimensional model of telomerase at MUN

- Elizabeth Blackburn's seminars:   Telomeres and Telomerase

- Telomerase at the US National Library of Medicine Medical Subject Headings (MeSH)

# Chapter 12

# Telomerase reverse transcriptase

**Telomerase reverse transcriptase** (abbreviated to **TERT**, or **hTERT** in humans) is a catalytic subunit of the enzyme telomerase, which, together with the telomerase RNA component (TERC), comprises the most important unit of the telomerase complex.[4][5]

Telomerases are part of a distinct subgroup of RNA-dependent polymerases. Telomerase lengthens telomeres in DNA strands, thereby allowing senescent cells that would otherwise become postmitotic and undergo apoptosis to exceed the Hayflick limit and become potentially immortal, as is often the case with cancerous cells. To be specific, TERT is responsible for catalyzing the addition of nucleotides in a TTAGGG sequence to the ends of a chromosome's telomeres.[6] This addition of repetitive DNA sequences prevents degradation of the chromosomal ends following multiple rounds of replication.[7]

hTERT absence (usually as a result of a chromosomal mutation) is associated with the disorder Cri du chat.[8][9]

## 12.1  Function

Telomerase is a ribonucleoprotein polymerase that maintains telomere ends by addition of the telomere repeat TTAGGG. The enzyme consists of a protein component with reverse transcriptase activity, encoded by this gene, and an RNA component that serves as a template for the telomere repeat. Telomerase expression plays a role in cellular senescence, as it is normally repressed in postnatal somatic cells, resulting in progressive shortening of telomeres. Studies in mice suggest that telomerase also participates in chromosomal repair, since de novo synthesis of telomere repeats may occur at double-stranded breaks. Alternatively spliced variants encoding different isoforms of telomerase reverse transcriptase have been identified; the full-length sequence of some variants has not been determined. Alternative splicing at this locus is thought to be one mechanism of regulation of telomerase activity.[10]

## 12.2  Regulation of hTERT

The hTERT gene, located on chromosome 5, consists of 16 exons and 15 introns spanning 35 kb. The core promoter of hTERT includes 330 base pairs upstream of the translation start site (AUG since it's RNA by using the words "exons" and "introns"), as well as 37 base pairs of exon 2 of the hTERT gene.[11][12][13] The hTERT promoter is GC-rich and lacks TATA and CAAT boxes but contains many sites for several transcription factors giving indication of a high level of regulation by multiple factors in many cellular contexts.[11] Transcription factors that can activate hTERT include many oncogenes (cancer-causing genes) such as c-Myc, Sp1, HIF-1, AP2, and many more, while many cancer suppressing genes such as p53, WT1, and Menin produce factors that suppress hTERT activity.[13][14] Another form of up-regulation is through demethylation of histones proximal to the promoter region, imitating the low density of trimethylated histones seen in embryonic stem cells.[15] This allows for the recruitment of histone acetyltransferase (HAT) to unwind the sequence allowing for transcription of the gene.[14]

Telomere deficiency is often linked to aging, cancers and the conditions dyskeratosis congenita (DKC) and Cri du chat. Meanwhile, over-expression of hTERT is often associated with cancers and tumor formation.[8][16][17][18] The regulation of hTERT is extremely important to the maintenance of stem and cancer cells and can be used in multiple ways in the field of regenerative medicine.

## 12.3  Stem cells

### 12.3.1  hTERT in stem cells

hTERT is often up-regulated in cells that divide rapidly, including both embryonic stem cells and adult stem cells.[17] It elongates the telomeres of stem cells, which, as a consequence, increases the lifespan of the stem cells by allowing for indefinite division without shortening of telomeres. Therefore, it is responsible for the self-renewal properties of stem cells. Telomerase are found specifically to target shorter telomere over longer telomere. due to various regulatory mechanisms inside the cells that reduce the affinity of telomerase to longer telomeres. This preferential affinity maintains a balance within the cell such that the telomeres are of sufficient length for their function and yet, at the same time, not contribute to aberrant telomere elongation [19]

High expression of hTERT is also often used as a landmark for pluripotency and multipotency state of embryonic and adult stem cells. Over-expression of hTERT was found to immortalize certain cell types as well as impart different interesting properties to different stem cells.[13][20]

### 12.3.2 Immortalization

hTERT immortalizes various normal cells in culture, thereby endowing the self-renewal properties of stem cells to non-stem cell cultures.[13][21] There are multiple ways in which immortalization of non-stem cells can be achieved, one of which being via the introduction of hTERT into the cells. Differentiated cells often express hTERC and TP1, a telomerase-associated protein that helps form the telomerase assembly, but does not express hTERT. Hence, hTERT acts as the limiting factor for telomerase activity in differentiated cells [13][22] However, with hTERT over-expression, active telomerase can be formed in differentiated cells. This method has been used to immortalize prostate epithelial and stromal-derived cells, which are typically difficult to culture *in vitro*. hTERT introduction allows *in vitro* culture of these cells and available for possible future research. hTERT introduction have an advantage over the use of viral protein for immortalization in that it does not involve the inactivation of tumor suppressor gene, which might lead to cancer formation.[21]

### 12.3.3 Enhancement of stem cell properties

Over-expression of hTERT in stem cells changes the properties of the cells.[20][23][24] hTERT over-expression increases the stem cell properties of human mesenchymal stem cells. The expression profile of mesenchymal stem cells converges towards embryonic stem cells, suggesting that these cells may have embryonic stem cell-like properties. However, it has been observed that mesenchymal stem cells undergo decreased levels of spontaneous differentiation.[20] This suggests that the differentiation capacity of adult stem cells may be dependent on telomerase activities. Therefore, over-expression of hTERT, which is akin to increasing telomerase activities, may create adult stem cells with a larger capacity for differentiation and

hence, a larger capacity for treatment.

Increasing the telomerase activities in stem cells gives different effects depending on the intrinsic nature of the different types of stem cells.[17] Hence, not all stem cells will have increased stem-cell properties. For example, research has shown that telomerase can be upregulated in CD34+ Umbilical Cord Blood Cells through hTERT over-expression. The survival of these stem cells was enhanced, although there was no increase in the amount of population doubling.[24]

## 12.4   Clinical significance

Deregulation of telomerase expression in somatic cells may be involved in oncogenesis.[10]

Genome-wide association studies suggest TERT is a susceptibility gene for development of many cancers,[25] including lung cancer.[26]

### 12.4.1   Role in cancer

Telomerase activity is associated with the number of times a cell can divide playing an important role in the immortality of cell lines, such as cancer cells. The enzyme complex acts through the addition of telomeric repeats to the ends of chromosomal DNA. This generates immortal cancer cells.[27] In fact, there is a strong correlation between telomerase activity and malignant tumors or cancerous cell lines.[28] Not all types of human cancer have increased telomerase activity. 90% of cancers are characterized by increased telomerase activity.[28] Lung cancer is the most well characterized type of cancer associated with telomerase.[29] There is a lack of substantial telomerase activity in some cell types such as primary human fibroblasts, which become senescent after about 30–50 population doublings.[28] There is also evidence that telomerase activity is increased in tissues, such as germ cell lines, that are self-renewing. Normal somatic cells, on the other hand, do not have detectable telomerase activity.[30] Since the catalytic component of telomerase is its reverse transcriptase, hTERT, and the RNA component hTERC. hTERT is an important gene to investigate in terms of cancer and tumorigenesis.

The hTERT gene has been examined for mutations and their association with the risk of contracting cancer. Over two hundred combinations of hTERT polymorphisms and cancer development have been found.[29] There were several different types of cancer involved, and the strength of the correlation between the polymorphism and developing cancer varied from weak to strong.[29] The regulation of hTERT has also been researched to determine possible mechanisms of telomerase activation in cancer cells. Glycogen synthase kinase 3 (GSK3) seems to be over-expressed in most cancer cells.[27] GSK3 is involved in promoter activation through controlling a network of transcription factors.[27] Leptin is also involved in increasing mRNA expression of hTERT via signal transducer and activation of transcription 3 (STAT3), proposing a mechanism for increased cancer incidence in obese individuals.[27] There are several other regulatory mechanisms that are altered or aberrant in cancer cells, including the Ras signaling pathway and other transcriptional regulators.[27] Phosphorylation is also a key process of post-transcriptional modification that regulates mRNA expression and cellular localization.[27] Clearly, there are many regulatory mechanisms of activation and repression of hTERT and telomerase activity in the cell, providing methods of immortalization in cancer cells.

### 12.4.2 Therapeutic potential

If increased telomerase activity is associated with malignancy, then possible cancer treatments could involve inhibiting its catalytic component, hTERT, to reduce the enzyme's activity and cause cell death. Since normal somatic cells do not express TERT, telomerase inhibition in cancer cells can cause senescence and apoptosis without affecting normal human cells.[27] It has been found that dominant-negative mutants of hTERT could reduce telomerase activity within the cell.[28] This led to apoptosis and cell death in cells with short telomere lengths, a promising result for cancer treatment.[28] Although cells with long telomeres did not experience apoptosis, they developed mortal characteristics and underwent telomere shortening.[28] Telomerase activity has also been found to be inhibited by phytochemicals such as isoprenoids, genistein, curcumin, etc.[27] These chemicals play a role in inhibiting the mTOR pathway via down-regulation of phosphorylation.[27] The mTOR pathway is very important in regulating protein synthesis and it interacts with telomerase to increase its expression.[27] Several other chemicals have been found to inhibit telomerase activity and are currently being tested as potential clinical treatment options such as nucleoside analogues, retinoic acid derivatives, quinolone antibiotics, and catechin derivatives.[30] There are also other molecular genetic-based methods of inhibiting telomerase, such as antisense therapy and RNA interference.[30]

hTERT peptide fragments have been shown to induce a cytotoxic T-cell reaction against telomerase-positive tumor cells *in vitro*.[31] The response is mediated by dendritic cells, which can display hTERT-associated antigens on MHC class I and II receptors following adenoviral transduction of an hTERT plasmid into dendritic cells, which mediate T-cell responses.[32] Dendritic cells are then able to present telomerase-associated antigens even with undetectable amounts of telomerase activity, as long as the hTERT plasmid is present.[33] Immunotherapy against telomerase-positive tumor cells is a promising field in cancer research that has been shown to be effective in *in vitro* and mouse model studies.[34]

## 12.5 Medical implications

### 12.5.1 iPS cells

Induced pluripotent stem cells (iPS cells) are somatic cells that have been reprogrammed into a stem cell-like state by the introduction of four factors (Oct3/4, Sox2, Klf4, and c-Myc).[35] iPS cells have the ability to self-renew indefinitely and contribute to all three germ layers when implanted into a blastocyst or use in teratoma formation.[35]

Early development of iPS cell lines were not efficient, as they yielded up to 5% of somatic cells successfully reprogrammed into a stem cell-like state.[36] By using immortalized somatic cells (differentiated cells with hTERT upregulated), iPS cell reprogramming was increased by twentyfold compared to reprogramming using mortal cells.[36]

The reactivation of hTERT, and subsequently telomerase, in human iPS cells has been used as an indication of pluripotency and reprogramming to an ES (embryonic stem) cell-like state when using mortal cells.[35] Reprogrammed cells that do not express sufficient hTERT levels enter a quiescent state following a number of replications depending on the length of the telomeres while maintaining stem cell-like abilities to differentiate.[36] Reactivation of TERT activity can be achieved using only three of the four reprogramming factors described by Takahashi and Yamanaka: To be specific, Oct3/4, Sox2 and Klf4 are essential, whereas c-Myc is not.[15] However, this study was done with cells containing endogenous levels of c-Myc that may have been sufficient for reprogramming.

Telomere length in healthy adult cells elongates and acquires epigenetic characteristics similar to those of ES cells when reprogrammed as iPS cells. Some epigenetic characteristics of ES cells include a low density of tri-methylated histones H3K9 and H4K20 at telomeres, as well as an increased detectable amount of TERT transcripts and protein activity.[15] Without the restoration of TERT and associated telomerase proteins, the efficiency of iPS cells would be drastically reduced. iPS cells would also lose the ability to self-renew and would eventually senesce.[15]

DKC (dyskeratosis congenita) patients are all characterized by the defective maintenance of telomeres leading to problems with stem cell regeneration.[16] iPS cells derived from DKC patients with a heterozygous mutation on the TERT gene display a 50% reduction in telomerase activity compared to wild type iPS cells.[37] Conversely, mutations on the TERC gene (RNA portion of telomerase complex) can be overcome by up-regulation due to reprogramming as long as the hTERT gene is intact and functional.[38] Lastly, iPS cells generated with DKC cells with a mutated dyskerin (DKC1) gene cannot assemble the hTERT/RNA complex and thus do not have functional telomerase.[37]

The functionality and efficiency of a reprogrammed iPS cell is determined by the ability of the cell to re-activate the telomerase complex and elongate its telomeres allowing for self-renewal. hTERT is a major limiting component of the telomerase complex and a deficiency of intact hTERT impedes the activity of telomerase, making iPS cells an unsuitable pathway towards therapy for telomere-deficient disorders.[37]

## 12.5.2   Androgen therapy

Although the mechanism is not fully understood, exposure of TERT-deficient hematopoietic cells to androgens resulted in an increased level of TERT activity.[39] Cells with a heterozygous TERT mutation, like those in DKC (dyskeratosis congenita) patients, which normally exhibit low baseline levels of TERT, could be restored to normal levels comparable to control cells. TERT mRNA levels are also increased with exposure to androgens.[39] Androgen therapy may become a suitable method for treating circulatory ailments such as bone marrow degeneration and low blood count linked with DKC and other telomerase-deficient conditions.[39]

## 12.5.3   Aging

As organisms age and cells proliferate, telomeres shorten with each round of replication. Cells restricted to a specific lineage are capable of division only a set number of times, set by the length of telomeres, before they senesce.[40] Depletion and uncapping of telomeres has been linked to organ degeneration, failure, and fibrosis due to progenitors' becoming quiescent and unable to differentiate.[19][40] Using an *in vivo* TERT deficient mouse model, reactivation of the TERT gene in quiescent populations in multiple organs reactivated telomerase and restored the cells' abilities to differentiate.[41] Reactivation of TERT down-regulates DNA damage signals associated with cellular mitotic checkpoints allowing for proliferation and elimination of a degenerative phenotype.[41] In another study, introducing the TERT gene into healthy one-year-old mice using an engineered adeno-associated virus led to a 24% increase in

lifespan, without any increase in cancer.[42]

## 12.5.4   Gene therapy

The hTERT gene has become a main focus for gene therapy involving cancer due to its expression in tumor cells but not somatic adult cells.[43] One method is to prevent the translation of hTERT mRNA through the introduction of siRNA, which are complimentary sequences that bind to the mRNA preventing processing of the gene post transcription.[44] This method does not completely eliminate telomerase activity, but it does lower telomerase activity and levels of hTERT mRNA seen in the cytoplasm.[44] Higher success rates were seen *in vitro* when combining the use of antisense hTERT sequences with the introduction of a tumor-suppressing plasmid by adenovirus infection such as PTEN.[45]

Another method that has been studied is manipulating the hTERT promoter to induce apoptosis in tumor cells. Plasmid DNA sequences can be manufactured using the hTERT promoter followed by genes encoding for specific proteins. The protein can be a toxin, an apoptotic factor, or a viral protein. Toxins such as diphtheria toxin interfere with cellular processes and eventually induce apoptosis.[43] Apoptotic death factors like FADD (Fas-Associated protein with Death Domain) can be used to force cells expressing hTERT to undergo apoptosis.[46] Viral proteins like viral thymidine kinase can be used for specific targeting of a drug.[47] By introducing a prodrug only activated by the viral enzyme, specific targeting of cells expressing hTERT can be achieved.[47] By using the hTERT promoter, only cells expressing hTERT will be affected and allows for specific targeting of tumor cells.[43][46][47]

Aside from cancer therapies, the hTERT gene has been used to promote the growth of hair follicles.[48]

A schematic animation for gene therapy is shown as follows.

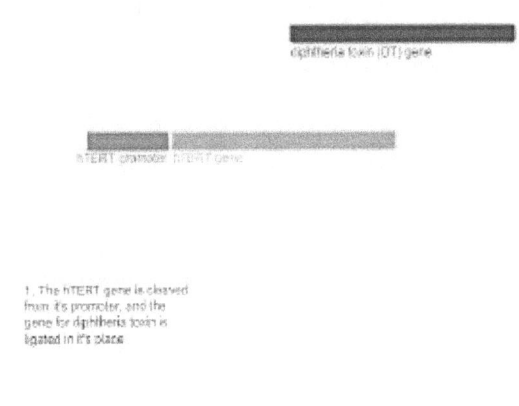

HTERT Final gif

## 12.6 Interactions

Telomerase reverse transcriptase has been shown to interact with:

- HSP90AA1,[49][50]

- Ku70,[51]

- Ku80,[51]

- MCRS1,[52]

- Nucleolin,[53]

- PINX1,[54] and

- YWHAQ.[55]

## 12.7 See also

- telomerase

- reverse transcriptase

## 12.8 References

[1] "Diseases that are genetically associated with TERT view/edit references on wikidata".

[2] "Human PubMed Reference:".

[3] "Mouse PubMed Reference:".

[4] Weinrich SL, Pruzan R, Ma L, Ouellette M, Tesmer VM, Holt SE, Bodnar AG, Lichtsteiner S, Kim NW, Trager JB, Taylor RD, Carlos R, Andrews WH, Wright WE, Shay JW, Harley CB, Morin GB (December 1997). "Reconstitution of human telomerase with the template RNA component hTR and the catalytic protein subunit hTRT". *Nat. Genet.* **17** (4): 498–502. doi:10.1038/ng1297-498. PMID 9398860.

[5] Kirkpatrick KL, Mokbel K (2001). "The significance of human telomerase reverse transcriptase (hTERT) in cancer". *Eur J Surg Oncol.* **27** (8): 754–60. doi:10.1053/ejso.2001.1151. PMID 11735173.

[6] Shampay J, Blackburn EH (January 1988). "Generation of telomere-length heterogeneity in Saccharomyces cerevisiae". *Proc. Natl. Acad. Sci. U.S.A.* **85** (2): 534–8. doi:10.1073/pnas.85.2.534. PMC 279585. PMID 3277178.

[7] Poole JC, Andrews LG, Tollefsbol TO (May 2001). "Activity, function, and gene regulation of the catalytic subunit of telomerase (hTERT)". *Gene.* **269** (1-2): 1–12. doi:10.1016/S0378-1119(01)00440-1. PMID 11376932.

[8] Zhang A, Zheng C, Hou M, Lindvall C, Li KJ, Erlandsson F, Björkholm M, Gruber A, Blennow E, Xu D (2003). "Deletion of the telomerase reverse transcriptase gene and haploinsufficiency of telomere maintenance in Cri du chat syndrome". *Am. J. Hum. Genet.* **72** (4): 940–8. doi:10.1086/374565. PMC 1180356. PMID 12629597.

[9] Cerruti Mainardi P (2006). "Cri du Chat syndrome". *Orphanet J Rare Dis.* **1**: 33. doi:10.1186/1750-1172-1-33. PMC 1574300. PMID 16953888.

[10] "Entrez Gene: TERT telomerase reverse transcriptase".

[11] Cong YS, Wen J, Bacchetti S (January 1999). "The human telomerase catalytic subunit hTERT: organization of the gene and characterization of the promoter". *Hum. Mol. Genet.* **8** (1): 137–42. doi:10.1093/hmg/8.1.137. PMID 9887342.

[12] Bryce LA, Morrison N, Hoare SF, Muir S, Keith WN (2000). "Mapping of the gene for the human telomerase reverse transcriptase, hTERT, to chromosome 5p15.33 by fluorescence in situ hybridization". *Neoplasia.* **2** (3): 197–201. doi:10.1038/sj.neo.7900092. PMC 1507564. PMID 10935505.

[13] Ćukušić A, Skrobot Vidaček N, Sopta M, Rubelj I (2008). "Telomerase regulation at the crossroads of cell fate". *Cytogenet. Genome Res.* **122** (3-4): 263–72. doi:10.1159/000167812. PMID 19188695.

[14] Kyo S, Takakura M, Fujiwara T, Inoue M (August 2008). "Understanding and exploiting hTERT promoter regulation for diagnosis and treatment of human cancers". *Cancer Sci.* **99** (8): 1528–38. doi:10.1111/j.1349-7006.2008.00878.x. PMID 18754863.

[15] Marion RM, Strati K, Li H, Tejera A, Schoeftner S, Ortega S, Serrano M, Blasco MA (February 2009). "Telomeres acquire embryonic stem cell characteristics in induced pluripotent stem cells". *Cell Stem Cell.* **4** (2): 141–54. doi:10.1016/j.stem.2008.12.010. PMID 19200803.

[16] Walne AJ, Dokal I (2009). "Advances in the understanding of dyskeratosis congenita". *British Journal of Haematology.* **145**: 164–172. doi:10.1111/j.1365-2141.2009.07598.x.

[17] Flores I, Benetti R, Blasco MA (June 2006). "Telomerase regulation and stem cell behaviour". *Current Opinion in Cell Biology.* **18** (3): 254–60. doi:10.1016/j.ceb.2006.03.003. PMID 16617011.

[18] Calado R, Young N (2012). "Telomeres in disease". *F1000 Med Rep.* **4**: 8. doi:10.3410/M4-8. PMC 3318193. PMID 22500192.

[19] Flores I, Blasco MA (September 2010). "The role of telomeres and telomerase in stem cell aging". *FEBS Lett.* **584** (17): 3826–30. doi:10.1016/j.febslet.2010.07.042. PMID 20674573.

[20] Tsai CC, Chen CL, Liu HC, Lee YT, Wang HW, Hou LT, Hung SC (2010). "Overexpression of hTERT increases stem-like properties and decreases spontaneous differentiation in human mesenchymal stem cell lines". *J. Biomed. Sci.* **17**: 64. doi:10.1186/1423-0127-17-64. PMC 2923118⊚. PMID 20670406.

[21] Kogan I, Goldfinger N, Milyavsky M, Cohen M, Shats I, Dobler G, Klocker H, Wasylyk B, Voller M, Aalders T, Schalken JA, Oren M, Rotter V (April 2006). "hTERT-immortalized prostate epithelial and stromal-derived cells: an authentic in vitro model for differentiation and carcinogenesis". *Cancer Res.* **66** (7): 3531–40. doi:10.1158/0008-5472.CAN-05-2183. PMID 16585177.

[22] Nakayama J, Tahara H, Tahara E, Saito M, Ito K, Nakamura H, Nakanishi T, Tahara E, Ide T, Ishikawa F (January 1998). "Telomerase activation by hTRT in human normal fibroblasts and hepatocellular carcinomas". *Nat. Genet.* **18** (1): 65–8. doi:10.1038/ng0198-65. PMID 9425903.

[23] Dashinimaev EB, Vishnyakova KS, Popov KV, Yegorov YE (2008). "Stable Culture of hTERT-transduced Human Embryonic Neural Stem Cells Holds All the Features of Primary Culture" (PDF). *Electronic Journal of Biology*. **4** (2): 93–97.

[24] Elwood NJ, Jiang XR, Chiu CP, Lebkowski JS, Smith CA (March 2004). "Enhanced long-term survival, but no increase in replicative capacity, following retroviral transduction of human cord blood CD34+ cells with human telomerase reverse transcriptase". *Haematologica*. **89** (3): 377–8. PMID 15020288.

[25] Baird DM (2010). "Variation at the TERT locus and predisposition for cancer". *Expert Rev Mol Med*. **12**: e16. doi:10.1017/S146239941000147X. PMID 20478107.

[26] McKay JD, Hung RJ, Gaborieau V, Boffetta P, Chabrier A, Byrnes G, Zaridze D, Mukeria A, Szeszenia-Dabrowska N, Lissowska J, et al. (December 2008). "Lung cancer susceptibility locus at 5p15.33". *Nat. Genet.* **40** (12): 1404–6. doi:10.1038/ng.254. PMC 2748187⊚. PMID 18978790.

[27] Sundin T, Hentosh P (2012). "InTERTesting association between telomerase, mTOR and phytochemicals". *Expert Rev Mol Med*. **14**: e8. doi:10.1017/erm.2012.1. PMID 22455872.

[28] Zhang X, Mar V, Zhou W, Harrington L, Robinson MO (September 1999). "Telomere shortening and apoptosis in telomerase-inhibited human tumor cells". *Genes Dev.* **13** (18): 2388–99. doi:10.1101/gad.13.18.2388. PMC 317024⊚. PMID 10500096.

[29] Mocellin S, Verdi D, Pooley KA, Landi MT, Egan KM, Baird DM, Prescott J, De Vivo I, Nitti D (June 2012). "Telomerase reverse transcriptase locus polymorphisms and cancer risk: a field synopsis and meta-analysis". *J. Natl. Cancer Inst.* **104** (11): 840–54. doi:10.1093/jnci/djs222. PMID 22523397.

[30] Glukhov AI, Svinareva LV, Severin SE, Shvets VI (2011). "Telomerase inhibitors as novel antitumour drugs". *Applied Biochemistry and Microbiology*. **47**: 655–660. doi:10.1134/S0003683811070039.

[31] Minev B, Hipp J, Firat H, Schmidt JD, Langlade-Demoyen P, Zanetti M (April 2000). "Cytotoxic T cell immunity against telomerase reverse transcriptase in humans". *Proc. Natl. Acad. Sci. U.S.A.* **97** (9): 4796–801. doi:10.1073/pnas.070560797. PMC 18312⊚. PMID 10759561.

[32] Frolkis M, Fischer MB, Wang Z, Lebkowski JS, Chiu CP, Majumdar AS (March 2003). "Dendritic cells reconstituted with human telomerase gene induce potent cytotoxic T-cell response against different types of tumors". *Cancer Gene Ther.* **10** (3): 239–49. doi:10.1038/sj.cgt.7700563. PMID 12637945.

[33] Vonderheide RH, Hahn WC, Schultze JL, Nadler LM (June 1999). "The telomerase catalytic subunit is a widely expressed tumor-associated antigen recognized by cytotoxic T lymphocytes". *Immunity*. **10** (6): 673–9. doi:10.1016/S1074-7613(00)80066-7. PMID 10403642.

[34] Rosenberg SA (March 1999). "A new era for cancer immunotherapy based on the genes that encode cancer antigens". *Immunity*. **10** (3): 281–7. doi:10.1016/S1074-7613(00)80028-X. PMID 10204484.

[35] Takahashi K, Tanabe K, Ohnuki M, Narita M, Ichisaka T, Tomoda K, Yamanaka S (November 2007). "Induction of pluripotent stem cells from adult human fibroblasts by defined factors". *Cell*. **131** (5): 861–72. doi:10.1016/j.cell.2007.11.019. PMID 18035408.

[36] Utikal J, Polo JM, Stadtfeld M, Maherali N, Kulalert W, Walsh RM, Khalil A, Rheinwald JG, Hochedlinger K (August 2009). "Immortalization eliminates a roadblock during cellular reprogramming into iPS cells". *Nature*. **460** (7259): 1145–8. doi:10.1038/nature08285. PMID 19668190.

[37] Batista LF, Pech MF, Zhong FL, Nguyen HN, Xie KT, Zaug AJ, Crary SM, Choi J, Sebastiano V, Cherry A, Giri N, Wernig M, Alter BP, Cech TR, Savage SA, Reijo Pera RA, Artandi SE (2011). "Telomere shortening and loss of self-renewal in dyskeratosis congenita induced pluripotent stem cells". *Nature*. **474** (7351): 399–402. doi:10.1038/nature10084. PMC 3155806⊚. PMID 21602826.

[38] Agarwal S, Loh YH, McLoughlin EM, Huang J, Park IH, Miller JD, Huo H, Okuka M, Dos Reis RM, Loewer S, Ng HH, Keefe DL, Goldman FD, Klingelhutz AJ, Liu L, Daley GQ (2010). "Telomere elongation in induced pluripotent stem cells from dyskeratosis congenita patients". *Nature*. **464** (7286): 292–296. doi:10.1038/nature08792. PMC 3058620⊚. PMID 20164838.

[39] Calado RT, Yewdell WT, Wilkerson KL, Regal JA, Kajigaya S, Stratakis CA, Young NS (September 2009). "Sex

hormones, acting on the TERT gene, increase telomerase activity in human primary hematopoietic cells". *Blood*. **114** (11): 2236–43. doi:10.1182/blood-2008-09-178871. PMC 2745844ⓈPMID 19561322.

[40] Sahin E, Depinho RA (March 2010). "Linking functional decline of telomeres, mitochondria and stem cells during ageing". *Nature*. **464** (7288): 520–8. doi:10.1038/nature08982. PMID 20336134.

[41] Jaskelioff M, Muller FL, Paik JH, Thomas E, Jiang S, Adams AC, Sahin E, Kost-Alimova M, Protopopov A, Cadiñanos J, Horner JW, Maratos-Flier E, Depinho RA (January 2011). "Telomerase reactivation reverses tissue degeneration in aged telomerase-deficient mice". *Nature*. **469** (7328): 102–6. doi:10.1038/nature09603. PMC 3057569ⓈPMID 21113150.

[42] Bernardes de Jesus B, Vera E, Schneeberger K, Tejera AM, Ayuso E, Bosch F, Blasco MA (August 2012). "Telomerase gene therapy in adult and old mice delays aging and increases longevity without increasing cancer". *EMBO Molecular Medicine*. **4** (8): 691–704. doi:10.1002/emmm.201200245. PMC 3494070ⓈPMID 22585399.

[43] Abdul-Ghani R, Ohana P, Matouk I, Ayesh S, Ayesh B, Laster M, Bibi O, Giladi H, Molnar-Kimber K, Sughayer MA, de Groot N, Hochberg A (December 2000). "Use of transcriptional regulatory sequences of telomerase (hTER and hTERT) for selective killing of cancer cells". *Mol. Ther.* **2** (6): 539–44. doi:10.1006/mthe.2000.0196. PMID 11124054.

[44] Zhang PH, Tu ZG, Yang MQ, Huang WF, Zou L, Zhou YL (June 2004). "[Experimental research of targeting hTERT gene inhibited in hepatocellular carcinoma therapy by RNA interference]". *Ai Zheng* (in Chinese). **23** (6): 619–25. PMID 15191658.

[45] You Y, Geng X, Zhao P, Fu Z, Wang C, Chao S, Liu N, Lu A, Gardner K, Pu P, Kong C, Ge Y, Judge SI, Li QQ (March 2007). "Evaluation of combination gene therapy with PTEN and antisense hTERT for malignant glioma in vitro and xenografts". *Cell. Mol. Life Sci.* **64** (5): 621–31. doi:10.1007/s00018-007-6424-4. PMID 17310280.

[46] Koga S, Hirohata S, Kondo Y, Komata T, Takakura M, Inoue M, Kyo S, Kondo S (2001). "FADD gene therapy using the human telomerase catalytic subunit (hTERT) gene promoter to restrict induction of apoptosis to tumors in vitro and in vivo". *Anticancer Res.* **21** (3B): 1937–43. PMID 11497281.

[47] Song JS, Kim HP, Yoon WS, Lee KW, Kim MH, Kim KT, Kim HS, Kim YT (November 2003). "Adenovirus-mediated suicide gene therapy using the human telomerase catalytic subunit (hTERT) gene promoter induced apoptosis of ovarian cancer cell line". *Biosci. Biotechnol. Biochem.* **67** (11): 2344–50. doi:10.1271/bbb.67.2344. PMID 14646192.

[48] Jan HM, Wei MF, Peng CL, Lin SJ, Lai PS, Shieh MJ (January 2012). "The use of polyethylenimine-DNA to topically deliver hTERT to promote hair growth". *Gene Ther.* **19** (1): 86–93. doi:10.1038/gt.2011.62. PMID 21593794.

[49] Haendeler J, Hoffmann J, Rahman S, Zeiher AM, Dimmeler S (February 2003). "Regulation of telomerase activity and anti-apoptotic function by protein-protein interaction and phosphorylation". *FEBS Lett.* **536** (1-3): 180–6. doi:10.1016/S0014-5793(03)00058-9. PMID 12586360.

[50] Kawauchi K, Ihjima K, Yamada O (May 2005). "IL-2 increases human telomerase reverse transcriptase activity transcriptionally and posttranslationally through phosphatidylinositol 3'-kinase/Akt, heat shock protein 90, and mammalian target of rapamycin in transformed NK cells". *J. Immunol.* **174** (9): 5261–9. doi:10.4049/jimmunol.174.9.5261. PMID 15843522.

[51] Chai W, Ford LP, Lenertz L, Wright WE, Shay JW (December 2002). "Human Ku70/80 associates physically with telomerase through interaction with hTERT". *J. Biol. Chem.* **277** (49): 47242–7. doi:10.1074/jbc.M208542200. PMID 12377759.

[52] Song H, Li Y, Chen G, Xing Z, Zhao J, Yokoyama KK, Li T, Zhao M (April 2004). "Human MCRS2, a cell-cycle-dependent protein, associates with LPTS/PinX1 and reduces the telomere length". *Biochem. Biophys. Res. Commun.* **316** (4): 1116–23. doi:10.1016/j.bbrc.2004.02.166. PMID 15044100.

[53] Khurts S, Masutomi K, Delgermaa L, Arai K, Oishi N, Mizuno H, Hayashi N, Hahn WC, Murakami S (December 2004). "Nucleolin interacts with telomerase". *J. Biol. Chem.* **279** (49): 51508–15. doi:10.1074/jbc.M407643200. PMID 15371412.

[54] Zhou XZ, Lu KP (November 2001). "The Pin2/TRF1-interacting protein PinX1 is a potent telomerase inhibitor". *Cell.* **107** (3): 347–59. doi:10.1016/S0092-8674(01)00538-4. PMID 11701125.

[55] Seimiya H, Sawada H, Muramatsu Y, Shimizu M, Ohko K, Yamane K, Tsuruo T (June 2000). "Involvement of 14-3-3 proteins in nuclear localization of telomerase". *EMBO J.* **19** (11): 2652–61. doi:10.1093/emboj/19.11.2652. PMC 212742ⓈPMID 10835362.

## 12.9  Further reading

- Mattson MP, Fu W, Zhang P (2001). "Emerging roles for telomerase in regulating cell differentiation and survival: a neuroscientist's perspective". *Mech. Ageing Dev.* **122** (7): 659–71. doi:10.1016/S0047-6374(01)00221-4. PMID 11322991.

- Castillo Ureta H, Barrera Saldaña HA, Martínez Rodríguez HG (2003). "[Telomerase: an enzyme with multiple applications in cancer research]". *Rev. Invest. Clin.* **54** (4): 342–8. PMID 12415959.

- Janknecht R (2004). "On the road to immortality: hTERT upregulation in cancer cells". *FEBS Lett.* **564** (1–2): 9–13. doi:10.1016/S0014-5793(04)00356-4. PMID 15094035.

- Cristofari G, Sikora K, Lingner J (2007). "Telomerase unplugged". *ACS Chem. Biol.* **2** (3): 155–8. doi:10.1021/cb700037c. PMID 17373762.

- Beliveau A, Yaswen P (2007). "Soothing the watchman: telomerase reduces the p53-dependent cellular stress response". *Cell Cycle.* **6** (11): 1284–7. doi:10.4161/cc.6.11.4298. PMID 17534147.

- Bellon M, Nicot C (2007). "Telomerase: a crucial player in HTLV-I-induced human T-cell leukemia". *Cancer genomics & proteomics.* **4** (1): 21–5. PMID 17726237.

## 12.10   External links

- GeneReviews/NCBI/NIH/UW entry on Dyskeratosis Congenita

- GeneReviews/NCBI/NIH/UW entry on Pulmonary Fibrosis, Familial

- TERT protein, human at the US National Library of Medicine Medical Subject Headings (MeSH)

# Chapter 13

# Reverse transcriptase

A **reverse transcriptase** (RT) is an enzyme used to generate complementary DNA (cDNA) from an RNA template, a process termed *reverse transcription*. It is mainly associated with retroviruses. However, non-retroviruses also use RT (for example, the hepatitis B virus, a member of the Hepadnaviridae, which are dsDNA-RT viruses, while retroviruses are ssRNA viruses). RT inhibitors are widely used as antiretroviral drugs. RT activities are also associated with the replication of chromosome ends (telomerase) and some mobile genetic elements (retrotransposons).

Retroviral RT has three sequential biochemical activities:

- (a) RNA-dependent DNA polymerase activity,
- (b) ribonuclease H, and
- (c) DNA-dependent DNA polymerase activity.

These activities are used by the retrovirus to convert single-stranded genomic RNA into double-stranded cDNA which can integrate into the host genome, potentially generating a long-term infection that can be very difficult to eradicate. The same sequence of reactions is widely used in the laboratory to convert RNA to DNA for use in molecular cloning, RNA sequencing, polymerase chain reaction (PCR), or genome analysis.

Well studied reverse transcriptases include:

- HIV-1 reverse transcriptase from human immunodeficiency virus type 1 (PDB: 1HMV) has two subunits, which have respective molecular weights of 66 and 51 kDa.[2]

- M-MLV reverse transcriptase from the Moloney murine leukemia virus is a single 75 kDa monomer.[3]

- AMV reverse transcriptase from the avian myeloblastosis virus also has two subunits, a 63 kDa subunit and a 95 kDa subunit.[3]

- Telomerase reverse transcriptase that maintains the telomeres of eukaryotic chromosomes.

## 13.1 History

Reverse transcriptases were discovered by Howard Temin at the University of Wisconsin–Madison in RSV virions[4] and independently isolated by David Baltimore in 1970 at MIT from two RNA tumour viruses: R-MLV and again RSV.[5] For their achievements, both shared the 1975 Nobel Prize in Physiology or Medicine (with Renato Dulbecco).

The idea of reverse transcription was very unpopular at first, as it contradicted the central dogma of molecular biology, which states that DNA is transcribed into RNA, which is then translated into proteins. However, in 1970, when the scientists Howard Temin and David Baltimore both independently discovered the enzyme responsible for reverse transcription, named reverse transcriptase, the possibility that genetic information could be passed on in this manner was finally accepted.[6]

## 13.2 Function in viruses

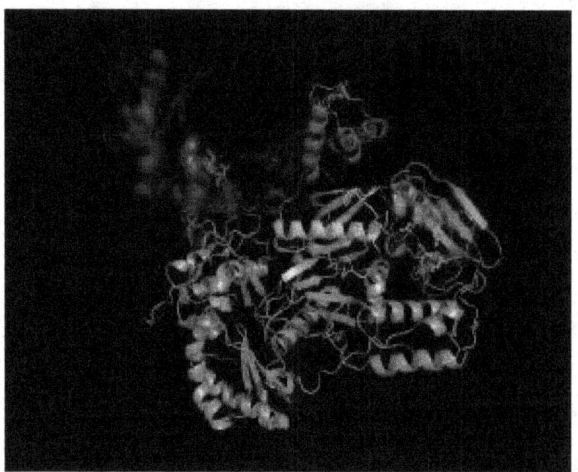

*Reverse transcriptase is shown with its finger, palm, and thumb regions. The catalytic amino acids of the RNase H active site and the polymerase active site are shown in ball-and-stick form.*

The enzymes are encoded and used by reverse-transcribing viruses, which use the enzyme during the process of replication. Reverse-transcribing RNA viruses, such as retroviruses, use the enzyme to reverse-transcribe their RNA genomes into DNA, which is then integrated into the host genome and replicated along with it. Reverse-transcribing DNA viruses, such as the hepadnaviruses, can allow RNA to serve as a template in assembling and making DNA strands. HIV infects humans with the use of this enzyme. Without reverse transcriptase, the viral genome would not be able to incorporate into the host cell, resulting in failure to replicate.

### 13.2.1   Process of reverse transcription

Reverse transcriptase creates single-stranded DNA from an RNA template.

In virus species with reverse transcriptase lacking DNA-dependent DNA polymerase activity, creation of double-stranded DNA can possibly be done by host-encoded DNA polymerase δ, mistaking the viral DNA-RNA for a primer and synthesizing a double-stranded DNA by similar mechanism as in primer removal, where the newly synthesized DNA displaces the original RNA template.

The process of reverse transcription is extremely error-prone, and it is during this step that mutations may occur. Such mutations may cause drug resistance.

### Retroviral reverse transcription

Retroviruses, also referred to as class VI ssRNA-RT viruses, are RNA reverse-transcribing viruses with a DNA intermediate. Their genomes consist of two molecules of positive-sense single-stranded RNA with a 5' cap and 3' polyadenylated tail. Examples of retroviruses include the human immunodeficiency virus (HIV) and the human T-lymphotropic virus (HTLV). Creation of double-stranded DNA occurs in the cytosol[7] as a series of these steps:

1. A specific cellular tRNA acts as a primer and hybridizes to a complementary part of the virus RNA genome called the primer binding site or PBS.

2. Complementary DNA then binds to the U5 (non-coding region) and R region (a direct repeat found at both ends of the RNA molecule) of the viral RNA.

3. A domain on the reverse transcriptase enzyme called RNAse H degrades the 5' end of the RNA which removes the U5 and R region.

4. The primer then "jumps" to the 3' end of the viral

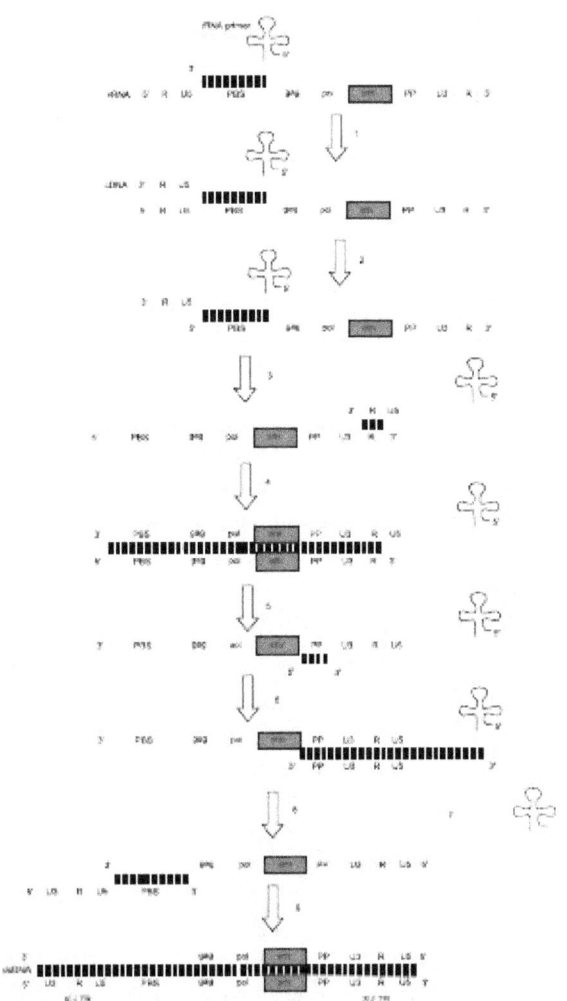

genome, and the newly synthesised DNA strands hybridizes to the complementary R region on the RNA.

5. The first strand of complementary DNA (cDNA) is extended, and the majority of viral RNA is degraded by RNAse H.

6. Once the strand is completed, second strand synthesis is initiated from the viral RNA.

7. There is then another "jump" where the PBS from the second strand hybridizes with the complementary PBS on the first strand.

8. Both strands are extended further and can be incorporated into the hosts genome by the enzyme integrase.

Creation of double-stranded DNA also involves *strand transfer*, in which there is a translocation of short DNA product from initial RNA-dependent DNA synthesis to acceptor template regions at the other end of the genome, which are later reached and processed by the reverse transcriptase for its DNA-dependent DNA activity.[8]

Retroviral RNA is arranged in 5' terminus to 3' terminus. The site where the primer is annealed to viral RNA is called the primer-binding site (PBS). The RNA 5'end to the PBS site is called U5, and the RNA 3' end to the PBS is called the leader. The tRNA primer is unwound between 14 and 22 nucleotides and forms a base-paired duplex with the viral RNA at PBS. The fact that the PBS is located near the 5' terminus of viral RNA is unusual because reverse transcriptase synthesize DNA from 3' end of the primer in the 5' to 3' direction (with respect to the RNA template). Therefore, the primer and reverse transcriptase must be relocated to 3' end of viral RNA. In order to accomplish this reposition, multiple steps and various enzymes including DNA polymerase, ribonuclease H(RNase H) and polynucleotide unwinding are needed.[9][10]

The HIV reverse transcriptase also has ribonuclease activity that degrades the viral RNA during the synthesis of cDNA, as well as DNA-dependent DNA polymerase activity that copies the sense cDNA strand into an *antisense* DNA to form a double-stranded viral DNA intermediate (vDNA).[11]

## 13.3 In eukaryotes

Self-replicating stretches of eukaryotic genomes known as retrotransposons utilize reverse transcriptase to move from one position in the genome to another via an RNA intermediate. They are found abundantly in the genomes of plants and animals. Telomerase is another reverse transcriptase found in many eukaryotes, including humans, which carries its own RNA template; this RNA is used as a template for DNA replication.[12]

## 13.4 In prokaryotes

Initial reports of reverse transcriptase in prokaryotes came as far back as 1971 (Beljanski et al., 1971a, 1972). These have since been broadly described as part of bacterial Retrons, distinct sequences that code for reverse transcriptase, and are used in the synthesis of msDNA. In order to initiate synthesis of DNA, a primer is needed. In bacteria, the primer is synthesized during replication.[13]

## 13.5 Evolutionary role

Valerian Dolja of Oregon State argues that viruses due to their diversity have played an evolutionary role in the development of cellular life, with reverse transcriptase playing a central role.[14]

## 13.6 Structure

Reverse transcriptase enzymes include an RNA-dependent DNA polymerase and a DNA-dependent DNA polymerase, which work together to perform transcription. In addition to the transcription function, retroviral reverse transcriptases have a domain belonging to the RNase H family, which is vital to their replication.

## 13.7 Replication fidelity

There are three different replication systems during the life cycle of a retrovirus. First of all, the reverse transcriptase synthesizes viral DNA from viral RNA, and then from newly made complementary DNA strand. The second replication process occurs when host cellular DNA polymerase replicates the integrated viral DNA. Lastly, RNA polymerase II transcribes the proviral DNA into RNA, which will be packed into virions. Therefore, mutation can occur during one or all of these replication steps.[15]

Reverse transcriptase has a high error rate when transcribing RNA into DNA since, unlike most other DNA polymerases, it has no proofreading ability. This high error rate allows mutations to accumulate at an accelerated rate relative to proofread forms of replication. The commercially available reverse transcriptases produced by Promega are quoted by their manuals as having error rates in the range of 1 in 17,000 bases for AMV and 1 in 30,000 bases for M-MLV.[16]

Other than creating single-nucleotide polymorphisms, reverse transcriptases have also been shown to be involved in processes such as transcript fusions, exon shuffling and creating artificial antisense transcripts.[17][18] It has been speculated that this *template switching* activity of reverse transcriptase, which can be demonstrated completely *in vivo*, may have been one of the causes for finding several thousand unannotated transcripts in the genomes of model organisms.[19]

## 13.8 Applications

### 13.8.1 Antiviral drugs

For more details on this topic, see Reverse-transcriptase inhibitor.

*The molecular structure of zidovudine (AZT), a drug used to inhibit HIV reverse transcriptase*

As HIV uses reverse transcriptase to copy its genetic material and generate new viruses (part of a retrovirus proliferation circle), specific drugs have been designed to disrupt the process and thereby suppress its growth. Collectively, these drugs are known as reverse-transcriptase inhibitors and include the nucleoside and nucleotide analogues zidovudine (trade name Retrovir), lamivudine (Epivir) and tenofovir (Viread), as well as non-nucleoside inhibitors, such as nevirapine (Viramune).

### 13.8.2 Molecular biology

For more details on this topic, see Reverse transcription polymerase chain reaction.

Reverse transcriptase is commonly used in research to apply the polymerase chain reaction technique to RNA in a technique called reverse transcription polymerase chain reaction (RT-PCR). The classical PCR technique can be applied only to DNA strands, but, with the help of reverse transcriptase, RNA can be transcribed into DNA, thus making PCR analysis of RNA molecules possible. Reverse transcriptase is used also to create cDNA libraries from mRNA. The commercial availability of reverse transcriptase greatly improved knowledge in the area of molecular biology, as, along with other enzymes, it allowed scientists to clone, sequence, and characterise RNA.

Reverse transcriptase has also been employed in insulin production. By inserting eukaryotic mRNA for insulin production along with reverse transcriptase into bacteria, the mRNA could be inserted into the prokaryote's genome. Large amounts of insulin can then be created, sidestepping the need to harvest pig pancreas and other such traditional sources. Directly inserting eukaryotic DNA into bacteria would not work because it carries introns, so would not translate successfully using the bacterial ribosomes. Processing in the eukaryotic cell during mRNA production removes these introns to provide a suitable template. Reverse transcriptase converted this edited RNA back into DNA so it could be incorporated in the genome.

## 13.9 See also

- cDNA library
- DNA polymerase
- msDNA
- Reverse transcribing virus
- RNA polymerase
- Telomerase
- Retrotransposon marker

## 13.10 References

[1] PDB: 3KLF; Tu X, Das K, Han Q, Bauman JD, Clark AD, Hou X, Frenkel YV, Gaffney BL, Jones RA, Boyer PL, Hughes SH, Sarafianos SG, Arnold E (September 2010). "Structural basis of HIV-1 resistance to AZT by excision.". *Nat. Struct. Mol. Biol.* **17** (10): 1202–9. doi:10.1038/nsmb.1908. PMC 2987654. PMID 20852643.

[2] Ferris, AL; Hizi, A; Showalter, SD; Pichuantes, S; Babe, L; Craik, CS; Hughes, SH (April 1990). "Immunologic and proteolytic analysis of HIV-1 reverse transcriptase structure." (PDF). *Virology.* **175** (2): 456–64. doi:10.1016/0042-6822(90)90430-y. PMID 1691562.

[3] Konishi A, Yasukawa K, Inouye K (2012). "Improving the thermal stability of avian myeloblastosis virus reverse transcriptase α-subunit by site-directed mutagenesis". *Biotechnol. Lett.* **34** (7): 1209–15. doi:10.1007/s10529-012-0904-9. PMID 22426840.

[4] Temin H. M., Mizutani S. (June 1970). "RNA-dependent DNA polymerase in virions of Rous sarcoma virus". *Nature.* **226** (5252): 1211–3. doi:10.1038/2261211a0. PMID 4316301.

[5] Baltimore D. (June 1970). "RNA-dependent DNA polymerase in virions of RNA tumour viruses". *Nature*. **226** (5252): 1209–11. doi:10.1038/2261209a0. PMID 4316300.

[6] "Central dogma reversed". *Nature*. **226** (5252): 1198–9. June 1970. doi:10.1038/2261198a0. PMID 5422595.

[7] Bio-Medicine.org - Retrovirus Retrieved on 17 Feb, 2009

[8] Telesnitsky A., Goff S. P. (1993). "Strong-stop strand transfer during reverse transcription". In Skalka, M. A.; Goff, S. P. *Reverse transcriptase* (1st ed.). New York: Cold Spring Harbor. p. 49. ISBN 0-87969-382-7.

[9] Bernstein A., Weiss R., Tooze J. (1985). "RNA tumor viruses". *Molecular Biology of Tumor Viruses* (2nd ed.). Cold Spring Harbor, N.Y.: Cold Spring Harbor Laboratory.

[10] Moelling, K; Broecker F. (2015) The reverse transcriptase–RNase H: from viruses to antiviral defense. Ann N Y Acad Sci. 1341:126-35. doi: 10.1111/nyas.12668.

[11] Doc Kaiser's Microbiology Home Page > IV. VIRUSES > F. ANIMAL VIRUS LIFE CYCLES > 3. The Life Cycle of HIV Community College of Baltimore County. Updated: Jan 2008.

[12] Krieger M, Scott MP, Matsudaira PT, Lodish HF, Darnell JE, Zipursky L, Kaiser C, Berk A (2004). *Molecular cell biology*. New York: W.H. Freeman and CO. ISBN 0-7167-4366-3.

[13] Hurwitz J., Leis J. P. (January 1972). "RNA-dependent DNA polymerase activity of RNA tumor viruses. I. Directing influence of DNA in the reaction". *J. Virol.* **9** (1): 116–29. PMC 356270. PMID 4333538.

[14] Arnold, Carrie (17 July 2014). "Could Giant Viruses Be the Origin of Life on Earth?". *news.nationalgeographic.com*. Retrieved 29 May 2016.

[15] Bbenek K., Kunkel A. T. (1993). "The fidelity of retroviral reverse transcriptases". In Skalka, M. A.; Goff, P. S. *Reverse transcriptase*. New York: Cold Spring Harbor Laboratory Press. p. 85. ISBN 0-87969-382-7.

[16] Promega kit instruction manual (1999)

[17] Houseley J., Tollervey D. (2010). "Apparent non-canonical trans-splicing is generated by reverse transcriptase in vitro". *PLoS ONE*. **5** (8): e12271. doi:10.1371/journal.pone.0012271. PMC 2923612. PMID 20805885.

[18] Zeng X. C., Wang S. X. (June 2002). "Evidence that BmTXK beta-BmKCT cDNA from Chinese scorpion Buthus martensii Karsch is an artifact generated in the reverse transcription process". *FEBS Lett.* **520** (1–3): 183–4; author reply 185. doi:10.1016/S0014-5793(02)02812-0. PMID 12044895.

[19] van Bakel H., Nislow C., Blencowe B. J., Hughes T. R. (2011). "Response to "The Reality of Pervasive Transcription"". *PLoS Biology*. **9** (7): e1001102. doi:10.1371/journal.pbio.1001102.

## 13.11 External links

- RNA Transcriptase at the US National Library of Medicine Medical Subject Headings (MeSH)

- animation of reverse transcriptase action and three reverse transcriptase inhibitors

- Molecule of the month (September 2002) at the RCSB PDB

- HIV Replication 3D Medical Animation. (Nov 2008). Video by Boehringer Ingelheim.

- Goodsell DS. "Molecule of the Month: Reverse Transcriptase (Sep 2002)". Research Collaboratory for Structural Bioinformatics (RCSB) Protein Data Bank (PDB). Retrieved 2013-01-13.

# Chapter 14

# Chromosome

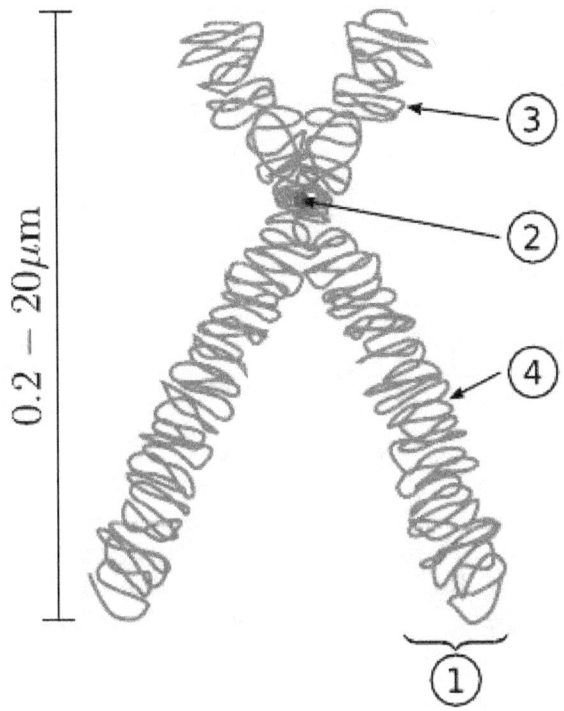

0.2 – 20 μm

*Diagram of a replicated and condensed metaphase eukaryotic chromosome. (1) Chromatid – one of the two identical parts of the chromosome after S phase. (2) Centromere – the point where the two chromatids touch. (3) Short strand. (4) Long strand.*

A **chromosome** is a packaged and organized structure containing most of the DNA of a living organism. Most eukaryotic cells have a set of chromosomes (46 in humans) with the genetic material spread among them.

During most of the duration of the cell cycle, a chromosome consists of one long double-helix DNA molecule (with associated proteins. During S phase, the chromosome gets replicated, resulting in an X-shaped structure called a metaphase chromosome. Both the original and the newly copied DNA are now called chromatids. The two "sister" chromatids are joined together at a protein junction called a centromere (forming the X-shaped structure).

Chromosomes are normally visible under a light microscope only when the cell is undergoing mitosis (cell division). Even then, the full chromosome containing both joined sister chromatids becomes visible only during a sequence of mitosis known as metaphase (when chromosomes align together, attached to the mitotic spindle and prepare to divide).[1] This DNA and its associated proteins and macromolecules is collectively known as chromatin, which is further packaged along with its associated molecules into a discrete structure called a nucleosome. Chromatin is present in most cells, with a few exceptions, for example, red blood cells. Occurring only in the nucleus of eukaryotic cells, chromatin composes the vast majority of all DNA, except for a small amount inherited maternally, which is found in mitochondria.

In prokaryotic cells, chromatin occurs free-floating in cytoplasm, as these cells lack organelles and a defined nucleus. Bacteria also lack histones. The main information-carrying macromolecule is a single piece of coiled double-helix DNA, containing many genes, regulatory elements and other noncoding DNA.[2] The DNA-bound macromolecules are proteins that serve to package the DNA and control its functions. Chromosomes vary widely between different organisms. Some species such as certain bacteria also contain plasmids or other extrachromosomal DNA. These are circular structures in the cytoplasm that contain cellular DNA and play a role in horizontal gene transfer.[1]

Compaction of the duplicated chromosomes during cell division (mitosis or meiosis) results either in a four-arm structure (pictured above) if the centromere is located in the middle of the chromosome or a two-arm structure if the centromere is located near one of the ends. Chromosomal recombination during meiosis and subsequent sexual reproduction plays a significant role in genetic diversity. If these structures are manipulated incorrectly, through processes known as chromosomal instability and translocation, the cell may undergo mitotic catastrophe and die, or it may unexpectedly evade apoptosis leading to the progression of cancer.

In prokaryotes (see nucleoids) and viruses,[2] the DNA is

often densely packed and organized: in the case of archaea, by homologs to eukaryotic histones, and in the case of bacteria, by histone-like proteins. Small circular genomes called plasmids are often found in bacteria and also in mitochondria and chloroplasts, reflecting their bacterial origins.

Some authors, as in this article, use the term chromosome in a wider sense, to refer to the individualized portions of chromatin in cells, either visible or not under light microscopy. However, others use the concept in a narrower sense, to refer to the individualized portions of chromatin during cell division, visible under light microscopy due to high condensation.

# 14.1 History of discovery

The word *chromosome* (/'krəʊməˌsəʊm, -ˌzəʊm/[3][4]) comes from the Greek χρῶμα (*chroma*, "colour") and σῶμα (*soma*, "body"), describing their strong staining by particular dyes.[5]

Schleiden,[1] Virchow and Bütschli were among the first scientists who recognized the structures now so familiar to everyone as chromosomes.[6] The term was coined by von Waldeyer-Hartz,[7] referring to the term chromatin, which was introduced by Walther Flemming.

In a series of experiments beginning in the mid-1880s, Theodor Boveri gave the definitive demonstration that chromosomes are the vectors of heredity. His two principles were the *continuity* of chromosomes and the *individuality* of chromosomes. It is the second of these principles that was so original. Wilhelm Roux suggested that each chromosome carries a different genetic load. Boveri was able to test and confirm this hypothesis. Aided by the rediscovery at the start of the 1900s of Gregor Mendel's earlier work, Boveri was able to point out the connection between the rules of inheritance and the behaviour of the chromosomes. Boveri influenced two generations of American cytologists: Edmund Beecher Wilson, Nettie Stevens, Walter Sutton and Theophilus Painter were all influenced by Boveri (Wilson, Stevens, and Painter actually worked with him).[8]

In his famous textbook *The Cell in Development and Heredity*, Wilson linked together the independent work of Boveri and Sutton (both around 1902) by naming the chromosome theory of inheritance the Boveri–Sutton chromosome theory (the names are sometimes reversed).[9] Ernst Mayr remarks that the theory was hotly contested by some famous geneticists: William Bateson, Wilhelm Johannsen, Richard Goldschmidt and T.H. Morgan, all of a rather dogmatic turn of mind. Eventually, complete proof came from chromosome maps in Morgan's own lab.[10]

The number of human chromosomes was published in 1923 by Theophilus Painter. By inspection through the microscope he counted 24 pairs, which would mean 48 chromosomes. His error was copied by others and it was not until 1956 that the true number, 46, was determined by Indonesia-born cytogeneticist Joe Hin Tjio.[11]

# 14.2 Prokaryotes

The prokaryotes – bacteria and archaea – typically have a single circular chromosome, but many variations exist.[12] The chromosomes of most bacteria, which some authors prefer to call genophores, can range in size from only 130,000 base pairs in the endosymbiotic bacteria *Candidatus Hodgkinia cicadicola*[13] and *Candidatus Tremblaya princeps*,[14] to more than 14,000,000 base pairs in the soil-dwelling bacterium *Sorangium cellulosum*.[15] Spirochaetes of the genus *Borrelia* are a notable exception to this arrangement, with bacteria such as *Borrelia burgdorferi*, the cause of Lyme disease, containing a single *linear* chromosome.[16]

## 14.2.1 Structure in sequences

Prokaryotic chromosomes have less sequence-based structure than eukaryotes. Bacteria typically have a one-point (the origin of replication) from which replication starts, whereas some archaea contain multiple replication origins.[17] The genes in prokaryotes are often organized in operons, and do not usually contain introns, unlike eukaryotes.

## 14.2.2 DNA packaging

Prokaryotes do not possess nuclei. Instead, their DNA is organized into a structure called the nucleoid.[18][19] The nucleoid is a distinct structure and occupies a defined region of the bacterial cell. This structure is, however, dynamic and is maintained and remodeled by the actions of a range of histone-like proteins, which associate with the bacterial chromosome.[20] In archaea, the DNA in chromosomes is even more organized, with the DNA packaged within structures similar to eukaryotic nucleosomes.[21][22]

Bacterial chromosomes tend to be tethered to the plasma membrane of the bacteria. In molecular biology application, this allows for its isolation from plasmid DNA by centrifugation of lysed bacteria and pelleting of the membranes (and the attached DNA).

Prokaryotic chromosomes and plasmids are, like eukaryotic DNA, generally supercoiled. The DNA must first be released into its relaxed state for access for transcription, regulation, and replication.

## 14.3 Eukaryotes

*Organization of DNA in a eukaryotic cell.*

See also: Eukaryotic chromosome fine structure

In eukaryotes, nuclear chromosomes are packaged by proteins into a condensed structure called chromatin. This allows the very long DNA molecules to fit into the cell nucleus. The structure of chromosomes and chromatin varies through the cell cycle. Chromosomes are even more condensed than chromatin and are an essential unit for cellular division. Chromosomes must be replicated, divided, and passed successfully to their daughter cells so as to ensure the genetic diversity and survival of their progeny. Chromosomes may exist as either duplicated or unduplicated. Unduplicated chromosomes are single double helixes, whereas duplicated chromosomes contain two identical copies (called chromatids or sister chromatids) joined by a centromere.

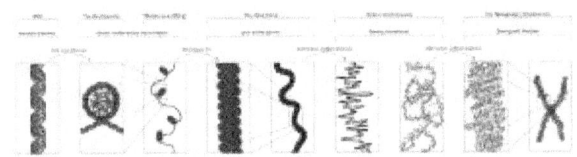

*The major structures in DNA compaction: DNA, the nucleosome, the 10 nm "beads-on-a-string" fibre, the 30 nm fibre and the metaphase chromosome.*

Eukaryotes (cells with nuclei such as those found in plants, fungi, and animals) possess multiple large linear chromosomes contained in the cell's nucleus. Each chromosome has one centromere, with one or two arms projecting from the centromere, although, under most circumstances, these arms are not visible as such. In addition, most eukaryotes have a small circular mitochondrial genome, and some

eukaryotes may have additional small circular or linear cytoplasmic chromosomes.

In the nuclear chromosomes of eukaryotes, the uncondensed DNA exists in a semi-ordered structure, where it is wrapped around histones (structural proteins), forming a composite material called chromatin.

### 14.3.1 Chromatin

Main article: Chromatin

Chromatin is the complex of DNA and protein found in the eukaryotic nucleus, which packages chromosomes. The structure of chromatin varies significantly between different stages of the cell cycle, according to the requirements of the DNA.

#### Interphase chromatin

During interphase (the period of the cell cycle where the cell is not dividing), two types of chromatin can be distinguished:

- Euchromatin, which consists of DNA that is active, e.g., being expressed as protein.

- Heterochromatin, which consists of mostly inactive DNA. It seems to serve structural purposes during the chromosomal stages. Heterochromatin can be further distinguished into two types:

  - *Constitutive heterochromatin*, which is never expressed. It is located around the centromere and usually contains repetitive sequences.
  - *Facultative heterochromatin*, which is sometimes expressed.

#### Metaphase chromatin and division

See also: mitosis and meiosis

In the early stages of mitosis or meiosis (cell division), the chromatin double helix become more and more condensed. They cease to function as accessible genetic material (transcription stops) and become a compact transportable form. This compact form makes the individual chromosomes visible, and they form the classic four arm structure, a pair of sister chromatids attached to each other at the centromere. The shorter arms are called *p arms* (from the French *petit*, small) and the longer arms are called *q arms* (q follows p in the Latin alphabet; q-g "grande"; alternatively it is sometimes said q is short for *queue* meaning tail in French[23]). This is the only natural context in

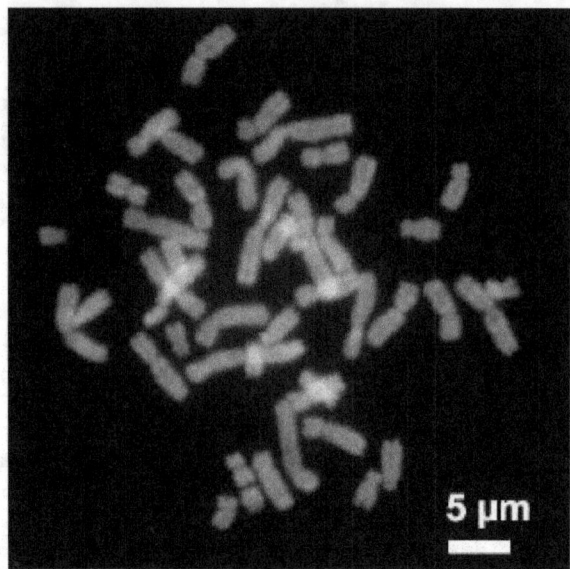

*Human chromosomes during metaphase*

Sequencing of the human genome has provided a great deal of information about each of the chromosomes. Below is a table compiling statistics for the chromosomes, based on the Sanger Institute's human genome information in the Vertebrate Genome Annotation (VEGA) database.[25] Number of genes is an estimate, as it is in part based on gene predictions. Total chromosome length is an estimate as well, based on the estimated size of unsequenced heterochromatin regions.

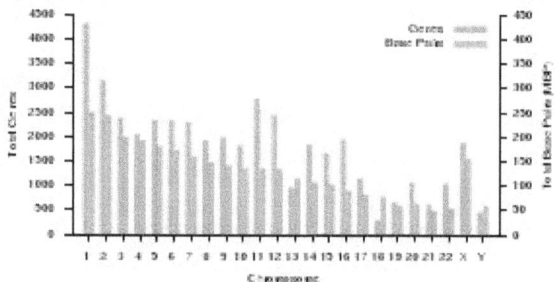

*Estimated number of genes and base pairs (in mega base pairs) on each human chromosome*

which individual chromosomes are visible with an optical microscope.

Mitotic metaphase chromosomes are best described by a linearly organized longitudinally compressed array of consecutive chromatin loops.[24]

During mitosis, microtubules grow from centrosomes located at opposite ends of the cell and also attach to the centromere at specialized structures called kinetochores, one of which is present on each sister chromatid. A special DNA base sequence in the region of the kinetochores provides, along with special proteins, longer-lasting attachment in this region. The microtubules then pull the chromatids apart toward the centrosomes, so that each daughter cell inherits one set of chromatids. Once the cells have divided, the chromatids are uncoiled and DNA can again be transcribed. In spite of their appearance, chromosomes are structurally highly condensed, which enables these giant DNA structures to be contained within a cell nucleus.

### 14.3.2 Human chromosomes

Chromosomes in humans can be divided into two types: autosomes (body chromosome(s)) and allosome (sex chromosome(s)). Certain genetic traits are linked to a person's sex and are passed on through the sex chromosomes. The autosomes contain the rest of the genetic hereditary information. All act in the same way during cell division. Human cells have 23 pairs of chromosomes (22 pairs of autosomes and one pair of sex chromosomes), giving a total of 46 per cell. In addition to these, human cells have many hundreds of copies of the mitochondrial genome.

# 14.4  Number in various organisms

Main article: List of organisms by chromosome count

### 14.4.1  In eukaryotes

These tables give the total number of chromosomes (including sex chromosomes) in a cell nucleus. For example, human cells are diploid and have 22 different types of autosome, each present as two copies, and two sex chromosomes. This gives 46 chromosomes in total. Other organisms have more than two copies of their chromosome types, such as bread wheat, which is *hexaploid* and has six copies of seven different chromosome types – 42 chromosomes in total.

Normal members of a particular eukaryotic species all have the same number of nuclear chromosomes (see the table). Other eukaryotic chromosomes, i.e., mitochondrial and plasmid-like small chromosomes, are much more variable in number, and there may be thousands of copies per cell.

Asexually reproducing species have one set of chromosomes that are the same in all body cells. However, asexual species can be either haploid or diploid.

Sexually reproducing species have somatic cells (body cells), which are diploid [2n] having two sets of chromo-

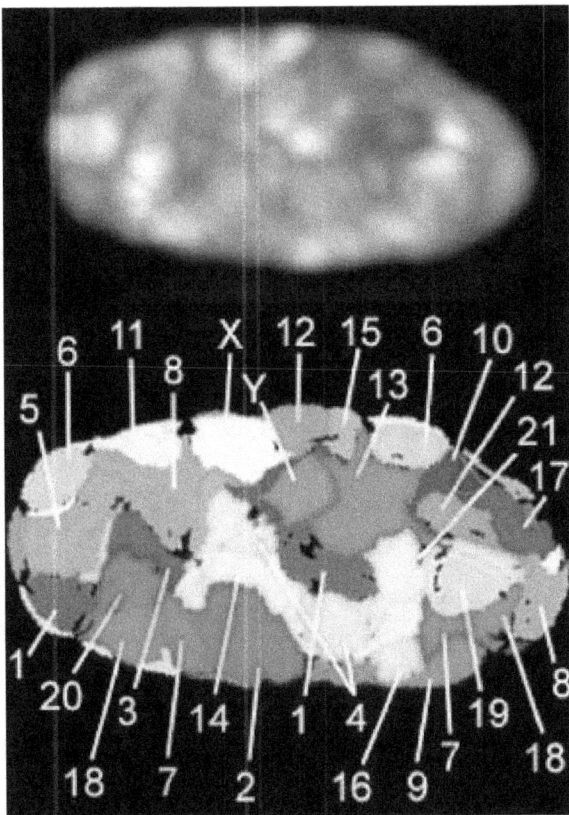

*The 23 human chromosome territories during prometaphase in fibroblast cells.*

somes (23 pairs in humans with one set of 23 chromosomes from each parent), one set from the mother and one from the father. Gametes, reproductive cells, are haploid [n]: They have one set of chromosomes. Gametes are produced by meiosis of a diploid germ line cell. During meiosis, the matching chromosomes of father and mother can exchange small parts of themselves (crossover), and thus create new chromosomes that are not inherited solely from either parent. When a male and a female gamete merge (fertilization), a new diploid organism is formed.

Some animal and plant species are polyploid [Xn]: They have more than two sets of homologous chromosomes. Plants important in agriculture such as tobacco or wheat are often polyploid, compared to their ancestral species. Wheat has a haploid number of seven chromosomes, still seen in some cultivars as well as the wild progenitors. The more-common pasta and bread wheats are polyploid, having 28 (tetraploid) and 42 (hexaploid) chromosomes, compared to the 14 (diploid) chromosomes in the wild wheat.[52]

### 14.4.2   In prokaryotes

Prokaryote species generally have one copy of each major chromosome, but most cells can easily survive with multiple copies.[53] For example, *Buchnera*, a symbiont of aphids has multiple copies of its chromosome, ranging from 10–400 copies per cell.[54] However, in some large bacteria, such as *Epulopiscium fishelsoni* up to 100,000 copies of the chromosome can be present.[55] Plasmids and plasmid-like small chromosomes are, as in eukaryotes, highly variable in copy number. The number of plasmids in the cell is almost entirely determined by the rate of division of the plasmid – fast division causes high copy number.

## 14.5   Karyotype

Main article: Karyotype
In general, the **karyotype** is the characteristic chromosome

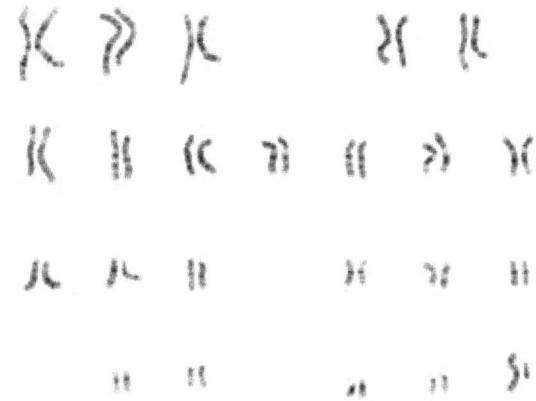

*Karyogram of a human male*

complement of a eukaryote species.[56] The preparation and study of karyotypes is part of cytogenetics.

Although the replication and transcription of DNA is highly standardized in eukaryotes, *the same cannot be said for their karyotypes*, which are often highly variable. There may be variation between species in chromosome number and in detailed organization. In some cases, there is significant variation within species. Often there is:

1. variation between the two sexes

2. variation between the germ-line and soma (between gametes and the rest of the body)

3. variation between members of a population, due to balanced genetic polymorphism

4. geographical variation between races

5. mosaics or otherwise abnormal individuals.

Also, variation in karyotype may occur during development from the fertilized egg.

The technique of determining the karyotype is usually called *karyotyping*. Cells can be locked part-way through division (in metaphase) in vitro (in a reaction vial) with colchicine. These cells are then stained, photographed, and arranged into a *karyogram*, with the set of chromosomes arranged, autosomes in order of length, and sex chromosomes (here X/Y) at the end.

Like many sexually reproducing species, humans have special gonosomes (sex chromosomes, in contrast to autosomes). These are XX in females and XY in males.

### 14.5.1 Historical note

Investigation into the human karyotype took many years to settle the most basic question: *How many chromosomes does a normal diploid human cell contain?* In 1912, Hans von Winiwarter reported 47 chromosomes in spermatogonia and 48 in oogonia, concluding an XX/XO sex determination mechanism.[57] Painter in 1922 was not certain whether the diploid number of man is 46 or 48, at first favouring 46.[58] He revised his opinion later from 46 to 48, and he correctly insisted on humans having an XX/XY system.[59]

New techniques were needed to definitively solve the problem:

1. Using cells in culture

2. Arresting mitosis in metaphase by a solution of colchicine

3. Pretreating cells in a hypotonic solution 0.075 M KCl, which swells them and spreads the chromosomes

4. Squashing the preparation on the slide forcing the chromosomes into a single plane

5. Cutting up a photomicrograph and arranging the result into an indisputable karyogram.

It took until 1954 before the human diploid number was confirmed as 46.[60][61] Considering the techniques of Winiwarter and Painter, their results were quite remarkable.[62] Chimpanzees (the closest living relatives to modern humans) have 48 chromosomes (as well as the other great apes: in humans two chromosomes fused to form chromosome 2).

(See Also: Argument from authority#Inaccurate chromosome number)

## 14.6  Aberrations

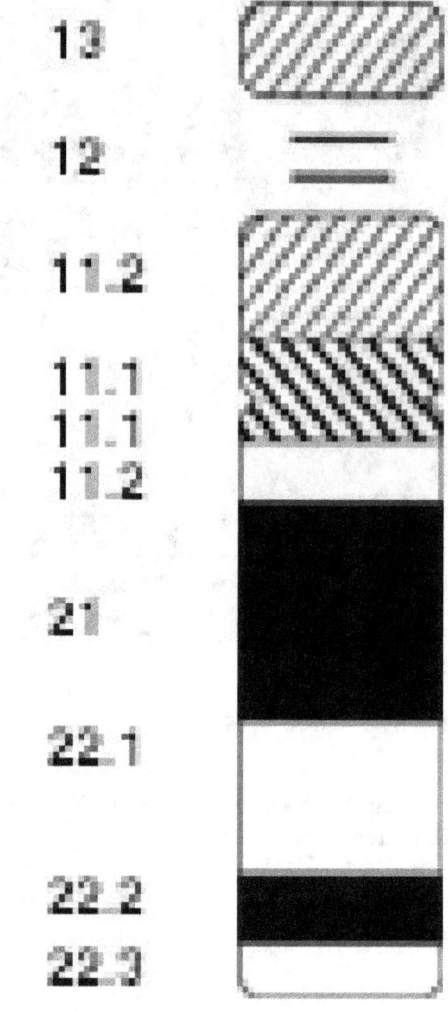

*In Down syndrome, there are three copies of chromosome 21*

Chromosomal aberrations are disruptions in the normal chromosomal content of a cell and are a major cause of genetic conditions in humans, such as Down syndrome, although most aberrations have little to no effect. Some chromosome abnormalities do not cause disease in carriers, such as translocations, or chromosomal inversions, although they may lead to a higher chance of bearing a child with a chromosome disorder. Abnormal numbers of chromosomes or chromosome sets, called aneuploidy, may be lethal or may give rise to genetic disorders. Genetic counseling is offered for families that may carry a chromosome rearrangement.

The gain or loss of DNA from chromosomes can lead to a variety of genetic disorders. Human examples include:

- Cri du chat, which is caused by the deletion of part of the short arm of chromosome 5. "Cri du chat" means "cry of the cat" in French; the condition was so-named because affected babies make high-pitched cries that sound like those of a cat. Affected individuals have wide-set eyes, a small head and jaw, moderate to severe mental health problems, and are very short.

- Down syndrome, the most common trisomy, usually caused by an extra copy of chromosome 21 (trisomy 21). Characteristics include decreased muscle tone, stockier build, asymmetrical skull, slanting eyes and mild to moderate developmental disability.[63]

- Edwards syndrome, or trisomy-18, the second most common trisomy.[64] Symptoms include motor retardation, developmental disability and numerous congenital anomalies causing serious health problems. Ninety percent of those affected die in infancy. They have characteristic clenched hands and overlapping fingers.

- Isodicentric 15, also called idic(15), partial tetrasomy 15q, or inverted duplication 15 (inv dup 15).

- Jacobsen syndrome, which is very rare. It is also called the terminal 11q deletion disorder.[65] Those affected have normal intelligence or mild developmental disability, with poor expressive language skills. Most have a bleeding disorder called Paris-Trousseau syndrome.

- Klinefelter syndrome (XXY). Men with Klinefelter syndrome are usually sterile and tend to be taller and have longer arms and legs than their peers. Boys with the syndrome are often shy and quiet and have a higher incidence of speech delay and dyslexia. Without testosterone treatment, some may develop gynecomastia during puberty.

- Patau Syndrome, also called D-Syndrome or trisomy-13. Symptoms are somewhat similar to those of trisomy-18, without the characteristic folded hand.

- Small supernumerary marker chromosome. This means there is an extra, abnormal chromosome. Features depend on the origin of the extra genetic material. Cat-eye syndrome and isodicentric chromosome 15 syndrome (or Idic15) are both caused by a supernumerary marker chromosome, as is Pallister–Killian syndrome.

- Triple-X syndrome (XXX). XXX girls tend to be tall and thin and have a higher incidence of dyslexia.

- Turner syndrome (X instead of XX or XY). In Turner syndrome, female sexual characteristics are present but underdeveloped. Females with Turner syndrome often have a short stature, low hairline, abnormal eye features and bone development and a "caved-in" appearance to the chest.

- Wolf–Hirschhorn syndrome, which is caused by partial deletion of the short arm of chromosome 4. It is characterized by growth retardation, delayed motor skills development, "Greek Helmet" facial features, and mild to profound mental health problems.

- XYY syndrome. XYY boys are usually taller than their siblings. Like XXY boys and XXX girls, they are more likely to have learning difficulties.

### 14.6.1  Sperm aneuploidy

Exposure of males to certain lifestyle, environmental and/or occupational hazards may increase the risk of aneuploid spermatozoa.[66] In particular, risk of aneuploidy is increased by tobacco smoking,[67][68] and occupational exposure to benzene,[69] insecticides,[70][71] and perfluorinated compounds.[72] Increased aneuploidy is often associated with increased DNA damage in spermatozoa.

## 14.7  See also

- Aneuploidy
- Chromosome segregation
- DNA
- Genetic deletion
- For information about chromosomes in genetic algorithms, see chromosome (genetic algorithm)
- Genetic genealogy
  - Genealogical DNA test
- Lampbrush chromosome
- List of number of chromosomes of various organisms
- Locus (explains gene location nomenclature)
- Maternal influence on sex determination
- Non-disjunction
- Sex-determination system
  - XY sex-determination system
    - X-chromosome
      - X-inactivation

- Y-chromosome
  - Y-chromosomal Aaron
  - Y-chromosomal Adam
- Polytene chromosome
- Neochromosome

## 14.8   Notes and references

[1] Schleyden, M. J. (1847). *Microscopical researches into the accordance in the structure and growth of animals and plants.*

[2] Johnson, J.; Chiu, W. (1 April 2000). "Structures of virus and virus-like particles". *Current Opinion in Structural Biology.* **10** (2): 229–235. doi:10.1016/S0959-440X(00)00073-7. PMID 10753814.

[3] Jones, Daniel (2003) [1917], Peter Roach, James Hartmann and Jane Setter, eds., *English Pronouncing Dictionary*, Cambridge: Cambridge University Press, ISBN 3-12-539683-2

[4] "Chromosome". *Merriam-Webster Dictionary.*

[5] Coxx, H. J. (1925). *Biological Stains - A Handbook on the Nature and Uses of the Dyes Employed in the Biological Laboratory.* Commission on Standardization of Biological Stains.

[6] Fokin S.I. (2013). "Otto Bütschli (1848–1920) Where we will genuflect?" (PDF). *Protistology.* **8** (1): 22–35.

[7] Waldeyer-Hartz (1888). "Über Karyokinese und ihre Beziehungen zu den Befruchtungsvorgängen". *Archiv für mikroskopische Anatomie und Entwicklungsmechanik.* **32**: 27.

[8] Carlson, Elof A. (2004). *Mendel's Legacy: The Origin of Classical Genetics* (PDF). Cold Spring Harbor, NY: Cold Spring Harbor Laboratory Press. p. 88. ISBN 978-087969675-7.

[9] Wilson, E.B. (1925). *The Cell in Development and Heredity,* Ed. 3. Macmillan, New York. p. 923.

[10] Mayr, E. (1982). *The growth of biological thought.* Harvard. p. 749.

[11] Matthews, Robert. "The bizarre case of the chromosome that never was" (PDF). Archived from the original (PDF) on 15 December 2013. Retrieved 13 July 2013.

[12] Thanbichler M; Shapiro L (2006). "Chromosome organization and segregation in bacteria". *J. Struct. Biol.* **156** (2): 292–303. doi:10.1016/j.jsb.2006.05.007. PMID 16860572.

[13] Van Leuven, JT; Meister, RC; Simon. C; McCutcheon, JP (11 September 2014). "Sympatric speciation in a bacterial endosymbiont results in two genomes with the functionality of one.". *Cell.* **158** (6): 1270–80. doi:10.1016/j.cell.2014.07.047. PMID 25175626.

[14] McCutcheon, JP; von Dohlen, CD (23 August 2011). "An interdependent metabolic patchwork in the nested symbiosis of mealybugs.". *Current Biology.* **21** (16): 1366–72. doi:10.1016/j.cub.2011.06.051. PMC 3169327. PMID 21835622.

[15] Han, K; Li, ZF; Peng, R; Zhu, LP; Zhou, T; Wang, LG; Li, SG; Zhang, XB; Hu, W; Wu, ZH; Qin, N; Li, YZ (2013). "Extraordinary expansion of a Sorangium cellulosum genome from an alkaline milieu.". *Scientific Reports.* **3**: 2101. doi:10.1038/srep02101. PMID 23812535.

[16] Hinnebusch J; Tilly K (1993). "Linear plasmids and chromosomes in bacteria". *Mol Microbiol.* **10** (5): 917–22. doi:10.1111/j.1365-2958.1993.tb00963.x. PMID 7934868.

[17] Kelman LM; Kelman Z (2004). "Multiple origins of replication in archaea". *Trends Microbiol.* **12** (9): 399–401. doi:10.1016/j.tim.2004.07.001. PMID 15337158.

[18] Thanbichler M; Wang SC; Shapiro L (2005). "The bacterial nucleoid: a highly organized and dynamic structure". *J. Cell. Biochem.* **96** (3): 506–21. doi:10.1002/jcb.20519. PMID 15988757.

[19] Le TB, Imakaev MV, Mirny LA, Laub MT (2013). "High-resolution mapping of the spatial organization of a bacterial chromosome". *Science.* **342** (6159): 731–4. doi:10.1126/science.1242059. PMC 3927313. PMID 24158908.

[20] Sandman K; Pereira SL; Reeve JN (1998). "Diversity of prokaryotic chromosomal proteins and the origin of the nucleosome". *Cell. Mol. Life Sci.* **54** (12): 1350–64. doi:10.1007/s000180050259. PMID 9893710.

[21] Sandman K; Reeve JN (2000). "Structure and functional relationships of archaeal and eukaryal histones and nucleosomes". *Arch. Microbiol.* **173** (3): 165–9. doi:10.1007/s002039900122. PMID 10763747.

[22] Pereira SL; Grayling RA; Lurz R; Reeve JN (1997). "Archaeal nucleosomes". *Proc. Natl. Acad. Sci. U.S.A.* **94** (23): 12633–7. Bibcode:1997PNAS...9412633P. doi:10.1073/pnas.94.23.12633. PMC 250063. PMID 9356501.

[23] "Chromosome Mapping: Idiograms" *Nature Education* - August 13, 2013

[24] Naumova N, Imakaev M, Fudenberg G, Zhan Y, Lajoie BR, Mirny LA, Dekker J (2013). "Organization of the mitotic chromosome". *Science.* **342** (6161): 948–53. doi:10.1126/science.1236083. PMC 4040465. PMID 24200812.

[25] Vega.sanger.ad.uk, all data in this table was derived from this database, November 11, 2008.

[26] "Ensembl genome browser 71: Homo sapiens – Chromosome summary – Chromosome 1: 1–1,000,000". *apr2013.archive.ensembl.org.* Retrieved 2016-04-11.

[27] Sequenced percentages are based on fraction of euchromatin portion, as the Human Genome Project goals called for determination of only the euchromatic portion of the genome. Telomeres, centromeres, and other heterochromatic regions have been left undetermined, as have a small number of unclonable gaps. See http://www.ncbi.nlm.nih.gov/genome/seq/ for more information on the Human Genome Project.

[28] Armstrong SJ; Jones GH (January 2003). "Meiotic cytology and chromosome behaviour in wild-type Arabidopsis thaliana". *J. Exp. Bot.* **54** (380): 1–10. doi:10.1093/jxb/54.380.1. PMID 12456750.

[29] Gill BS; Kimber G (April 1974). "The Giemsa C-Banded Karyotype of Rye". *Proc. Natl. Acad. Sci. U.S.A.* **71** (4): 1247–9. Bibcode:1974PNAS...71.1247G. doi:10.1073/pnas.71.4.1247. PMC 388202⊘. PMID 4133848.

[30] Dubcovsky J; Luo MC; Zhong GY; et al. (1996). "Genetic Map of Diploid Wheat, Triticum Monococcum L., and Its Comparison with Maps of Hordeum Vulgare L.". *Genetics.* **143** (2): 983–99. PMC 1207354⊘. PMID 8725244.

[31] Kato A; Lamb JC; Birchler JA (September 2004). "Chromosome painting using repetitive DNA sequences as probes for somatic chromosome identification in maize". *Proc. Natl. Acad. Sci. U.S.A.* **101** (37): 13554–9. Bibcode:2004PNAS..10113554K. doi:10.1073/pnas.0403659101. PMC 518793⊘. PMID 15342909.

[32] Leitch IJ; Soltis DE; Soltis PS; Bennett MD (2005). "Evolution of DNA amounts across land plants (embryophyta)". *Ann. Bot.* **95** (1): 207–17. doi:10.1093/aob/mci014. PMID 15596468.

[33] Ambarish, C.N. Sridhar, K.R. (2014). "Cytological and karyological observations of two endemic pill-millipedes Arthrosphaera (Pocock, 1895) (Diplopoda: Sphaerotheriida) of the Western Ghats of India". *Caryologia.* **66** (1). doi:10.1080/00087114.

[34] Vitturi R; Colomba MS; Pirrone AM; Mandrioli M (2002). "rDNA (18S-28S and 5S) colocalization and linkage between ribosomal genes and (TTAGGG)(n) telomeric sequence in the earthworm, Octodrilus complanatus (Annelida: Oligochaeta: Lumbricidae), revealed by single- and double-color FISH". *J. Hered.* **93** (4): 279–82. doi:10.1093/jhered/93.4.279. PMID 12407215.

[35] Nie W; Wang J; O'Brien PC; et al. (2002). "The genome phylogeny of domestic cat, red panda and five mustelid species revealed by comparative chromosome painting and G-banding". *Chromosome Res.* **10** (3): 209–22. doi:10.1023/A:1015292005631. PMID 12067210.

[36] Romanenko, Svetlana A.; Perelman, Polina L.; Serdukova, Natalya A.; Trifonov, Vladimir A.; Biltueva, Larisa S.; Wang, Jinhuan; Li, Tangliang; Nie, Wenhui; O'Brien, Patricia C.M.; Volobouev, Vitaly T.; Stanyon, Roscoe;

Ferguson-Smith, Malcolm A.; Yang, Fengtang; Graphodatsky, Alexander S. (2006). "Reciprocal chromosome painting between three laboratory rodent species". *Mammalian Genome.* **17** (12): 1183–92. doi:10.1007/s00335-006-0081-z. PMID 17143584.

[37] Painter, TS (1928). "A Comparison of the Chromosomes of the Rat and Mouse with Reference to the Question of Chromosome Homology in Mammals". *Genetics.* **13** (2): 180–9. PMC 1200977⊘. PMID 17246549.

[38] Hayes, H.; Rogel-Gaillard, C.; Zijlstra, C.; De Haan, N.A.; Urien, C.; Bourgeaux, N.; Bertaud, M.; Bosma, A.A. (2002). "Establishment of an R-banded rabbit karyotype nomenclature by FISH localization of 23 chromosome-specific genes on both G- and R-banded chromosomes". *Cytogenetic and Genome Research.* **98** (2–3): 199–205. doi:10.1159/000069807. PMID 12698004.

[39] "The Genetics of the Popular Aquarium Pet - Guppy Fish". Retrieved 2009-12-06.

[40] De Grouchy J (1987). "Chromosome phylogenies of man, great apes, and Old World monkeys". *Genetica.* **73** (1–2): 37–52. doi:10.1007/bf00057436. PMID 3333352.

[41] T.J. Robinson; F. Yang; W.R. Harrison (2002). "Chromosome painting refines the history of genome evolution in hares and rabbits (order Lagomorpha)". *Cytogenetic and Genome Research.* **96** (1–4): 223–227. doi:10.1159/000063034. PMID 12438803.

[42] "section 4.W4". *Rabbits, Hares and Pikas. Status Survey and Conservation Action Plan,* pp. 61–94

[43] Vitturi R; Libertini A; Sineo L; et al. (2005). "Cytogenetics of the land snails Cantareus aspersus and C. mazzullii (Mollusca: Gastropoda: Pulmonata)". *Micron.* **36** (4): 351–7. doi:10.1016/j.micron.2004.12.010. PMID 15857774.

[44] Yasukochi Y; Ashakumary LA; Baba K; Yoshido A; Sahara K (2006). "A Second-Generation Integrated Map of the Silkworm Reveals Synteny and Conserved Gene Order Between Lepidopteran Insects". *Genetics.* **173** (3): 1319–28. doi:10.1534/genetics.106.055541. PMC 1526672⊘. PMID 16547103.

[45] Houck, M.L.; Kumamoto, A.T.; Gallagher, D.S.; Benirschke, K. (2001). "Comparative cytogenetics of the African elephant *(Loxodonta africana)* and Asiatic elephant *(Elephas maximus)*". *Cytogenetic and Genome Research.* **93** (3–4): 249–52. doi:10.1159/000056992. PMID 11528120.

[46] Umeko Semba; Yasuko Umeda; Yoko Shibuya; Hiroaki Okabe; Sumio Tanase & Tetsuro Yamamoto (2004). "Primary structures of guinea pig high- and low-molecular-weight kininogens". *International Immunopharmacology.* **4** (10–11): 1391–1400. doi:10.1016/j.intimp.2004.06.003. PMID 15313436.

[47] Wayne RK; Ostrander EA (1999). "Origin, genetic diversity, and genome structure of the domestic dog". *BioEssays*. **21** (3): 247–57. doi:10.1002/(SICI)1521-1878(199903)21:3<247::AID-BIES9>3.0.CO;2-Z. PMID 10333734.

[48] Ciudad J; Cid E; Velasco A; Lara JM; Aijón J; Orfao A (2002). "Flow cytometry measurement of the DNA contents of G0/G1 diploid cells from three different teleost fish species". *Cytometry*. **48** (1): 20–5. doi:10.1002/cyto.10100. PMID 12116377.

[49] Burt DW (2002). "Origin and evolution of avian microchromosomes". *Cytogenet. Genome Res.* **96** (1–4): 97–112. doi:10.1159/000063018. PMID 12438785.

[50] Itoh, Masahiro; Ikeuchi, Tatsuro; Shimba, Hachiro; Mori, Michiko; Sasaki, Motomichi; Makino, Sajiro (1969). "A Comparative Karyotype Study in Fourteen Species of Birds". *The Japanese journal of genetics*. **44** (3): 163–170. doi:10.1266/jjg.44.163.

[51] Smith J; Burt DW (1998). "Parameters of the chicken genome (Gallus gallus)". *Anim. Genet.* **29** (4): 290–4. doi:10.1046/j.1365-2052.1998.00334.x. PMID 9745667.

[52] Sakamura, Tetsu (1918). "Kurze Mitteilung über die Chromosomenzahlen und die Verwandtschaftsverhältnisse der Triticum-Arten". *Shokubutsugaku Zasshi*. **32** (379): 150–3. doi:10.15281/jplantres1887.32.379_150.

[53] Charlebois R.L. (ed) 1999. *Organization of the prokaryote genome*. ASM Press, Washington DC.

[54] Komaki K; Ishikawa H (March 2000). "Genomic copy number of intracellular bacterial symbionts of aphids varies in response to developmental stage and morph of their host". *Insect Biochem. Mol. Biol.* **30** (3): 253–8. doi:10.1016/S0965-1748(99)00125-3. PMID 10732993.

[55] Mendell JE; Clements KD; Choat JH; Angert ER (May 2008). "Extreme polyploidy in a large bacterium". *Proc. Natl. Acad. Sci. U.S.A.* **105** (18): 6730–4. Bibcode:2008PNAS..105.6730M. doi:10.1073/pnas.0707522105. PMC 2373351. PMID 18445653.

[56] White, M. J. D. (1973). *The chromosomes* (6th ed.). London: Chapman and Hall, distributed by Halsted Press. New York. p. 28. ISBN 0-412-11930-7.

[57] von Winiwarter H (1912). "Études sur la spermatogénèse humaine". *Archives de Biologie*. **27** (93): 147–9.

[58] Painter TS (1922). "The spermatogenesis of man". *Anat. Res.* **23**: 129.

[59] Painter TS (1923). "Studies in mammalian spermatogenesis II. The spermatogenesis of man". *J. Exp. Zoology*. **37** (3): 291–336. doi:10.1002/jez.1400370303.

[60] Tjio JH; Levan A (1956). "The chromosome number of man". *Hereditas*. **42** (1-2): 1–6. doi:10.1111/j.1601-5223.1956.tb03010.x.

[61] Ford C.E; Hamerton J.L (1956). "The Chromosomes of Man". *Nature*. **178** (4541): 1020–1023. Bibcode:1956Natur.178.1020F. doi:10.1038/1781020a0. PMID 13378517.

[62] Hsu T.C. *Human and mammalian cytogenetics: a historical perspective*. Springer-Verlag, N.Y. p10: "It's amazing that he [Painter] even came close!"

[63] Miller, Kenneth R. (2000). "Chapter 9-3". *Biology* (5th ed.). Upper Saddle River, New Jersey: Prentice Hall. pp. 194–5. ISBN 0-13-436265-9.

[64] "What is Trisomy 18?". *Trisomy 18 Foundation*. Retrieved 4 February 2017.

[65] European Chromosome 11 Network

[66] Templado C, Uroz L, Estop A (2013). "New insights on the origin and relevance of aneuploidy in human spermatozoa". *Mol. Hum. Reprod.* **19** (10): 634–43. doi:10.1093/molehr/gat039. PMID 23720770.

[67] Shi Q, Ko E, Barclay L, Hoang T, Rademaker A, Martin R (2001). "Cigarette smoking and aneuploidy in human sperm". *Mol. Reprod. Dev.* **59** (4): 417–21. doi:10.1002/mrd.1048. PMID 11468778.

[68] Rubes J, Lowe X, Moore D, Perreault S, Slott V, Evenson D, Selevan SG, Wyrobek AJ (1998). "Smoking cigarettes is associated with increased sperm disomy in teenage men". *Fertil. Steril.* **70** (4): 715–23. PMID 9797104.

[69] Xing C, Marchetti F, Li G, Weldon RH, Kurtovich E, Young S, Schmid TE, Zhang L, Rappaport S, Waidyanatha S, Wyrobek AJ, Eskenazi B (2010). "Benzene exposure near the U.S. permissible limit is associated with sperm aneuploidy". *Environ. Health Perspect.* **118** (6): 833–9. doi:10.1289/ehp.0901531. PMC 2898861. PMID 20418200.

[70] Xia Y, Bian Q, Xu L, Cheng S, Song L, Liu J, Wu W, Wang S, Wang X (2004). "Genotoxic effects on human spermatozoa among pesticide factory workers exposed to fenvalerate". *Toxicology*. **203** (1-3): 49–60. doi:10.1016/j.tox.2004.05.018. PMID 15363581.

[71] Xia Y, Cheng S, Bian Q, Xu L, Collins MD, Chang HC, Song L, Liu J, Wang S, Wang X (2005). "Genotoxic effects on spermatozoa of carbaryl-exposed workers". *Toxicol. Sci.* **85** (1): 615–23. doi:10.1093/toxsci/kfi066. PMID 15615886.

[72] Governini L, Guerranti C, De Leo V, Boschi L, Luddi A, Gori M, Orvieto R, Piomboni P (2015). "Chromosomal aneuploidies and DNA fragmentation of human spermatozoa from patients exposed to perfluorinated compounds". *Andrologia*. **47** (9): 1012–9. doi:10.1111/and.12371. PMID 25382683.

## 14.9   External links

- An Introduction to DNA and Chromosomes from HOPES: Huntington's Outreach Project for Education at Stanford

- Chromosome Abnormalities at AtlasGeneticsOncology

- On-line exhibition on chromosomes and genome (SIB)

- What Can Our Chromosomes Tell Us?, from the University of Utah's Genetic Science Learning Center

- Try making a karyotype yourself, from the University of Utah's Genetic Science Learning Center

- Kimballs Chromosome pages

- Chromosome News from Genome News Network

- Eurochromnet, European network for Rare Chromosome Disorders on the Internet

- Ensembl.org, Ensembl project, presenting chromosomes, their genes and syntenic loci graphically via the web

- Genographic Project

- Home reference on Chromosomes from the U.S. National Library of Medicine

- Visualisation of human chromosomes and comparison to other species

- Unique - The Rare Chromosome Disorder Support Group Support for people with rare chromosome disorders

# Chapter 15

# Nucleotide

*This nucleotide contains the five-carbon sugar deoxyribose (at center), a nitrogenous base called adenine (upper right), and one phosphate group (left). The sugar joined with the base forms a nucleoside called deoxyadenosine. The whole structure is a nucleotide constituent of DNA, with the name deoxyadenosine monophosphate.*

**Nucleotides** are organic molecules that serve as the monomer units for forming the nucleic acid polymers DNA (deoxyribonucleic acid) and RNA (ribonucleic acid), both of which are essential biomolecules in all life-forms on Earth. Nucleotides are the building blocks of nucleic acids; they are composed of three subunit molecules: a nitrogenous base, a five-carbon sugar (ribose or deoxyribose), and at least one phosphate group. They are also known as *phosphate* nucleotides.

A nucleoside is a nitrogenuous base and a 5-carbon sugar. Thus a nucleoside plus a phosphate group yields a nucleotide.

Nucleotides also play a central role in life-form metabolism at the fundamental, cellular level. They carry packets of chemical energy—in the form of the nucleoside triphosphates ATP, GTP, CTP and UTP—throughout the cell to the many cellular functions that demand energy, which include synthesizing amino acids, proteins and cell membranes and parts; moving the cell and moving cell parts, both internally and intercellularly; dividing the cell, etc.[1] In addition, nucleotides participate in cell signaling (cGMP and cAMP), and are incorporated into important cofactors of enzymatic reactions (e.g. coenzyme A, FAD, FMN, NAD, and NADP+).

In experimental biochemistry, nucleotides can be radiolabeled with radionuclides to yield radionucleotides.

## 15.1 Structure

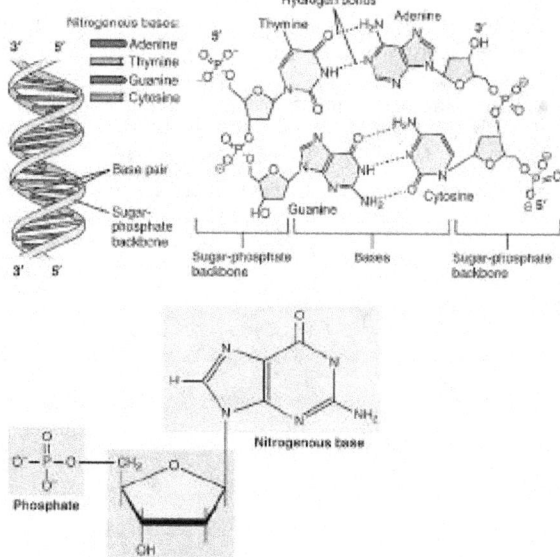

*Showing the arrangement of nucleotides within the structure of nucleic acids: At lower left, a monophosphate nucleotide; its nitrogenous base represents one side of a base-pair. At upper right, four nucleotides form two base-pairs: thymine and adenine (connected by double hydrogen bonds) and guanine and cytosine (connected by triple hydrogen bonds). The individual nucleotide monomers are chain-joined at their sugar and phosphate molecules, forming two 'backbones' (a double helix) of a nucleic acid, shown at upper left.*

A nucleotide is composed of three distinctive chemical subunits: a five-carbon sugar molecule, a nitrogenous base—which two together are called a nucleoside—and one phosphate group. With all three joined, a nucleotide is also termed a "nucleoside *mono*phosphate". The chemistry sources ACS Style Guide[2] and IUPAC Gold Book[3] prescribe that a nucleotide should contain only one phosphate group, but common usage in molecular biology textbooks often extends the definition to include molecules with two, or with three, phosphates.[1][4][5][6] Thus, the terms "nucleoside *di*phosphate" or "nucleoside *tri*phosphate" may also indicate nucleotides.

Nucleotides contain either a purine or a pyrimidine base—i.e., the nitrogenous base molecule, also known as a nucleobase—and are termed *ribo*nucleotides if the sugar is ribose, or *deoxyribo*nucleotides if the sugar is deoxyribose. Individual phosphate molecules repetitively connect the sugar-ring molecules in two adjacent nucleotide monomers, thereby connecting the nucleotide monomers of a nucleic acid end-to-end into a long chain. These chain-joins of sugar and phosphate molecules create a 'backbone' strand for a single- or double helix. In any one strand, the chemical orientation (directionality) of the chain-joins runs from the 5'-end to the 3'-end (*read*: 5 prime-end to 3 prime-end)—referring to the five carbon sites on sugar molecules in adjacent nucleotides. In a double helix, the two strands are oriented in opposite directions, which permits base pairing and complementarity between the base-pairs, all which is essential for replicating or transcribing the encoded information found in DNA.

Unlike in nucleic acid nucleotides, singular cyclic nucleotides are formed when the phosphate group is bound twice to the same sugar molecule, i.e., at the corners of the sugar hydroxyl groups.[1] These individual nucleotides function in cell metabolism rather than the nucleic acid structures of long-chain molecules.

Nucleic acids then are polymeric macromolecules assembled from nucleotides, the monomer-units of nucleic acids. The purine bases adenine and guanine and pyrimidine base cytosine occur in both DNA and RNA, while the pyrimidine bases thymine (in DNA) and uracil (in RNA)in just one. Adenine forms a base pair with thymine with two hydrogen bonds, while guanine pairs with cytosine with three hydrogen bonds.

## 15.2 Synthesis

Nucleotides can be synthesized by a variety of means both in vitro and in vivo.

In vivo, nucleotides can be synthesized de novo or recycled through salvage pathways.[7] The components used in

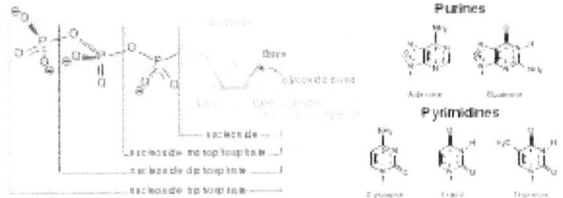

*Structural elements of three nucleotides—each in turn is attached to the nucleoside (in yellow, blue, green) at center; 1st, the nucleotide termed as a nucleoside monophosphate nucleotide is formed by adding a phosphate group (in red); 2nd, adding a second phosphate group forms a nucleoside diphosphate nucleotide; 3rd, adding a third phosphate group results in a nucleoside triphosphate nucleotide. + The nitrogenous base (nucleobase) is indicated by "Base" and "glycosidic bond" (sugar bond). All five primary, or canonical, bases—the purines and pyrimidines—are sketched at right (in blue).*

de novo nucleotide synthesis are derived from biosynthetic precursors of carbohydrate and amino acid metabolism, and from ammonia and carbon dioxide. The liver is the major organ of de novo synthesis of all four nucleotides. De novo synthesis of pyrimidines and purines follows two different pathways. Pyrimidines are synthesized first from aspartate and carbamoyl-phosphate in the cytoplasm to the common precursor ring structure orotic acid, onto which a phosphorylated ribosyl unit is covalently linked. Purines, however, are first synthesized from the sugar template onto which the ring synthesis occurs. For reference, the syntheses of the purine and pyrimidine nucleotides are carried out by several enzymes in the cytoplasm of the cell, not within a specific organelle. Nucleotides undergo breakdown such that useful parts can be reused in synthesis reactions to create new nucleotides.

In vitro, protecting groups may be used during laboratory production of nucleotides. A purified nucleoside is protected to create a phosphoramidite, which can then be used to obtain analogues not found in nature and/or to synthesize an oligonucleotide.

### 15.2.1 Pyrimidine ribonucleotide synthesis

Main article: Pyrimidine metabolism

The synthesis of the pyrimidines CTP and UTP occurs in the cytoplasm and starts with the formation of carbamoyl phosphate from glutamine and $CO_2$. Next, aspartate carbamoyltransferase catalyzes a condensation reaction between aspartate and carbamoyl phosphate to form carbamoyl aspartic acid, which is cyclized into 4,5-dihydroorotic acid by dihydroorotase. The latter is converted to orotate by dihydroorotate oxidase. The net re-

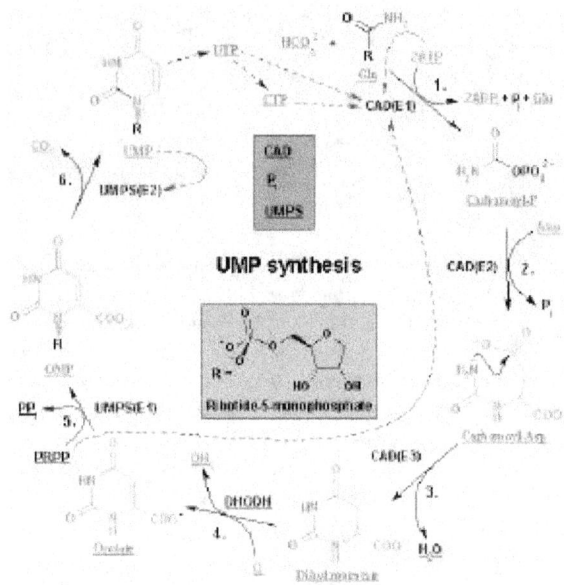

**The synthesis of UMP.**
*The color scheme is as follows: enzymes, coenzymes, substrate names, inorganic molecules*

action is:

$$(S)\text{-Dihydroorotate} + O_2 \rightarrow \text{Orotate} + H_2O_2$$

Orotate is covalently linked with a phosphorylated ribosyl unit. The covalent linkage between the ribose and pyrimidine occurs at position $C_1$[8] of the ribose unit, which contains a pyrophosphate, and $N_1$ of the pyrimidine ring. Orotate phosphoribosyltransferase (PRPP transferase) catalyzes the net reaction yielding orotidine monophosphate (OMP):

Orotate + 5-Phospho-α-D-ribose 1-diphosphate (PRPP) → Orotidine 5'-phosphate + Pyrophosphate

Orotidine 5'-monophosphate is decarboxylated by orotidine-5'-phosphate decarboxylase to form uridine monophosphate (UMP). PRPP transferase catalyzes both the ribosylation and decarboxylation reactions, forming UMP from orotic acid in the presence of PRPP. It is from UMP that other pyrimidine nucleotides are derived. UMP is phosphorylated by two kinases to uridine triphosphate (UTP) via two sequential reactions with ATP. First the diphosphate form UDP is produced, which in turn is phosphorylated to UTP. Both steps are fueled by ATP hydrolysis:

$$ATP + UMP \rightarrow ADP + UDP$$

$$UDP + ATP \rightarrow UTP + ADP$$

CTP is subsequently formed by amination of UTP by the catalytic activity of CTP synthetase. Glutamine is the $NH_3$ donor and the reaction is fueled by ATP hydrolysis, too:

$$UTP + \text{Glutamine} + ATP + H_2O \rightarrow CTP + ADP + P_i$$

Cytidine monophosphate (CMP) is derived from cytidine triphosphate (CTP) with subsequent loss of two phosphates.[9] [10]

## 15.2.2 Purine ribonucleotide synthesis

Main article: Purine metabolism

The atoms which are used to build the purine nucleotides come from a variety of sources:

The synthesis of IMP. The color scheme is as follows: enzymes, coenzymes, substrate names, metal ions, inorganic molecules

The de novo synthesis of purine nucleotides by which these precursors are incorporated into the purine ring proceeds by a 10-step pathway to the branch-point intermediate IMP, the nucleotide of the base hypoxanthine. AMP and GMP are subsequently synthesized from this intermediate via separate, two-step pathways. Thus, purine moieties are initially formed as part of the ribonucleotides rather than as free bases.

Six enzymes take part in IMP synthesis. Three of them are multifunctional:

- GART (reactions 2, 3, and 5)
- PAICS (reactions 6, and 7)

- ATIC (reactions 9, and 10)

The pathway starts with the formation of PRPP. PRPS1 is the enzyme that activates R5P, which is formed primarily by the pentose phosphate pathway, to PRPP by reacting it with ATP. The reaction is unusual in that a pyrophosphoryl group is directly transferred from ATP to $C_1$ of R5P and that the product has the α configuration about C1. This reaction is also shared with the pathways for the synthesis of Trp, His, and the pyrimidine nucleotides. Being on a major metabolic crossroad and requiring much energy, this reaction is highly regulated.

In the first reaction unique to purine nucleotide biosynthesis, PPAT catalyzes the displacement of PRPP's pyrophosphate group ($PP_i$) by an amide nitrogen donated from either glutamine (N), glycine (N&C), aspartate (N), folic acid ($C_1$), or $CO_2$. This is the committed step in purine synthesis. The reaction occurs with the inversion of configuration about ribose $C_1$, thereby forming β-5-phosphorybosylamine (5-PRA) and establishing the anomeric form of the future nucleotide.

Next, a glycine is incorporated fueled by ATP hydrolysis and the carboxyl group forms an amine bond to the $NH_2$ previously introduced. A one-carbon unit from folic acid coenzyme $N_{10}$-formyl-THF is then added to the amino group of the substituted glycine followed by the closure of the imidazole ring. Next, a second $NH_2$ group is transferred from a glutamine to the first carbon of the glycine unit. A carboxylation of the second carbon of the glycin unit is concomitantly added. This new carbon is modified by the additional of a third $NH_2$ unit, this time transferred from an aspartate residue. Finally, a second one-carbon unit from formyl-THF is added to the nitrogen group and the ring covalently closed to form the common purine precursor inosine monophosphate (IMP).

Inosine monophosphate is converted to adenosine monophosphate in two steps. First, GTP hydrolysis fuels the addition of aspartate to IMP by adenylosuccinate synthase, substituting the carbonyl oxygen for a nitrogen and forming the intermediate adenylosuccinate. Fumarate is then cleaved off forming adenosine monophosphate. This step is catalyzed by adenylosuccinate lyase.

Inosine monophosphate is converted to guanosine monophosphate by the oxidation of IMP forming xanthylate, followed by the insertion of an amino group at $C_2$. $NAD^+$ is the electron acceptor in the oxidation reaction. The amide group transfer from glutamine is fueled by ATP hydrolysis.

### 15.2.3  Pyrimidine and purine degradation

In humans, pyrimidine rings (C, T, U) can be degraded completely to $CO_2$ and $NH_3$ (urea excretion). That having been said, purine rings (G, A) cannot. Instead they are degraded to the metabolically inert uric acid which is then excreted from the body. Uric acid is formed when GMP is split into the base guanine and ribose. Guanine is deaminated to xanthine which in turn is oxidized to uric acid. This last reaction is irreversible. Similarly, uric acid can be formed when AMP is deaminated to IMP from which the ribose unit is removed to form hypoxanthine. Hypoxanthine is oxidized to xanthine and finally to uric acid. Instead of uric acid secretion, guanine and IMP can be used for recycling purposes and nucleic acid synthesis in the presence of PRPP and aspartate ($NH_3$ donor).

## 15.3  Unnatural base pair (UBP)

Main article: Base pair § Unnatural base pair (UBP)

An unnatural base pair (UBP) is a designed subunit (or nucleobase) of DNA which is created in a laboratory and does not occur in nature. In 2012, a group of American scientists led by Floyd Romesberg, a chemical biologist at the Scripps Research Institute in San Diego, California, published that his team designed an unnatural base pair (UBP).[11] The two new artificial nucleotides or *Unnatural Base Pair* (UBP) were named **d5SICS** and **dNaM**. More technically, these artificial nucleotides bearing hydrophobic nucleobases, feature two fused aromatic rings that form a (d5SICS–dNaM) complex or base pair in DNA.[12][13] In 2014 the same team from the Scripps Research Institute reported that they synthesized a stretch of circular DNA known as a plasmid containing natural T-A and C-G base pairs along with the best-performing UBP Romesberg's laboratory had designed, and inserted it into cells of the common bacterium *E. coli* that successfully replicated the unnatural base pairs through multiple generations.[14] This is the first known example of a living organism passing along an expanded genetic code to subsequent generations.[12][15] This was in part achieved by the addition of a supportive algal gene that expresses a nucleotide triphosphate transporter which efficiently imports the triphosphates of both d5SICSTP and dNaMTP into *E. coli* bacteria.[12] Then, the natural bacterial replication pathways use them to accurately replicate the plasmid containing d5SICS–dNaM.

The successful incorporation of a third base pair is a significant breakthrough toward the goal of greatly expanding the number of amino acids which can be encoded by DNA, from the existing 20 amino acids to a theoretically possible 172, thereby expanding the potential for living organisms to produce novel proteins.[14] The artificial strings of DNA do not encode for anything yet, but scientists speculate they could be designed to manufacture new proteins which could have industrial or pharmaceutical uses.[16]

## 15.4 Length unit

Nucleotide (abbreviated "nt") is a common unit of length for single-stranded nucleic acids, similar to how base pair is a unit of length for double-stranded nucleic acids.

## 15.5 Nucleotide supplements

A study done by the Department of Sports Science at the University of Hull in Hull, UK has shown that nucleotides have significant impact on cortisol levels in saliva. Post exercise, the experimental nucleotide group had lower cortisol levels in their blood than the control or the placebo. Additionally, post supplement values of Immunoglobulin A were significantly higher than either the placebo or the control. The study concluded, "nucleotide supplementation blunts the response of the hormones associated with physiological stress."[17]

Another study conducted in 2013 looked at the impact nucleotide supplementation had on the immune system in athletes. In the study, all athletes were male and were highly skilled in taekwondo. Out of the twenty athletes tested, half received a placebo and half received 480mg per day of nucleotide supplement. After thirty days, the study concluded that nucleotide supplementation may counteract the impairment of the body's immune function after heavy exercise.[18]

## 15.6 Abbreviation codes for degenerate bases

Main article: Nucleic acid notation

The IUPAC has designated the symbols for nucleotides.[19] Apart from the five (A, G, C, T/U) bases, often degenerate bases are used especially for designing PCR primers. These nucleotide codes are listed here. Some primer sequences may also include the character "I", which codes for the non-standard nucleotide inosine. Inosine occurs in tRNAs, and will pair with adenine, cytosine, or thymine. This character does not appear in the following table however, because it does not represent a degeneracy. While inosine can serve a similar function as the degeneracy "D", it is an

actual nucleotide, rather than a representation of a mix of nucleotides that covers each possible pairing needed.

## 15.7 See also

- Biology
- Chromosome
- Gene
- Genetics
- Nucleic acid analogues
- Nucleic acid sequence
- Nucleobase

## 15.8 References

[1] Alberts B, Johnson A, Lewis J, Raff M, Roberts K & Walter P (2002). *Molecular Biology of the Cell* (4th ed.). Garland Science. ISBN 0-8153-3218-1. pp. 120–121.

[2] Coghill, Anne M.; Garson, Lorrin R., eds. (2006). *The ACS style guide: effective communication of scientific information* (3rd ed.). Washington, D.C.: American Chemical Society. p. 244. ISBN 978-0-8412-3999-9.

[3] "Nucleotides". *IUPAC Gold Book*. International Union of Pure and Applied Chemists. doi:10.1351/goldbook.N04255. Retrieved 30 June 2014.

[4] Lehninger, Albert L. (1975). *Biochemistry: the molecular basis of cell structure and function*. New York: Worth Publishers Inc. doi:10.1002/jobm.19770170116.

[5] Stryer, Lubert (1988). *Biochemistry* (3rd ed.). New York: W. H. Freeman. ISBN 9780716719205.

[6] Garrett, Reginald H.; Grisham, Charles M. (2007). *Biochemistry* (4th ed.). Belmont, California: Brooks/Cole, Cengage Learning.

[7] Zaharevitz, DW; Anerson, LW; Manlinowski, NM; Hyman, R; Strong, JM; Cysyk, RL. "Contribution of de-novo and salvage synthesis to the uracil nucleotide pool in mouse tissues and tumors in vivo".

[8] See IUPAC nomenclature of organic chemistry for details on carbon residue numbering

[9] Jones, M. E. (1980). "Pyrimidine nucleotide biosynthesis in animals: Genes, enzymes, and regulation of UMP biosynthesis". *Annu. Rev. Biochem.* **49** (1): 253–79. doi:10.1146/annurev.bi.49.070180.001345. PMID 6105839.

[10] McMurry. JE; Begley. TP (2005). *The organic chemistry of biological pathways*. Roberts & Company. ISBN 978-0-9747077-1-6.

[11] Malyshev, Denis A.; Dhami. Kirandeep; Quach, Henry T.; Lavergne, Thomas; Ordoukhanian, Phillip (24 July 2012). "Efficient and sequence-independent replication of DNA containing a third base pair establishes a functional six-letter genetic alphabet". *Proceedings of the National Academy of Sciences of the United States of America*. **109** (30): 12005–12010. Bibcode:2012PNAS..10912005M. doi:10.1073/pnas.1205176109. Retrieved 2014-05-11.

[12] Malyshev, Denis A.; Dhami, Kirandeep; Lavergne, Thomas; Chen, Tingjian; Dai, Nan; Foster, Jeremy M.; Corrêa, Ivan R.; Romesberg, Floyd E. (May 7, 2014). "A semi-synthetic organism with an expanded genetic alphabet". *Nature*. **509**: 385–8. doi:10.1038/nature13314. PMC 4058825. PMID 24805238. Retrieved May 7, 2014.

[13] Callaway, Ewan (May 7, 2014). "Scientists Create First Living Organism With 'Artificial' DNA". *Nature News*. Huffington Post. Retrieved 8 May 2014.

[14] Fikes. Bradley J. (May 8, 2014). "Life engineered with expanded genetic code". *San Diego Union Tribune*. Retrieved 8 May 2014.

[15] Sample, Ian (May 7, 2014). "First life forms to pass on artificial DNA engineered by US scientists". *The Guardian*. Retrieved 8 May 2014.

[16] Pollack, Andrew (May 7, 2014). "Scientists Add Letters to DNA's Alphabet, Raising Hope and Fear". *New York Times*. Retrieved 8 May 2014.

[17] Mc Naughton. L.; Bentley, D.; Koeppel. P. (2007-03-01). "The effects of a nucleotide supplement on the immune and metabolic response to short term, high intensity exercise performance in trained male subjects". *The Journal of Sports Medicine and Physical Fitness*. **47** (1): 112–118. ISSN 0022-4707. PMID 17369807.

[18] Riera, Joan; Pons, Victoria; Martinez-Puig, Daniel; Chetrit. Carlos; Tur, Josep A.; Pons, Antoni; Drobnic, Franchek (2013-04-08). "Dietary nucleotide improves markers of immune response to strenuous exercise under a cold environment". *Journal of the International Society of Sports Nutrition*. **10** (1): 20. doi:10.1186/1550-2783-10-20. PMC 3626726. PMID 23566489.

[19] Nomenclature Committee of the International Union of Biochemistry (NC-IUB) (1984). "Nomenclature for Incompletely Specified Bases in Nucleic Acid Sequences". Retrieved 2008-02-04.

## 15.9  External links

- Abbreviations and Symbols for Nucleic Acids. Polynucleotides and their Constituents (IUPAC)
- Provisional Recommendations 2004 (IUPAC)
- Chemistry explanation of nucleotide structure

# Chapter 16

# DNA

For a non-technical introduction to the topic, see Introduction to genetics. For other uses, see DNA (disambiguation).

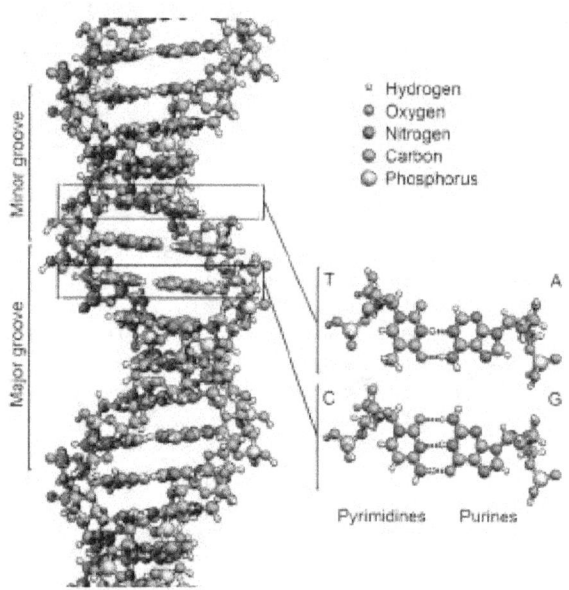

*The structure of the DNA double helix. The atoms in the structure are colour-coded by element and the detailed structure of two base pairs are shown in the bottom right.*

**Deoxyribonucleic acid** (🔊/diˈɒksiˌraɪboʊnjuːˌkliːɪk, -ˌkleɪk/;[1] **DNA**) is a molecule that carries the genetic instructions used in the growth, development, functioning and reproduction of all known living organisms and many viruses. DNA and RNA are nucleic acids; alongside proteins, lipids and complex carbohydrates (polysaccharides), they are one of the four major types of macromolecules that are essential for all known forms of life. Most DNA molecules consist of two biopolymer strands coiled around each other to form a double helix.

The two DNA strands are termed polynucleotides since they are composed of simpler monomer units called nucleotides.[2][3] Each nucleotide is composed of one of

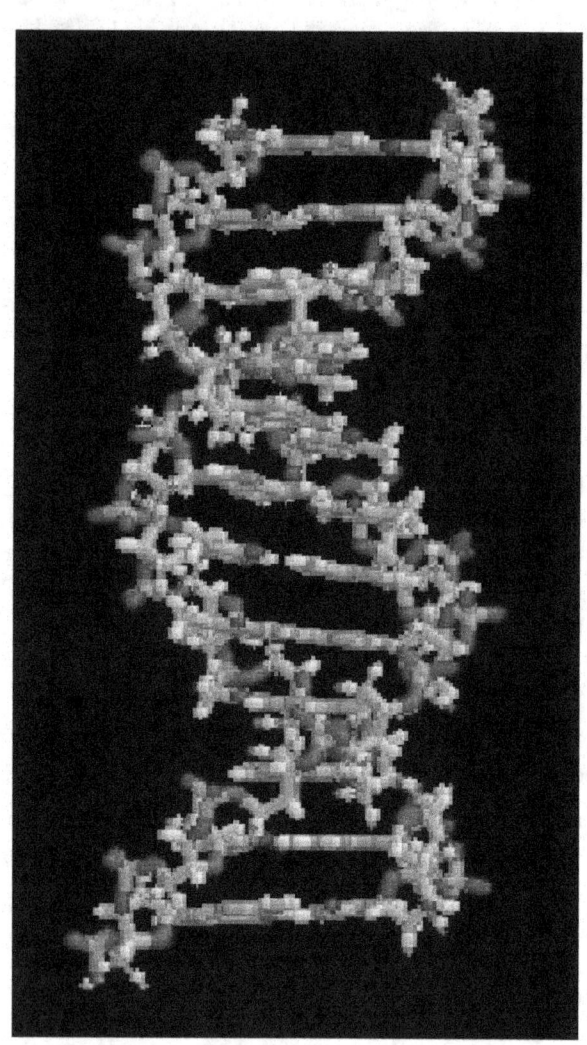

*The structure of part of a DNA double helix*

four nitrogen-containing nucleobases—either cytosine (C), guanine (G), adenine (A), or thymine (T)—and a sugar called deoxyribose and a phosphate group. The nucleotides are joined to one another in a chain by covalent bonds between the sugar of one nucleotide and the phosphate of the next, resulting in an alternating sugar-phosphate backbone.

The nitrogenous bases of the two separate polynucleotide strands are bound together (according to base pairing rules (A with T, and C with G) with hydrogen bonds to make double-stranded DNA. The total amount of related DNA base pairs on Earth is estimated at $5.0 \times 10^{37}$ and weighs 50 billion tonnes.[4] In comparison, the total mass of the biosphere has been estimated to be as much as 4 trillion tons of carbon (TtC).[5]

DNA stores biological information. The DNA backbone is resistant to cleavage, and both strands of the double-stranded structure store the same biological information. This information is replicated as and when the two strands separate. A large part of DNA (more than 98% for humans) is non-coding, meaning that these sections do not serve as patterns for protein sequences.

The two strands of DNA run in opposite directions to each other and are thus antiparallel. Attached to each sugar is one of four types of nucleobases (informally, *bases*). It is the sequence of these four nucleobases along the backbone that encodes biological information. RNA strands are created using DNA strands as a template in a process called transcription. Under the genetic code, these RNA strands are translated to specify the sequence of amino acids within proteins in a process called translation.

Within eukaryotic cells, DNA is organized into long structures called chromosomes. During cell division these chromosomes are duplicated in the process of DNA replication, providing each cell its own complete set of chromosomes. Eukaryotic organisms (animals, plants, fungi, and protists) store most of their DNA inside the cell nucleus and some of their DNA in organelles, such as mitochondria or chloroplasts.[6] In contrast, prokaryotes (bacteria and archaea) store their DNA only in the cytoplasm. Within the eukaryotic chromosomes, chromatin proteins such as histones compact and organize DNA. These compact structures guide the interactions between DNA and other proteins, helping control which parts of the DNA are transcribed.

DNA was first isolated by Friedrich Miescher in 1869. Its molecular structure was identified by James Watson and Francis Crick in 1953, whose model-building efforts were guided by X-ray diffraction data acquired by Rosalind Franklin. DNA is used by researchers as a molecular tool to explore physical laws and theories, such as the ergodic theorem and the theory of elasticity. The unique material properties of DNA have made it an attractive molecule for material scientists and engineers interested in micro- and nano-fabrication. Among notable advances in this field are DNA origami and DNA-based hybrid materials.[7]

# 16.1   Properties

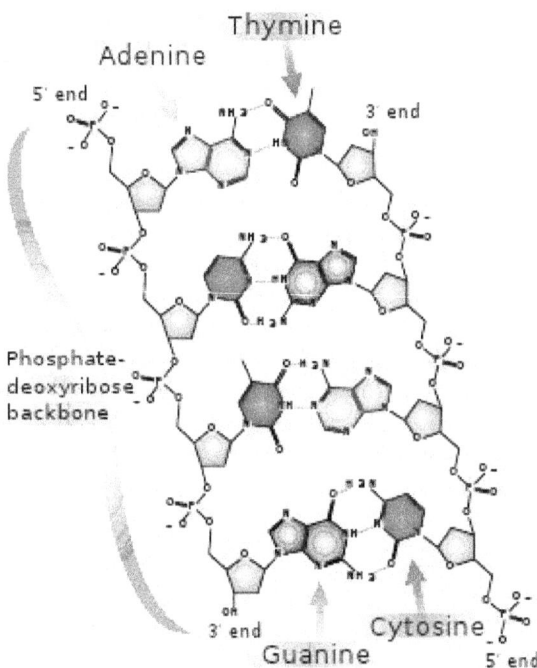

*Chemical structure of DNA; hydrogen bonds shown as dotted lines*

DNA is a long polymer made from repeating units called nucleotides.[8][9] The structure of DNA is dynamic along its length, being capable of coiling into tight loops, and other shapes.[10] In all species it is composed of two helical chains, bound to each other by hydrogen bonds. Both chains are coiled round the same axis, and have the same pitch of 34 ångströms (3.4 nanometres). The pair of chains has a radius of 10 ångströms (1.0 nanometre).[11] According to another study, when measured in a different solution, the DNA chain measured 22 to 26 ångströms wide (2.2 to 2.6 nanometres), and one nucleotide unit measured 3.3 Å (0.33 nm) long.[12] Although each individual nucleotide repeating unit is very small, DNA polymers can be very large molecules containing millions to hundreds of millions of nucleotides. For instance, the DNA in the largest human chromosome, chromosome number 1, consists of approximately 220 million base pairs[13] and would be 85 mm long if straightened.

In living organisms, DNA does not usually exist as a single molecule, but instead as a pair of molecules that are held tightly together.[14][15] These two long strands entwine like vines, in the shape of a double helix. The nucleotide contains both a segment of the backbone of the molecule (which holds the chain together) and a nucleobase (which interacts with the other DNA strand in the helix). A nucleobase linked to a sugar is called a nucleoside and a base linked to a sugar and one or more phosphate groups is called a nucleotide. A polymer comprising multiple linked nucleotides (as in DNA) is called a polynucleotide.[16]

The backbone of the DNA strand is made from alternating phosphate and sugar residues.[17] The sugar in DNA is 2-deoxyribose, which is a pentose (five-carbon) sugar. The sugars are joined together by phosphate groups that form phosphodiester bonds between the third and fifth carbon atoms of adjacent sugar rings. These asymmetric bonds mean a strand of DNA has a direction. In a double helix, the direction of the nucleotides in one strand is opposite to their direction in the other strand: the strands are *antiparallel*. The asymmetric ends of DNA strands are said to have a directionality of *five prime* (5′) and *three prime* (3′), with the 5′ end having a terminal phosphate group and the 3′ end a terminal hydroxyl group. One major difference between DNA and RNA is the sugar, with the 2-deoxyribose in DNA being replaced by the alternative pentose sugar ribose in RNA.[15]

The DNA double helix is stabilized primarily by two forces: hydrogen bonds between nucleotides and base-stacking interactions among aromatic nucleobases.[19] In the aqueous environment of the cell, the conjugated π bonds of nucleotide bases align perpendicular to the axis of the DNA molecule, minimizing their interaction with the solvation shell. The four bases found in DNA are adenine (A), cytosine (C), guanine (G) and thymine (T). These four bases are attached to the sugar-phosphate to form the complete nucleotide, as shown for adenosine monophosphate. Adenine pairs with thymine and guanine pairs with cytosine. It was represented by A-T base pairs and G-C base pairs.[20][21]

### 16.1.1 Nucleobase classification

The nucleobases are classified into two types: the purines, A and G, being fused five- and six-membered heterocyclic compounds, and the pyrimidines, the six-membered rings C and T.[15] A fifth pyrimidine nucleobase, uracil (U), usually takes the place of thymine in RNA and differs from thymine by lacking a methyl group on its ring. In addition to RNA and DNA, many artificial nucleic acid analogues have been created to study the properties of nucleic acids, or for use in biotechnology.[22]

Uracil is not usually found in DNA, occurring only as a breakdown product of cytosine. However, in several bacteriophages, *Bacillus subtilis* bacteriophages PBS1 and PBS2 and *Yersinia* bacteriophage piR1-37, thymine has been replaced by uracil.[23] Another phage - Staphylococcal phage S6 - has been identified with a genome where thymine has been replaced by uracil.[24]

Base J (beta-d-glucopyranosyloxymethyluracil), a modified

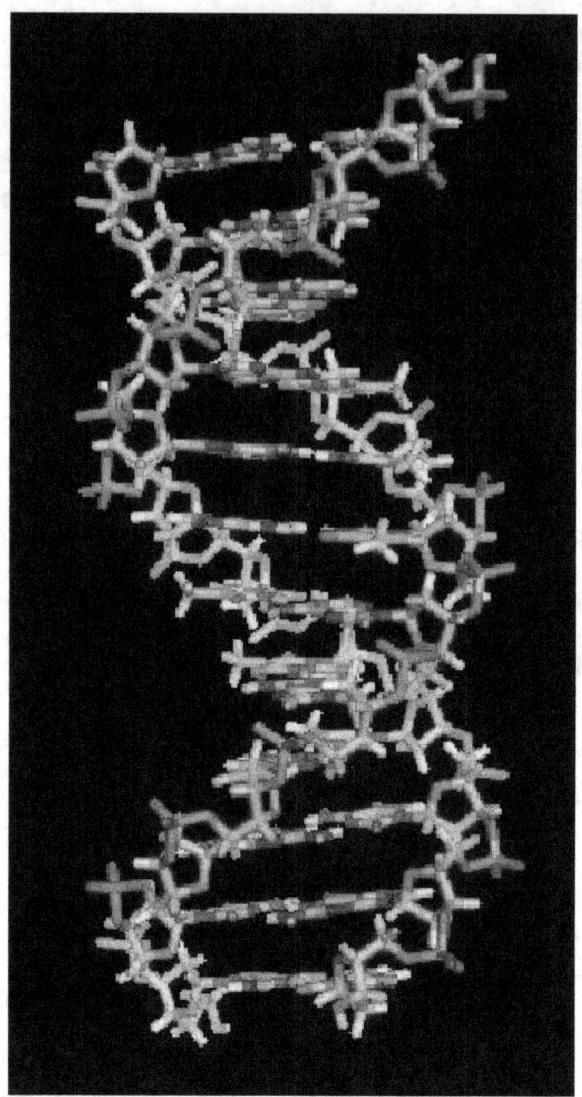

*A section of DNA. The bases lie horizontally between the two spiraling strands.[18] (animated version).*

form of uracil, is also found in several organisms: the flagellates *Diplonema* and *Euglena*, and all the kinetoplastid genera.[25] Biosynthesis of J occurs in two steps: in the first step, a specific thymidine in DNA is converted into hydroxymethyldeoxyuridine; in the second, HOMedU is glycosylated to form J.[26] Proteins that bind specifically to this base have been identified.[27][28][29] These proteins appear to be distant relatives of the Tet1 oncogene that is involved in the pathogenesis of acute myeloid leukemia.[30] J appears to act as a termination signal for RNA polymerase II.[31][32]

### 16.1.2 Grooves

Twin helical strands form the DNA backbone. Another double helix may be found tracing the spaces, or grooves,

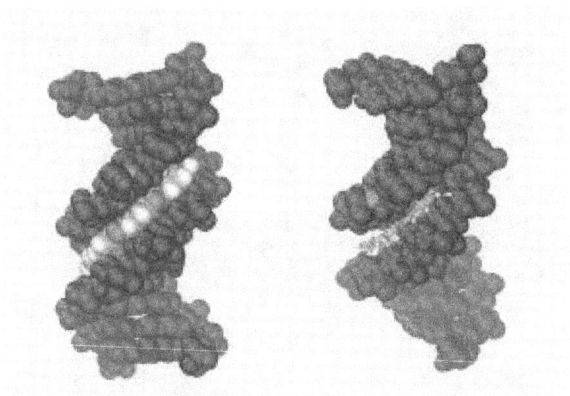

*DNA major and minor grooves. The latter is a binding site for the Hoechst stain dye 33258.*

between the strands. These voids are adjacent to the base pairs and may provide a binding site. As the strands are not symmetrically located with respect to each other, the grooves are unequally sized. One groove, the major groove, is 22 Å wide and the other, the minor groove, is 12 Å wide.[33] The width of the major groove means that the edges of the bases are more accessible in the major groove than in the minor groove. As a result, proteins such as transcription factors that can bind to specific sequences in double-stranded DNA usually make contact with the sides of the bases exposed in the major groove.[34] This situation varies in unusual conformations of DNA within the cell *(see below)*, but the major and minor grooves are always named to reflect the differences in size that would be seen if the DNA is twisted back into the ordinary B form.

## 16.1.3   Base pairing

Further information: Base pair

In a DNA double helix, each type of nucleobase on one strand bonds with just one type of nucleobase on the other strand. This is called complementary base pairing. Here, purines form hydrogen bonds to pyrimidines, with adenine bonding only to thymine in two hydrogen bonds, and cytosine bonding only to guanine in three hydrogen bonds. This arrangement of two nucleotides binding together across the double helix is called a base pair. As hydrogen bonds are not covalent, they can be broken and rejoined relatively easily. The two strands of DNA in a double helix can thus be pulled apart like a zipper, either by a mechanical force or high temperature.[35] As a result of this base pair complementarity, all the information in the double-stranded sequence of a DNA helix is duplicated on each strand, which is vital in DNA replication. This reversible and specific interaction between complementary base pairs is critical for

all the functions of DNA in living organisms.[9]

Top, a **GC** base pair with three hydrogen bonds. Bottom, an **AT** base pair with two hydrogen bonds. Non-covalent hydrogen bonds between the pairs are shown as dashed lines.

The two types of base pairs form different numbers of hydrogen bonds, AT forming two hydrogen bonds, and GC forming three hydrogen bonds (see figures, right). DNA with high GC-content is more stable than DNA with low GC-content.

As noted above, most DNA molecules are actually two polymer strands, bound together in a helical fashion by non-covalent bonds; this double stranded structure (**dsDNA**) is maintained largely by the intrastrand base stacking interactions, which are strongest for G,C stacks. The two strands can come apart – a process known as melting – to form two single-stranded DNA molecules (**ssDNA**) molecules. Melting occurs at high temperature, low salt and high pH (low pH also melts DNA, but since DNA is unstable due to acid depurination, low pH is rarely used).

The stability of the dsDNA form depends not only on the GC-content (% G,C basepairs) but also on sequence (since stacking is sequence specific) and also length (longer molecules are more stable). The stability can be measured in various ways; a common way is the "melting temperature", which is the temperature at which 50% of the ds molecules are converted to ss molecules; melting temperature is dependent on ionic strength and the concentration of DNA. As a result, it is both the percentage of GC base pairs and the overall length of a DNA double helix that determines the strength of the association between the two strands of DNA. Long DNA helices with a high GC-content have stronger-interacting strands, while short helices with high AT content have weaker-interacting strands.[36] In biology, parts of the DNA double helix that need to separate easily, such as the TATAAT Pribnow box in some promoters, tend to have a high AT content, making the strands easier to pull apart.[37]

In the laboratory, the strength of this interaction can be measured by finding the temperature necessary to break the hydrogen bonds, their melting temperature (also called *Tm* value). When all the base pairs in a DNA double helix melt, the strands separate and exist in solution as two entirely independent molecules. These single-stranded DNA molecules have no single common shape, but some conformations are more stable than others.[38]

### 16.1.4 Sense and antisense

Further information: Sense (molecular biology)

A DNA sequence is called "sense" if its sequence is the same as that of a messenger RNA copy that is translated into protein.[39] The sequence on the opposite strand is called the "antisense" sequence. Both sense and antisense sequences can exist on different parts of the same strand of DNA (i.e. both strands can contain both sense and antisense sequences). In both prokaryotes and eukaryotes, antisense RNA sequences are produced, but the functions of these RNAs are not entirely clear.[40] One proposal is that antisense RNAs are involved in regulating gene expression through RNA-RNA base pairing.[41]

A few DNA sequences in prokaryotes and eukaryotes, and more in plasmids and viruses, blur the distinction between sense and antisense strands by having overlapping genes.[42] In these cases, some DNA sequences do double duty, encoding one protein when read along one strand, and a second protein when read in the opposite direction along the other strand. In bacteria, this overlap may be involved in the regulation of gene transcription,[43] while in viruses, overlapping genes increase the amount of information that can be encoded within the small viral genome.[44]

### 16.1.5 Supercoiling

Further information: DNA supercoil

DNA can be twisted like a rope in a process called DNA supercoiling. With DNA in its "relaxed" state, a strand usually circles the axis of the double helix once every 10.4 base pairs, but if the DNA is twisted the strands become more tightly or more loosely wound.[45] If the DNA is twisted in the direction of the helix, this is positive supercoiling, and the bases are held more tightly together. If they are twisted in the opposite direction, this is negative supercoiling, and the bases come apart more easily. In nature, most DNA has slight negative supercoiling that is introduced by enzymes called topoisomerases.[46] These enzymes are also needed to relieve the twisting stresses introduced into DNA strands during processes such as transcription and DNA replication.[47]

### 16.1.6 Alternative DNA structures

Further information: Molecular Structure of Nucleic Acids: A Structure for Deoxyribose Nucleic Acid, Molecular models of DNA, and DNA structure

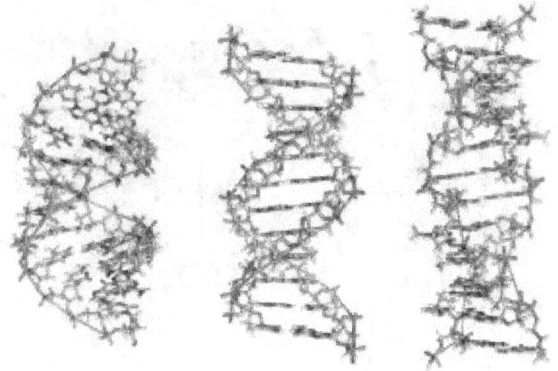

*From left to right, the structures of A, B and Z DNA*

DNA exists in many possible conformations that include A-DNA, B-DNA, and Z-DNA forms, although, only B-DNA and Z-DNA have been directly observed in functional organisms.[17] The conformation that DNA adopts depends on the hydration level, DNA sequence, the amount and direction of supercoiling, chemical modifications of the bases, the type and concentration of metal ions, and the presence of polyamines in solution.[48]

The first published reports of A-DNA X-ray diffraction patterns—and also B-DNA—used analyses based on Patterson transforms that provided only a limited amount of structural information for oriented fibers of DNA.[49][50] An alternative analysis was then proposed by Wilkins *et al.*, in 1953, for the *in vivo* B-DNA X-ray diffraction-scattering patterns of highly hydrated DNA fibers in terms of squares of Bessel functions.[51] In the same journal, James Watson and Francis Crick presented their molecular modeling analysis of the DNA X-ray diffraction patterns to suggest that the structure was a double-helix.[11]

Although the *B-DNA form* is most common under the conditions found in cells,[52] it is not a well-defined conformation but a family of related DNA conformations[53] that occur at the high hydration levels present in living cells. Their corresponding X-ray diffraction and scattering patterns are characteristic of molecular paracrystals with a significant degree of disorder.[54][55]

Compared to B-DNA, the A-DNA form is a wider right-handed spiral, with a shallow, wide minor groove and a narrower, deeper major groove. The A form occurs under non-physiological conditions in partly dehydrated samples of DNA, while in the cell it may be produced in hybrid pairings of DNA and RNA strands, and in enzyme-DNA complexes.[56][57] Segments of DNA where the bases have been chemically modified by methylation may undergo a larger change in conformation and adopt the Z form. Here, the strands turn about the helical axis in a left-handed spiral, the opposite of the more common B form.[58] These unusual structures can be recognized by specific Z-DNA binding proteins and may be involved in the regulation of transcription.[59]

### 16.1.7   Alternative DNA chemistry

For many years exobiologists have proposed the existence of a shadow biosphere, a postulated microbial biosphere of Earth that uses radically different biochemical and molecular processes than currently known life. One of the proposals was the existence of lifeforms that use arsenic instead of phosphorus in DNA. A report in 2010 of the possibility in the bacterium GFAJ-1, was announced,[60][60][61] though the research was disputed,[61][62] and evidence suggests the bacterium actively prevents the incorporation of arsenic into the DNA backbone and other biomolecules.[63]

### 16.1.8   Quadruplex structures

Further information: G-quadruplex

At the ends of the linear chromosomes are specialized regions of DNA called telomeres. The main function of these regions is to allow the cell to replicate chromosome ends using the enzyme telomerase, as the enzymes that normally replicate DNA cannot copy the extreme 3′ ends of chromosomes.[64] These specialized chromosome caps also help protect the DNA ends, and stop the DNA repair systems in the cell from treating them as damage to be corrected.[65] In human cells, telomeres are usually lengths of single-stranded DNA containing several thousand repeats of a simple TTAGGG sequence.[66]

These guanine-rich sequences may stabilize chromosome ends by forming structures of stacked sets of four-base units, rather than the usual base pairs found in other DNA molecules. Here, four guanine bases form a flat plate and these flat four-base units then stack on top of each other, to form a stable G-quadruplex structure.[68] These structures are stabilized by hydrogen bonding between the edges of the bases and chelation of a metal ion in the centre of each four-base unit.[69] Other structures can also be formed, with the central set of four bases coming from either a single strand folded around the bases, or several different parallel strands, each contributing one base to the central structure.

In addition to these stacked structures, telomeres also form large loop structures called telomere loops, or T-loops. Here, the single-stranded DNA curls around in a long circle stabilized by telomere-binding proteins.[70] At the very end of the T-loop, the single-stranded telomere DNA is held onto a region of double-stranded DNA by the telomere strand disrupting the double-helical DNA and base pairing

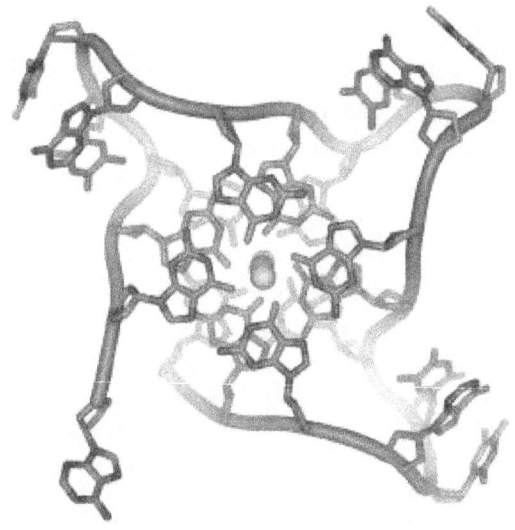

DNA quadruplex formed by telomere repeats. The looped conformation of the DNA backbone is very different from the typical DNA helix. The green spheres in the center represent potassium ions.[67]

to one of the two strands. This triple-stranded structure is called a displacement loop or D-loop.[68]

Branched DNA can form networks containing multiple branches.

### 16.1.9   Branched DNA

Further information: Branched DNA and DNA nanotechnology

In DNA, fraying occurs when non-complementary regions exist at the end of an otherwise complementary double-strand of DNA. However, branched DNA can occur if a third strand of DNA is introduced and contains adjoining regions able to hybridize with the frayed regions of the pre-existing double-strand. Although the simplest example of branched DNA involves only three strands of DNA, complexes involving additional strands and multiple branches are also possible.[71] Branched DNA can be used in nanotechnology to construct geometric shapes, see the section on uses in technology below.

## 16.2   Chemical modifications and altered DNA packaging

Structure of cytosine with and without the 5-methyl group. Deamination converts 5-methylcytosine into thymine.

### 16.2.1 Base modifications and DNA packaging

Further information: DNA methylation and Chromatin remodeling

The expression of genes is influenced by how the DNA is packaged in chromosomes, in a structure called chromatin. Base modifications can be involved in packaging, with regions that have low or no gene expression usually containing high levels of methylation of cytosine bases. DNA packaging and its influence on gene expression can also occur by covalent modifications of the histone protein core around which DNA is wrapped in the chromatin structure or else by remodeling carried out by chromatin remodeling complexes (see Chromatin remodeling). There is, further, crosstalk between DNA methylation and histone modification, so they can coordinately affect chromatin and gene expression.[72]

For one example, cytosine methylation produces 5-methylcytosine, which is important for X-inactivation of chromosomes.[73] The average level of methylation varies between organisms – the worm *Caenorhabditis elegans* lacks cytosine methylation, while vertebrates have higher levels, with up to 1% of their DNA containing 5-methylcytosine.[74] Despite the importance of 5-methylcytosine, it can deaminate to leave a thymine base, so methylated cytosines are particularly prone to mutations.[75] Other base modifications include adenine methylation in bacteria, the presence of 5-hydroxymethylcytosine in the brain,[76] and the glycosylation of uracil to produce the "J-base" in kinetoplastids.[77][78]

### 16.2.2 Damage

Further information: DNA damage (naturally occurring), Mutation, and DNA damage theory of aging

DNA can be damaged by many sorts of mutagens, which change the DNA sequence. Mutagens include oxidizing agents, alkylating agents and also high-energy electromagnetic radiation such as ultraviolet light and X-rays. The type of DNA damage produced depends on the type of mutagen. For example, UV light can damage DNA by producing thymine dimers, which are cross-links between pyrimidine bases.[80] On the other hand, oxidants such as free radicals or hydrogen peroxide produce multiple forms of damage, including base modifications, particularly of guanosine, and double-strand breaks.[81] A typical human cell contains about 150,000 bases that

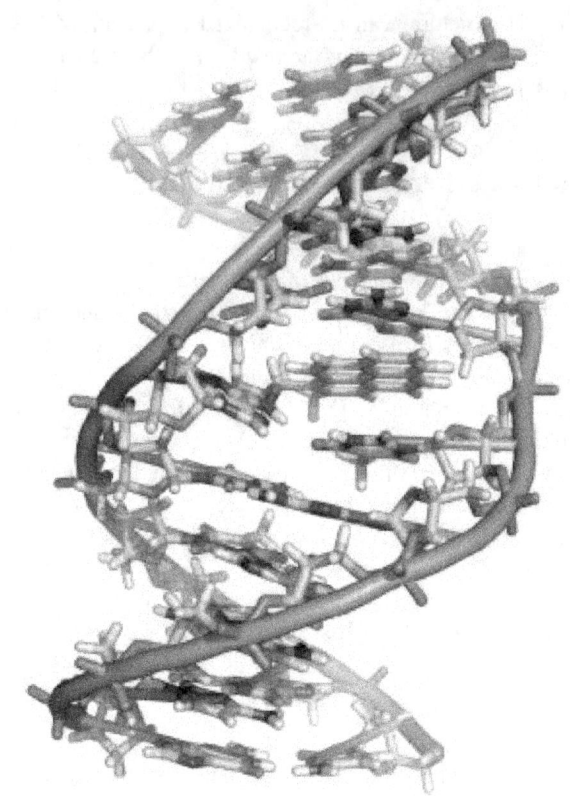

*A covalent adduct between a metabolically activated form of benzo[a]pyrene, the major mutagen in tobacco smoke, and DNA[79]*

have suffered oxidative damage.[82] Of these oxidative lesions, the most dangerous are double-strand breaks, as these are difficult to repair and can produce point mutations, insertions, deletions from the DNA sequence, and chromosomal translocations.[83] These mutations can cause cancer. Because of inherent limits in the DNA repair mechanisms, if humans lived long enough, they would all eventually develop cancer.[84][85] DNA damages that are naturally occurring, due to normal cellular processes that produce reactive oxygen species, the hydrolytic activities of cellular water, etc., also occur frequently. Although most of these damages are repaired, in any cell some DNA damage may remain despite the action of repair processes. These remaining DNA damages accumulate with age in mammalian postmitotic tissues. This accumulation appears to be an important underlying cause of aging.[86][87][88]

Many mutagens fit into the space between two adjacent base pairs, this is called *intercalation*. Most intercalators are aromatic and planar molecules; examples include ethidium bromide, acridines, daunomycin, and doxorubicin. For an intercalator to fit between base pairs, the bases must separate, distorting the DNA strands by unwinding of the double helix. This inhibits both transcription and DNA replication, causing toxicity and mutations.[89] As a result, DNA intercalators may be carcinogens, and in the case of thalidomide, a teratogen.[90] Others such as benzo[*a*]pyrene diol epoxide and aflatoxin form DNA adducts that induce errors in replication.[91] Nevertheless, due to their ability to inhibit DNA transcription and replication, other similar toxins are also used in chemotherapy to inhibit rapidly growing cancer cells.[92]

## 16.3  Biological functions

*Location of eukaryote nuclear DNA within the chromosomes.*

DNA usually occurs as linear chromosomes in eukaryotes, and circular chromosomes in prokaryotes. The set of chromosomes in a cell makes up its genome; the human genome has approximately 3 billion base pairs of DNA arranged into 46 chromosomes.[93] The information carried by DNA is held in the sequence of pieces of DNA called genes. Transmission of genetic information in genes is achieved via complementary base pairing. For example, in transcription, when a cell uses the information in a gene, the DNA sequence is copied into a complementary RNA sequence through the attraction between the DNA and the correct RNA nucleotides. Usually, this RNA copy is then used to make a matching protein sequence in a process called translation, which depends on the same interaction between RNA nucleotides. In alternative fashion, a cell may simply copy its genetic information in a process called DNA replication. The details of these functions are covered in other articles; here the focus is on the interactions between DNA and other molecules that mediate the function of the genome.

### 16.3.1  Genes and genomes

Further information:   Cell nucleus,  Chromatin, Chromosome, Gene, and Noncoding DNA

Genomic DNA is tightly and orderly packed in the process called DNA condensation, to fit the small available volumes of the cell. In eukaryotes, DNA is located in the cell nucleus, with small amounts in mitochondria and chloroplasts. In prokaryotes, the DNA is held within an irregularly shaped body in the cytoplasm called the nucleoid.[94] The genetic information in a genome is held within genes, and the complete set of this information in an organism is called its genotype. A gene is a unit of heredity and is a region of DNA that influences a particular characteristic in an organism. Genes contain an open reading frame that can be transcribed, and regulatory sequences such as promoters and enhancers, which control transcription of the open reading frame.

In many species, only a small fraction of the total sequence of the genome encodes protein. For example, only about 1.5% of the human genome consists of protein-coding exons, with over 50% of human DNA consisting of non-coding repetitive sequences.[95] The reasons for the presence of so much noncoding DNA in eukaryotic genomes and the extraordinary differences in genome size, or *C-value*, among species, represent a long-standing puzzle known as the "C-value enigma".[96] However, some DNA sequences that do not code protein may still encode functional non-coding RNA molecules, which are involved in the regulation of gene expression.[97]

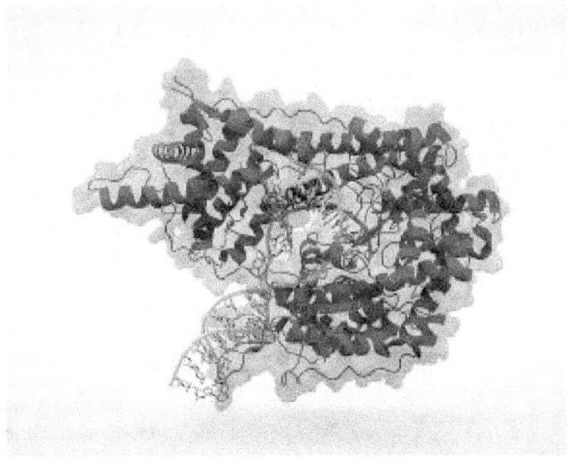

*T7 RNA polymerase (blue) producing an mRNA (green) from a DNA template (orange).*[98]

Some noncoding DNA sequences play structural roles in chromosomes. Telomeres and centromeres typically contain few genes but are important for the function and stability of chromosomes.[65][99] An abundant form of noncoding DNA in humans are pseudogenes, which are copies of genes that have been disabled by mutation.[100] These sequences are usually just molecular fossils, although they can occasionally serve as raw genetic material for the creation of new genes through the process of gene duplication and divergence.[101]

## 16.3.2 Transcription and translation

Further information: Genetic code, Transcription (genetics), and Protein biosynthesis

A gene is a sequence of DNA that contains genetic information and can influence the phenotype of an organism. Within a gene, the sequence of bases along a DNA strand defines a messenger RNA sequence, which then defines one or more protein sequences. The relationship between the nucleotide sequences of genes and the amino-acid sequences of proteins is determined by the rules of translation, known collectively as the genetic code. The genetic code consists of three-letter 'words' called *codons* formed from a sequence of three nucleotides (e.g. ACT, CAG, TTT).

In transcription, the codons of a gene are copied into messenger RNA by RNA polymerase. This RNA copy is then decoded by a ribosome that reads the RNA sequence by base-pairing the messenger RNA to transfer RNA, which carries amino acids. Since there are 4 bases in 3-letter combinations, there are 64 possible codons ($4^3$ combinations). These encode the twenty standard amino acids, giving most amino acids more than one possible codon. There are also three 'stop' or 'nonsense' codons signifying the end of the coding region; these are the TAA, TGA, and TAG codons.

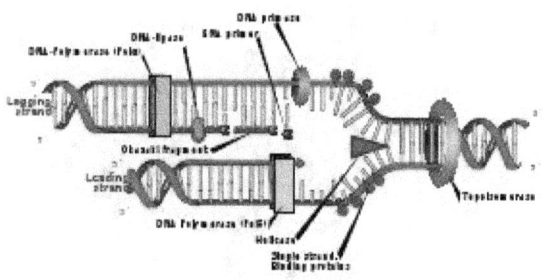

*DNA replication. The double helix is unwound by a helicase and topoisomerase. Next, one DNA polymerase produces the leading strand copy. Another DNA polymerase binds to the lagging strand. This enzyme makes discontinuous segments (called Okazaki fragments) before DNA ligase joins them together.*

## 16.3.3 Replication

Further information: DNA replication

Cell division is essential for an organism to grow, but, when a cell divides, it must replicate the DNA in its genome so that the two daughter cells have the same genetic information as their parent. The double-stranded structure of DNA provides a simple mechanism for DNA replication. Here, the two strands are separated and then each strand's complementary DNA sequence is recreated by an enzyme called DNA polymerase. This enzyme makes the complementary strand by finding the correct base through complementary base pairing and bonding it onto the original strand. As DNA polymerases can only extend a DNA strand in a 5' to 3' direction, different mechanisms are used to copy the antiparallel strands of the double helix.[102] In this way, the base on the old strand dictates which base appears on the new strand, and the cell ends up with a perfect copy of its DNA.

## 16.3.4 Extracellular nucleic acids

Naked extracellular DNA (eDNA), most of it released by cell death, is nearly ubiquitous in the environment. Its concentration in soil may be as high as 2 µg/L, and its concentration in natural aquatic environments may be as high at 88 µg/L.[103] Various possible functions have been proposed for eDNA: it may be involved in horizontal gene transfer;[104] it may provide nutrients;[105] and it may act as a buffer to recruit or titrate ions or antibiotics.[106] Extracellular DNA acts as a functional extracellular matrix component in the biofilms of several bacterial species. It may act as a recognition factor to regulate the attachment and dispersal of specific cell types in the biofilm;[107] it may contribute to biofilm formation;[108] and it may contribute to the biofilm's physical strength and resistance to biological stress.[109]

Cell-free fetal DNA is found in the blood of the mother, and can be sequenced to determine a great deal of information about the developing fetus.[110]

## 16.4 Interactions with proteins

All the functions of DNA depend on interactions with proteins. These protein interactions can be non-specific, or the protein can bind specifically to a single DNA sequence. Enzymes can also bind to DNA and of these, the polymerases that copy the DNA base sequence in transcription and DNA replication are particularly important.

## 16.4.1 DNA-binding proteins

Further information: DNA-binding protein
Interaction of DNA (in orange) with histones (in blue). These proteins' basic amino acids bind to the acidic phosphate groups on DNA.

Structural proteins that bind DNA are well-understood examples of non-specific DNA-protein interactions. Within chromosomes, DNA is held in complexes with structural proteins. These proteins organize the DNA into a compact structure called chromatin. In eukaryotes, this structure involves DNA binding to a complex of small basic proteins called histones, while in prokaryotes multiple types of proteins are involved.[111][112] The histones form a disk-shaped complex called a nucleosome, which contains two complete turns of double-stranded DNA wrapped around its surface. These non-specific interactions are formed through basic residues in the histones, making ionic bonds to the acidic sugar-phosphate backbone of the DNA, and are thus largely independent of the base sequence.[113] Chemical modifications of these basic amino acid residues include methylation, phosphorylation, and acetylation.[114] These chemical changes alter the strength of the interaction between the DNA and the histones, making the DNA more or less accessible to transcription factors and changing the rate of transcription.[115] Other non-specific DNA-binding proteins in chromatin include the high-mobility group proteins, which bind to bent or distorted DNA.[116] These proteins are important in bending arrays of nucleosomes and arranging them into the larger structures that make up chromosomes.[117]

A distinct group of DNA-binding proteins is the DNA-binding proteins that specifically bind single-stranded DNA. In humans, replication protein A is the best-understood member of this family and is used in processes where the double helix is separated, including DNA replication, recombination, and DNA repair.[118] These binding proteins seem to stabilize single-stranded DNA and protect it from forming stem-loops or being degraded by nucleases.

In contrast, other proteins have evolved to bind to particular DNA sequences. The most intensively studied of these are the various transcription factors, which are proteins that regulate transcription. Each transcription factor binds to one particular set of DNA sequences and activates or inhibits the transcription of genes that have these sequences close to their promoters. The transcription factors do this in two ways. Firstly, they can bind the RNA polymerase responsible for transcription, either directly or through other mediator proteins; this locates the polymerase at the promoter and allows it to begin transcription.[120] Alternatively, transcription factors can bind enzymes that modify the histones at the promoter. This changes the accessibility of the DNA template to the polymerase.[121]

As these DNA targets can occur throughout an organism's genome, changes in the activity of one type of transcription factor can affect thousands of genes.[122] Consequently, these proteins are often the targets of the signal transduction processes that control responses to environmental changes or cellular differentiation and development. The specificity of these transcription factors' interactions with DNA come

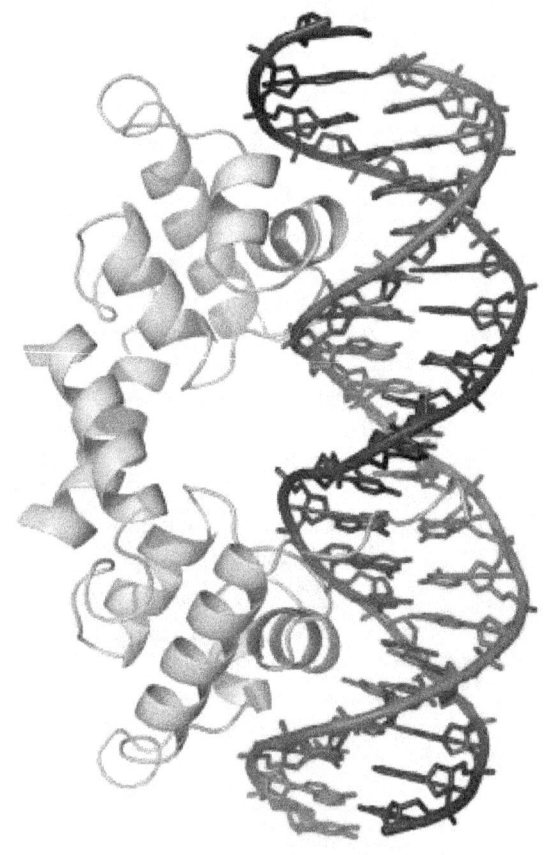

*The lambda repressor helix-turn-helix transcription factor bound to its DNA target*[119]

from the proteins making multiple contacts to the edges of the DNA bases, allowing them to "read" the DNA sequence. Most of these base-interactions are made in the major groove, where the bases are most accessible.[34]

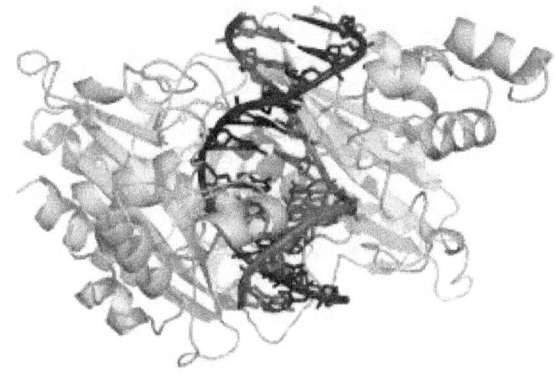

*The restriction enzyme EcoRV (green) in a complex with its substrate DNA*[123]

### 16.4.2    DNA-modifying enzymes

**Nucleases and ligases**

Nucleases are enzymes that cut DNA strands by catalyzing the hydrolysis of the phosphodiester bonds. Nucleases that hydrolyse nucleotides from the ends of DNA strands are called exonucleases, while endonucleases cut within strands. The most frequently used nucleases in molecular biology are the restriction endonucleases, which cut DNA at specific sequences. For instance, the EcoRV enzyme shown to the left recognizes the 6-base sequence 5′-GATATC-3′ and makes a cut at the horizontal line. In nature, these enzymes protect bacteria against phage infection by digesting the phage DNA when it enters the bacterial cell, acting as part of the restriction modification system.[124] In technology, these sequence-specific nucleases are used in molecular cloning and DNA fingerprinting.

Enzymes called DNA ligases can rejoin cut or broken DNA strands.[125] Ligases are particularly important in lagging strand DNA replication, as they join together the short segments of DNA produced at the replication fork into a complete copy of the DNA template. They are also used in DNA repair and genetic recombination.[125]

**Topoisomerases and helicases**

Topoisomerases are enzymes with both nuclease and ligase activity. These proteins change the amount of supercoiling in DNA. Some of these enzymes work by cutting the DNA helix, and allowing one section to rotate, thereby reducing its level of supercoiling; the enzyme then seals the DNA break.[46] Other types of these enzymes are capable of cutting one DNA helix and then passing a second strand of DNA through this break, before rejoining the helix.[126] Topoisomerases are required for many processes involving DNA, such as DNA replication and transcription.[47]

Helicases are proteins that are a type of molecular motor. They use the chemical energy in nucleoside triphosphates, predominantly adenosine triphosphate (ATP), to break hydrogen bonds between bases and unwind the DNA double helix into single strands.[127] These enzymes are essential for most processes where enzymes need to access the DNA bases.

**Polymerases**

Polymerases are enzymes that synthesize polynucleotide chains from nucleoside triphosphates. The sequence of their products is created based on existing polynucleotide chains—which are called *templates*. These enzymes function by repeatedly adding a nucleotide to the 3′ hydroxyl group at the end of the growing polynucleotide chain. As a consequence, all polymerases work in a 5′ to 3′ direction.[128] In the active site of these enzymes, the incoming nucleoside triphosphate base-pairs to the template: this allows polymerases to accurately synthesize the complementary strand of their template. Polymerases are classified according to the type of template that they use.

In DNA replication, DNA-dependent DNA polymerases make copies of DNA polynucleotide chains. To preserve biological information, it is essential that the sequence of bases in each copy are precisely complementary to the sequence of bases in the template strand. Many DNA polymerases have a proofreading activity. Here, the polymerase recognizes the occasional mistakes in the synthesis reaction by the lack of base pairing between the mismatched nucleotides. If a mismatch is detected, a 3′ to 5′ exonuclease activity is activated and the incorrect base removed.[129] In most organisms, DNA polymerases function in a large complex called the replisome that contains multiple accessory subunits, such as the DNA clamp or helicases.[130]

RNA-dependent DNA polymerases are a specialized class of polymerases that copy the sequence of an RNA strand into DNA. They include reverse transcriptase, which is a viral enzyme involved in the infection of cells by retroviruses, and telomerase, which is required for the replication of telomeres.[64][131] Telomerase is an unusual polymerase because it contains its own RNA template as part of its structure.[65]

Transcription is carried out by a DNA-dependent RNA polymerase that copies the sequence of a DNA strand into RNA. To begin transcribing a gene, the RNA polymerase binds to a sequence of DNA called a promoter and separates the DNA strands. It then copies the gene sequence into a messenger RNA transcript until it reaches a region of DNA called the terminator, where it halts and detaches from the DNA. As with human DNA-dependent DNA polymerases, RNA polymerase II, the enzyme that transcribes most of the genes in the human genome, operates as part of a large protein complex with multiple regulatory and accessory subunits.[132]

## 16.5    Genetic recombination

Structure of the Holliday junction intermediate in genetic recombination.    The four separate DNA strands are coloured red, blue, green and yellow.[133]

Further information: Genetic recombination

A DNA helix usually does not interact with other segments of DNA, and in human cells, the different chromosomes even occupy separate areas in the nucleus called "chromosome territories".[134] This physical separation of

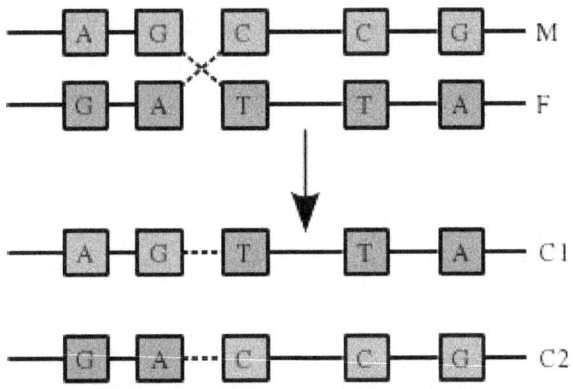

*Recombination involves the breaking and rejoining of two chromosomes (M and F) to produce two rearranged chromosomes (C1 and C2).*

## 16.6   Evolution

Further information: RNA world hypothesis

DNA contains the genetic information that allows all modern living things to function, grow and reproduce. However, it is unclear how long in the 4-billion-year history of life DNA has performed this function, as it has been proposed that the earliest forms of life may have used RNA as their genetic material.[140][141] RNA may have acted as the central part of early cell metabolism as it can both transmit genetic information and carry out catalysis as part of ribozymes.[142] This ancient RNA world where nucleic acid would have been used for both catalysis and genetics may have influenced the evolution of the current genetic code based on four nucleotide bases. This would occur, since the number of different bases in such an organism is a trade-off between a small number of bases increasing replication accuracy and a large number of bases increasing the catalytic efficiency of ribozymes.[143] However, there is no direct evidence of ancient genetic systems, as recovery of DNA from most fossils is impossible because DNA survives in the environment for less than one million years, and slowly degrades into short fragments in solution.[144] Claims for older DNA have been made, most notably a report of the isolation of a viable bacterium from a salt crystal 250 million years old,[145] but these claims are controversial.[146][147]

Building blocks of DNA (adenine, guanine, and related organic molecules) may have been formed extraterrestrially in outer space.[148][149][150] Complex DNA and RNA organic compounds of life, including uracil, cytosine, and thymine, have also been formed in the laboratory under conditions mimicking those found in outer space, using starting chemicals, such as pyrimidine, found in meteorites. Pyrimidine, like polycyclic aromatic hydrocarbons (PAHs), the most carbon-rich chemical found in the universe, may have been formed in red giants or in interstellar cosmic dust and gas clouds.[151]

different chromosomes is important for the ability of DNA to function as a stable repository for information, as one of the few times chromosomes interact is in chromosomal crossover which occurs during sexual reproduction, when genetic recombination occurs. Chromosomal crossover is when two DNA helices break, swap a section and then rejoin.

Recombination allows chromosomes to exchange genetic information and produces new combinations of genes, which increases the efficiency of natural selection and can be important in the rapid evolution of new proteins.[135] Genetic recombination can also be involved in DNA repair, particularly in the cell's response to double-strand breaks.[136]

The most common form of chromosomal crossover is homologous recombination, where the two chromosomes involved share very similar sequences. Non-homologous recombination can be damaging to cells, as it can produce chromosomal translocations and genetic abnormalities. The recombination reaction is catalyzed by enzymes known as recombinases, such as RAD51.[137] The first step in recombination is a double-stranded break caused by either an endonuclease or damage to the DNA.[138] A series of steps catalyzed in part by the recombinase then leads to joining of the two helices by at least one Holliday junction, in which a segment of a single strand in each helix is annealed to the complementary strand in the other helix. The Holliday junction is a tetrahedral junction structure that can be moved along the pair of chromosomes, swapping one strand for another. The recombination reaction is then halted by cleavage of the junction and re-ligation of the released DNA.[139]

## 16.7   Uses in technology

### 16.7.1   Genetic engineering

Further information: Molecular biology, Nucleic acid methods, and Genetic engineering

Methods have been developed to purify DNA from organisms, such as phenol-chloroform extraction, and to manipulate it in the laboratory, such as restriction digests and the polymerase chain reaction. Modern biology and biochemistry make intensive use of these techniques in recombinant DNA technology. Recombinant DNA is a man-made DNA sequence that has been assembled from other DNA sequences. They can be transformed into organisms in the form of plasmids or in the appropriate format, by using a viral vector.[152] The genetically modified organisms produced can be used to produce products such as recombinant proteins, used in medical research,[153] or be grown in agriculture.[154][155]

## 16.7.2 DNA profiling

Further information: DNA profiling

Forensic scientists can use DNA in blood, semen, skin, saliva or hair found at a crime scene to identify a matching DNA of an individual, such as a perpetrator. This process is formally termed DNA profiling, but may also be called "genetic fingerprinting". In DNA profiling, the lengths of variable sections of repetitive DNA, such as short tandem repeats and minisatellites, are compared between people. This method is usually an extremely reliable technique for identifying a matching DNA.[156] However, identification can be complicated if the scene is contaminated with DNA from several people.[157] DNA profiling was developed in 1984 by British geneticist Sir Alec Jeffreys,[158] and first used in forensic science to convict Colin Pitchfork in the 1988 Enderby murders case.[159]

The development of forensic science and the ability to now obtain genetic matching on minute samples of blood, skin, saliva, or hair has led to re-examining many cases. Evidence can now be uncovered that was scientifically impossible at the time of the original examination. Combined with the removal of the double jeopardy law in some places, this can allow cases to be reopened where prior trials have failed to produce sufficient evidence to convince a jury. People charged with serious crimes may be required to provide a sample of DNA for matching purposes. The most obvious defense to DNA matches obtained forensically is to claim that cross-contamination of evidence has occurred. This has resulted in meticulous strict handling procedures with new cases of serious crime. DNA profiling is also used successfully to positively identify victims of mass casualty incidents,[160] bodies or body parts in serious accidents, and individual victims in mass war graves, via matching to family members.

DNA profiling is also used in DNA paternity testing to determine if someone is the biological parent or grandparent of a child with the probability of parentage is typically 99.99% when the alleged parent is biologically related to the child. Normal DNA sequencing methods happen after birth, but there are new methods to test paternity while a mother is still pregnant.[161]

## 16.7.3 DNA enzymes or catalytic DNA

Further information: Deoxyribozyme

Deoxyribozymes, also called DNAzymes or catalytic DNA, are first discovered in 1994.[162] They are mostly single stranded DNA sequences isolated from a large pool of random DNA sequences through a combinatorial approach called in vitro selection or systematic evolution of ligands by exponential enrichment (SELEX). DNAzymes catalyze variety of chemical reactions including RNA-DNA cleavage, RNA-DNA ligation, amino acids phosphorylation-dephosphorylation, carbon-carbon bond formation, and etc. DNAzymes can enhance catalytic rate of chemical reactions up to 100,000,000,000-fold over the uncatalyzed reaction.[163] The most extensively studied class of DNAzymes is RNA-cleaving types which have been used to detect different metal ions and designing therapeutic agents. Several metal-specific DNAzymes have been reported including the GR-5 DNAzyme (lead-specific),[162] the CA1-3 DNAzymes (copper-specific),[164] the 39E DNAzyme (uranyl-specific) and the NaA43 DNAzyme (sodium-specific).[165] The NaA43 DNAzyme, which is reported to be more than 10,000-fold selective for sodium over other metal ions, was used to make a real-time sodium sensor in living cells.

## 16.7.4 Bioinformatics

Further information: Bioinformatics

Bioinformatics involves the development of techniques to store, data mine, search and manipulate biological data, including DNA nucleic acid sequence data. These have led to widely applied advances in computer science, especially string searching algorithms, machine learning, and database theory.[166] String searching or matching algorithms, which find an occurrence of a sequence of letters inside a larger sequence of letters, were developed to search for specific sequences of nucleotides.[167] The DNA sequence may be aligned with other DNA sequences to identify homologous sequences and locate the specific mutations that make them distinct. These techniques, especially multiple sequence alignment, are used in studying phylogenetic relationships and protein function.[168] Data sets representing entire genomes' worth of DNA sequences, such as those produced by the Human Genome Project, are difficult to use without the annotations that identify the locations of genes and regulatory elements on each chromosome. Regions of DNA sequence that have the characteristic patterns associated with protein- or RNA-coding genes can be identified by gene finding algorithms, which allow researchers to predict the presence of particular gene products and their possible functions in an organism even before they have been isolated experimentally.[169] Entire genomes may also be compared, which can shed light on the evolutionary history of particular organism and permit the examination of complex evolutionary events.

### 16.7.5   DNA nanotechnology

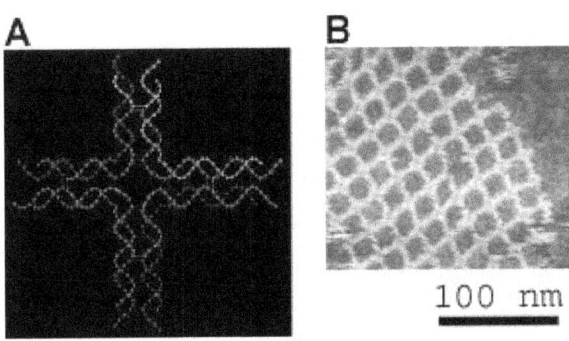

*The DNA structure at left (schematic shown) will self-assemble into the structure visualized by atomic force microscopy at right. DNA nanotechnology is the field that seeks to design nanoscale structures using the molecular recognition properties of DNA molecules. Image from Strong, 2004.*

Further information: DNA nanotechnology

DNA nanotechnology uses the unique molecular recognition properties of DNA and other nucleic acids to create self-assembling branched DNA complexes with useful properties.[170] DNA is thus used as a structural material rather than as a carrier of biological information. This has led to the creation of two-dimensional periodic lattices (both tile-based and using the *DNA origami* method) and three-dimensional structures in the shapes of polyhedra.[171] Nanomechanical devices and algorithmic self-assembly have also been demonstrated,[172] and these DNA structures have been used to template the arrangement of other molecules such as gold nanoparticles and streptavidin proteins.[173]

### 16.7.6   History and anthropology

Further information: Phylogenetics and Genetic genealogy

Because DNA collects mutations over time, which are then inherited, it contains historical information, and, by comparing DNA sequences, geneticists can infer the evolutionary history of organisms, their phylogeny.[174] This field of phylogenetics is a powerful tool in evolutionary biology. If DNA sequences within a species are compared, population geneticists can learn the history of particular populations. This can be used in studies ranging from ecological genetics to anthropology; For example, DNA evidence is being used to try to identify the Ten Lost Tribes of Israel.[175][176]

### 16.7.7   Information storage

Main article: DNA digital data storage

In a paper published in *Nature* in January 2013, scientists from the European Bioinformatics Institute and Agilent Technologies proposed a mechanism to use DNA's ability to code information as a means of digital data storage. The group was able to encode 739 kilobytes of data into DNA code, synthesize the actual DNA, then sequence the DNA and decode the information back to its original form, with a reported 100% accuracy. The encoded information consisted of text files and audio files. A prior experiment was published in August 2012. It was conducted by researchers at Harvard University, where the text of a 54,000-word book was encoded in DNA.[177][178]

Moreover, in living cells, the storage can be turned active by enzymes. Light-gated protein domains fused to DNA processing enzymes are suitable for that task *in vitro*.[179][180] Fluorescent exonucleases can transmit the output according to the nucleotide they have read.[181]

## 16.8   History of DNA research

Further information: History of molecular biology
DNA was first isolated by the Swiss physician Friedrich

*James Watson and Francis Crick (right), co-originators of the double-helix model, with Maclyn McCarty (left).*

*Pencil sketch of the DNA double helix by Francis Crick in 1953*

Miescher who, in 1869, discovered a microscopic substance in the pus of discarded surgical bandages. As it resided in the nuclei of cells, he called it "nuclein".[182][183] In 1878, Albrecht Kossel isolated the non-protein component of "nuclein", nucleic acid, and later isolated its five primary nucleobases.[184][185] In 1919, Phoebus Levene identified the base, sugar, and phosphate nucleotide unit.[186] Levene suggested that DNA consisted of a string of nucleotide units linked together through the phosphate groups. Levene thought the chain was short and the bases repeated in a fixed order. In 1937, William Astbury produced the first X-ray diffraction patterns that showed that DNA had a regular structure.[187]

In 1927, Nikolai Koltsov proposed that inherited traits would be inherited via a "giant hereditary molecule" made up of "two mirror strands that would replicate in a semi-conservative fashion using each strand as a template".[188][189] In 1928, Frederick Griffith in his experiment discovered that traits of the "smooth" form of *Pneumococcus* could be transferred to the "rough" form of the same bacteria by mixing killed "smooth" bacteria with the live "rough" form.[190][191] This system provided the first clear suggestion that DNA carries genetic information— the Avery–MacLeod–McCarty experiment—when Oswald Avery, along with coworkers Colin MacLeod and Maclyn McCarty, identified DNA as the transforming principle in 1943.[192] DNA's role in heredity was confirmed in 1952

when Alfred Hershey and Martha Chase in the Hershey–Chase experiment showed that DNA is the genetic material of the T2 phage.[193]

In 1953, James Watson and Francis Crick suggested what is now accepted as the first correct double-helix model of DNA structure in the journal *Nature*.[11] Their double-helix, molecular model of DNA was then based on one X-ray diffraction image (labeled as "Photo 51")[194] taken by Rosalind Franklin and Raymond Gosling in May 1952, and the information that the DNA bases are paired.

Experimental evidence supporting the Watson and Crick model was published in a series of five articles in the same issue of *Nature*.[195] Of these, Franklin and Gosling's paper was the first publication of their own X-ray diffraction data and original analysis method that partly supported the Watson and Crick model;[50][196] this issue also contained an article on DNA structure by Maurice Wilkins and two of his colleagues, whose analysis and *in vivo* B-DNA X-ray patterns also supported the presence *in vivo* of the double-helical DNA configurations as proposed by Crick and Watson for their double-helix molecular model of DNA in the prior two pages of *Nature*.[51] In 1962, after Franklin's death, Watson, Crick, and Wilkins jointly received the Nobel Prize in Physiology or Medicine.[197] Nobel Prizes are awarded only to living recipients. A debate continues about who should receive credit for the discovery.[198]

In an influential presentation in 1957, Crick laid out the central dogma of molecular biology, which foretold the relationship between DNA, RNA, and proteins, and articulated the "adaptor hypothesis".[199] Final confirmation of the replication mechanism that was implied by the double-helical structure followed in 1958 through the Meselson–Stahl experiment.[200] Further work by Crick and coworkers showed that the genetic code was based on non-overlapping triplets of bases, called codons, allowing Har Gobind Khorana, Robert W. Holley, and Marshall Warren Nirenberg to decipher the genetic code.[201] These findings represent the birth of molecular biology.

## 16.9  See also

- Autosome
- Crystallography
- DNA-encoded chemical library
- DNA microarray
- DNA sequencing
- Macromolecule
- Genetic disorder

- Haplotype

- Comparison of nucleic acid simulation software

- Meiosis

- Mitochondrial DNA

- Nuclear DNA

- Nucleic acid double helix

- Nucleic acid notation

- Nucleic acid sequence

- Pangenesis

- Phosphoramidite

- Ribosomal DNA

- Southern blot

- X-ray scattering techniques

- Xeno nucleic acid

- RNA

- Deoxyribozyme

## 16.10   References

[1] "deoxyribonucleic acid". *Merriam-Webster Dictionary*.

[2] Alberts B, Johnson A, Lewis J, Raff M, Roberts K, Walter P (2014). *Molecular Biology of the Cell* (6th ed.). Garland. p. Chapter 4: DNA, Chromosomes and Genomes. ISBN 9780815344322.

[3] Purcell A. "DNA". *Basic Biology*.

[4] Nuwer R (18 July 2015). "Counting All the DNA on Earth". *The New York Times*. New York: The New York Times Company. ISSN 0362-4331. Retrieved 2015-07-18.

[5] "The Biosphere: Diversity of Life". *Aspen Global Change Institute*. Basalt, CO. Retrieved 2015-07-19.

[6] Russell P (2001). *iGenetics*. New York: Benjamin Cummings. ISBN 0-8053-4553-1.

[7] Mashaghi A, Katan A (2013). "A physicist's view of DNA". *De Physicus*. **24c** (3): 59–61. arXiv:1311.2545v1. Bibcode:2013arXiv1311.2545M.

[8] Saenger W (1984). *Principles of Nucleic Acid Structure*. New York: Springer-Verlag. ISBN 0-387-90762-9.

[9] Alberts B, Johnson A, Lewis J, Raff M, Roberts K, Peter W (2002). *Molecular Biology of the Cell* (Fourth ed.). New York and London: Garland Science. ISBN 0-8153-3218-1. OCLC 145080076.

[10] Irobalieva RN, Fogg JM, Catanese DJ, Catanese DJ, Sutthibutpong T, Chen M, Barker AK, Ludtke SJ, Harris SA, Schmid MF, Chiu W, Zechiedrich L (October 2015). "Structural diversity of supercoiled DNA". *Nature Communications*. **6**: 8440. doi:10.1038/ncomms9440. PMC 4608029. PMID 26455586.

[11] Watson JD, Crick FH (April 1953). "Molecular structure of nucleic acids; a structure for deoxyribose nucleic acid" (PDF). *Nature*. **171** (4356): 737–8. Bibcode:1953Natur.171..737W. doi:10.1038/171737a0. PMID 13054692.

[12] Mandelkern M, Elias JG, Eden D, Crothers DM (October 1981). "The dimensions of DNA in solution". *Journal of Molecular Biology*. **152** (1): 153–61. doi:10.1016/0022-2836(81)90099-1. PMID 7338906.

[13] Gregory SG, Barlow KF, McLay KE, Kaul R, Swarbreck D, Dunham A, et al. (May 2006). "The DNA sequence and biological annotation of human chromosome 1". *Nature*. **441** (7091): 315–21. Bibcode:2006Natur.441..315G. doi:10.1038/nature04727. PMID 16710414.

[14] Watson JD, Crick FH (April 1953). "Molecular structure of nucleic acids; a structure for deoxyribose nucleic acid" (PDF). *Nature*. **171** (4356): 737–8. Bibcode:1953Natur.171..737W. doi:10.1038/171737a0. PMID 13054692.

[15] Berg J., Tymoczko J. and Stryer L. (2002) *Biochemistry*. W. H. Freeman and Company ISBN 0-7167-4955-6

[16] Abbreviations and Symbols for Nucleic Acids, Polynucleotides and their Constituents IUPAC-IUB Commission on Biochemical Nomenclature (CBN). Retrieved 3 January 2006.

[17] Ghosh A, Bansal M (April 2003). "A glossary of DNA structures from A to Z". *Acta Crystallographica. Section D, Biological Crystallography*. **59** (Pt 4): 620–6. doi:10.1107/S0907444903003251. PMID 12657780.

[18] Created from PDB 1D65

[19] Yakovchuk P, Protozanova E, Frank-Kamenetskii MD (2006). "Base-stacking and base-pairing contributions into thermal stability of the DNA double helix". *Nucleic Acids Research*. **34** (2): 564–74. doi:10.1093/nar/gkj454. PMC 1360284. PMID 16449200.

[20] Burton E. Tropp - *"Molecular Biology"*- Jones and Barlett Learning. ISBN 978-0-7637-8663-2

[21] "Watson-Crick Structure of DNA - 1953". *Steven Carr*. Memorial University of Newfoundland. Retrieved 13 July 2016.

[22] Verma S, Eckstein F (1998). "Modified oligonucleotides: synthesis and strategy for users". Annual Review of Biochemistry. 67: 99–134. doi:10.1146/annurev.biochem.67.1.99. PMID 9759484.

[23] Kiljunen S, Hakala K, Pinta E, Huttunen S, Pluta P, Gador A, Lönnberg H, Skurnik M (December 2005). "Yersiniophage phiR1-37 is a tailed bacteriophage having a 270 kb DNA genome with thymidine replaced by deoxyuridine". Microbiology. 151 (Pt 12): 4093–102. doi:10.1099/mic.0.28265-0. PMID 16339954.

[24] Uchiyama J, Takemura-Uchiyama I, Sakaguchi Y, Gamoh K, Kato S, Daibata M, Ujihara T, Misawa N, Matsuzaki S (September 2014). "Intragenus generalized transduction in Staphylococcus spp. by a novel giant phage". The ISME Journal. 8 (9): 1949–52. doi:10.1038/ismej.2014.29. PMC 4139722. PMID 24599069.

[25] Simpson L (March 1998). "A base called J". Proceedings of the National Academy of Sciences of the United States of America. 95 (5): 2037–8. Bibcode:1998PNAS...95.2037S. doi:10.1073/pnas.95.5.2037. PMC 33841. PMID 9482833.

[26] Borst P, Sabatini R (2008). "Base J: discovery, biosynthesis, and possible functions". Annual Review of Microbiology. 62: 235–51. doi:10.1146/annurev.micro.62.081307.162750. PMID 18729733.

[27] Cross M, Kieft R, Sabatini R, Wilm M, de Kort M, van der Marel GA, van Boom JH, van Leeuwen F, Borst P (November 1999). "The modified base J is the target for a novel DNA-binding protein in kinetoplastid protozoans". The EMBO Journal. 18 (22): 6573–81. doi:10.1093/emboj/18.22.6573. PMC 1171720. PMID 10562569.

[28] DiPaolo C, Kieft R, Cross M, Sabatini R (February 2005). "Regulation of trypanosome DNA glycosylation by a SWI2/SNF2-like protein". Molecular Cell. 17 (3): 441–51. doi:10.1016/j.molcel.2004.12.022. PMID 15694344.

[29] Vainio S, Genest PA, ter Riet B, van Luenen H, Borst P (April 2009). "Evidence that J-binding protein 2 is a thymidine hydroxylase catalyzing the first step in the biosynthesis of DNA base J". Molecular and Biochemical Parasitology. 164 (2): 157–61. doi:10.1016/j.molbiopara.2008.12.001. PMID 19114062.

[30] Iyer LM, Tahiliani M, Rao A, Aravind L (June 2009). "Prediction of novel families of enzymes involved in oxidative and other complex modifications of bases in nucleic acids". Cell Cycle. 8 (11): 1698–710. doi:10.4161/cc.8.11.8580. PMC 2995806. PMID 19411852.

[31] van Luenen HG, Farris C, Jan S, Genest PA, Tripathi P, Velds A, Kerkhoven RM, Nieuwland M, Haydock A, Ramasamy G, Vainio S, Heidebrecht T, Perrakis A, Pagie L, van Steensel B, Myler PJ, Borst P (August 2012). "Glucosylated hydroxymethyluracil, DNA base J, prevents transcriptional readthrough in Leishmania". Cell. 150 (5): 909–21. doi:10.1016/j.cell.2012.07.030. PMC 3684241. PMID 22939620.

[32] Hazelbaker DZ, Buratowski S (November 2012). "Transcription: base J blocks the way". Current Biology. 22 (22): R960–2. doi:10.1016/j.cub.2012.10.010. PMC 3648658. PMID 23174300.

[33] Wing R, Drew H, Takano T, Broka C, Tanaka S, Itakura K, Dickerson RE (October 1980). "Crystal structure analysis of a complete turn of B-DNA". Nature. 287 (5784): 755–8. Bibcode:1980Natur.287..755W. doi:10.1038/287755a0. PMID 7432492.

[34] Pabo CO, Sauer RT (1984). "Protein-DNA recognition". Annual Review of Biochemistry. 53: 293–321. doi:10.1146/annurev.bi.53.070184.001453. PMID 6236744.

[35] Clausen-Schaumann H, Rief M, Tolksdorf C, Gaub HE (April 2000). "Mechanical stability of single DNA molecules". Biophysical Journal. 78 (4): 1997–2007. Bibcode:2000BpJ....78.1997C. doi:10.1016/S0006-3495(00)76747-6. PMC 1300792. PMID 10733978.

[36] Chalikian TV, Völker J, Plum GE, Breslauer KJ (July 1999). "A more unified picture for the thermodynamics of nucleic acid duplex melting: a characterization by calorimetric and volumetric techniques". Proceedings of the National Academy of Sciences of the United States of America. 96 (14): 7853–8. Bibcode:1999PNAS...96.7853C. doi:10.1073/pnas.96.14.7853. PMC 22151. PMID 10393911.

[37] deHaseth PL, Helmann JD (June 1995). "Open complex formation by Escherichia coli RNA polymerase: the mechanism of polymerase-induced strand separation of double helical DNA". Molecular Microbiology. 16 (5): 817–24. doi:10.1111/j.1365-2958.1995.tb02309.x. PMID 7476180.

[38] Isaksson J, Acharya S, Barman J, Cheruku P, Chattopadhyaya J (December 2004). "Single-stranded adenine-rich DNA and RNA retain structural characteristics of their respective double-stranded conformations and show directional differences in stacking pattern". Biochemistry. 43 (51): 15996–6010. doi:10.1021/bi048221v. PMID 15609994.

[39] Designation of the two strands of DNA JCBN/NC-IUB Newsletter 1989. Retrieved 7 May 2008

[40] Hüttenhofer A, Schattner P, Polacek N (May 2005). "Non-coding RNAs: hope or hype?". Trends in Genetics. 21 (5): 289–97. doi:10.1016/j.tig.2005.03.007. PMID 15851066.

[41] Munroe SH (November 2004). "Diversity of antisense regulation in eukaryotes: multiple mechanisms, emerging patterns". *Journal of Cellular Biochemistry.* **93** (4): 664–71. doi:10.1002/jcb.20252. PMID 15389973.

[42] Makalowska I, Lin CF, Makalowski W (February 2005). "Overlapping genes in vertebrate genomes". *Computational Biology and Chemistry.* **29** (1): 1–12. doi:10.1016/j.compbiolchem.2004.12.006. PMID 15680581.

[43] Johnson ZI, Chisholm SW (November 2004). "Properties of overlapping genes are conserved across microbial genomes". *Genome Research.* **14** (11): 2268–72. doi:10.1101/gr.2433104. PMC 525685. PMID 15520290.

[44] Lamb RA, Horvath CM (August 1991). "Diversity of coding strategies in influenza viruses". *Trends in Genetics.* **7** (8): 261–6. doi:10.1016/0168-9525(91)90326-L. PMID 1771674.

[45] Benham CJ, Mielke SP (2005). "DNA mechanics". *Annual Review of Biomedical Engineering.* **7**: 21–53. doi:10.1146/annurev.bioeng.6.062403.132016. PMID 16004565.

[46] Champoux JJ (2001). "DNA topoisomerases: structure, function, and mechanism". *Annual Review of Biochemistry.* **70**: 369–413. doi:10.1146/annurev.biochem.70.1.369. PMID 11395412.

[47] Wang JC (June 2002). "Cellular roles of DNA topoisomerases: a molecular perspective". *Nature Reviews. Molecular Cell Biology.* **3** (6): 430–40. doi:10.1038/nrm831. PMID 12042765.

[48] Basu HS, Feuerstein BG, Zarling DA, Shafer RH, Marton LJ (October 1988). "Recognition of Z-RNA and Z-DNA determinants by polyamines in solution: experimental and theoretical studies". *Journal of Biomolecular Structure & Dynamics.* **6** (2): 299–309. doi:10.1080/07391102.1988.10507714. PMID 2482766.

[49] Franklin RE, Gosling RG (6 March 1953). "The Structure of Sodium Thymonucleate Fibres I. The Influence of Water Content" (PDF). *Acta Crystallogr.* **6** (8–9): 673–7. doi:10.1107/S0365110X53001939.
Franklin RE, Gosling RG (1953). "The structure of sodium thymonucleate fibres. II. The cylindrically symmetrical Patterson function". *Acta Crystallogr.* **6** (8–9): 678–85. doi:10.1107/S0365110X53001940.

[50] Franklin RE, Gosling RG (April 1953). "Molecular configuration in sodium thymonucleate" (PDF). *Nature.* **171** (4356): 740–1. Bibcode:1953Natur.171..740F. doi:10.1038/171740a0. PMID 13054694.

[51] Wilkins MH, Stokes AR, Wilson HR (April 1953). "Molecular structure of deoxypentose nucleic acids" (PDF). *Nature.* **171** (4356): 738–40. Bibcode:1953Natur.171..738W. doi:10.1038/171738a0. PMID 13054693.

[52] Leslie AG, Arnott S, Chandrasekaran R, Ratliff RL (October 1980). "Polymorphism of DNA double helices". *Journal of Molecular Biology.* **143** (1): 49–72. doi:10.1016/0022-2836(80)90124-2. PMID 7441761.

[53] Baianu, I.C. (1980). "Structural Order and Partial Disorder in Biological systems". *Bull. Math. Biol.* **42** (4): 137–141. doi:10.1007/BF02462372. http://cogprints.org/3822/

[54] Hosemann R., Bagchi R.N., *Direct analysis of diffraction by matter,* North-Holland Publs., Amsterdam – New York, 1962.

[55] Baianu, I.C. (1978). "X-ray scattering by partially disordered membrane systems". *Acta Crystallogr A.* **34** (5): 751–753. Bibcode:1978AcCrA..34..751B. doi:10.1107/S0567739478001540.

[56] Wahl MC, Sundaralingam M (1997). "Crystal structures of A-DNA duplexes". *Biopolymers.* **44** (1): 45–63. doi:10.1002/(SICI)1097-0282(1997)44:1<45::AID-BIP4>3.0.CO;2-#. PMID 9097733.

[57] Lu XJ, Shakked Z, Olson WK (July 2000). "A-form conformational motifs in ligand-bound DNA structures". *Journal of Molecular Biology.* **300** (4): 819–40. doi:10.1006/jmbi.2000.3690. PMID 10891271.

[58] Rothenburg S, Koch-Nolte F, Haag F (December 2001). "DNA methylation and Z-DNA formation as mediators of quantitative differences in the expression of alleles". *Immunological Reviews.* **184**: 286–98. doi:10.1034/j.1600-065x.2001.1840125.x. PMID 12086319.

[59] Oh DB, Kim YG, Rich A (December 2002). "Z-DNA-binding proteins can act as potent effectors of gene expression in vivo". *Proceedings of the National Academy of Sciences of the United States of America.* **99** (26): 16666–71. Bibcode:2002PNAS...9916666O. doi:10.1073/pnas.262672699. PMC 139201. PMID 12486233.

[60] Palmer J (2 December 2010). "Arsenic-loving bacteria may help in hunt for alien life". *BBC News.* Retrieved 2 December 2010.

[61] Bortman, Henry (2 December 2010). "Arsenic-Eating Bacteria Opens New Possibilities for Alien Life". *Space.com.* Retrieved 2 December 2010.

[62] Katsnelson A (2 December 2010). "Arsenic-eating microbe may redefine chemistry of life". *Nature News.* doi:10.1038/news.2010.645.

[63] Cressey D (3 October 2012). "'Arsenic-life' Bacterium Prefers Phosphorus after all". *Nature News.* doi:10.1038/nature.2012.11520.

[64] Greider CW, Blackburn EH (December 1985). "Identification of a specific telomere terminal transferase activity in Tetrahymena extracts". *Cell.* **43** (2 Pt 1): 405–13. doi:10.1016/0092-8674(85)90170-9. PMID 3907856.

[65] Nugent CI, Lundblad V (April 1998). "The telomerase reverse transcriptase: components and regulation". *Genes & Development*. **12** (8): 1073–85. doi:10.1101/gad.12.8.1073. PMID 9553037.

[66] Wright WE, Tesmer VM, Huffman KE, Levene SD, Shay JW (November 1997). "Normal human chromosomes have long G-rich telomeric overhangs at one end". *Genes & Development*. **11** (21): 2801–9. doi:10.1101/gad.11.21.2801. PMC 316649. PMID 9353250.

[67] Created from

[68] Burge S, Parkinson GN, Hazel P, Todd AK, Neidle S (2006). "Quadruplex DNA: sequence, topology and structure". *Nucleic Acids Research*. **34** (19): 5402–15. doi:10.1093/nar/gkl655. PMC 1636468. PMID 17012276.

[69] Parkinson GN, Lee MP, Neidle S (June 2002). "Crystal structure of parallel quadruplexes from human telomeric DNA". *Nature*. **417** (6891): 876–80. Bibcode:2002Natur.417..876P. doi:10.1038/nature755. PMID 12050675.

[70] Griffith JD, Comeau L, Rosenfield S, Stansel RM, Bianchi A, Moss H, de Lange T (May 1999). "Mammalian telomeres end in a large duplex loop". *Cell*. **97** (4): 503–14. doi:10.1016/S0092-8674(00)80760-6. PMID 10338214.

[71] Seeman NC (November 2005). "DNA enables nanoscale control of the structure of matter". *Quarterly Reviews of Biophysics*. **38** (4): 363–71. doi:10.1017/S0033583505004087. PMC 3478329. PMID 16515737.

[72] Hu Q, Rosenfeld MG (2012). "Epigenetic regulation of human embryonic stem cells". *Frontiers in Genetics*. **3**: 238. doi:10.3389/fgene.2012.00238. PMC 3488762. PMID 23133442.

[73] Klose RJ, Bird AP (February 2006). "Genomic DNA methylation: the mark and its mediators". *Trends in Biochemical Sciences*. **31** (2): 89–97. doi:10.1016/j.tibs.2005.12.008. PMID 16403636.

[74] Bird A (January 2002). "DNA methylation patterns and epigenetic memory". *Genes & Development*. **16** (1): 6–21. doi:10.1101/gad.947102. PMID 11782440.

[75] Walsh CP, Xu GL (2006). "Cytosine methylation and DNA repair". *Current Topics in Microbiology and Immunology*. Current Topics in Microbiology and Immunology. **301**: 283–315. doi:10.1007/3-540-31390-7_11. ISBN 3-540-29114-8. PMID 16570853.

[76] Kriaucionis S, Heintz N (May 2009). "The nuclear DNA base 5-hydroxymethylcytosine is present in Purkinje neurons and the brain". *Science*. **324** (5929): 929–30. Bibcode:2009Sci...324..929K. doi:10.1126/science.1169786. PMC 3263819. PMID 19372393.

[77] Ratel D, Ravanat JL, Berger F, Wion D (March 2006). "N6-methyladenine: the other methylated base of DNA". *BioEssays*. **28** (3): 309–15. doi:10.1002/bies.20342. PMC 2754416. PMID 16479578.

[78] Gommers-Ampt JH, Van Leeuwen F, de Beer AL, Vliegenthart JF, Dizdaroglu M, Kowalak JA, Crain PF, Borst P (December 1993). "beta-D-glucosyl-hydroxymethyluracil: a novel modified base present in the DNA of the parasitic protozoan T. brucei". *Cell*. **75** (6): 1129–36. doi:10.1016/0092-8674(93)90322-H. PMID 8261512.

[79] Created from PDB 1JDG

[80] Douki T, Reynaud-Angelin A, Cadet J, Sage E (August 2003). "Bipyrimidine photoproducts rather than oxidative lesions are the main type of DNA damage involved in the genotoxic effect of solar UVA radiation". *Biochemistry*. **42** (30): 9221–6. doi:10.1021/bi034593c. PMID 12885257.

[81] Cadet J, Delatour T, Douki T, Gasparutto D, Pouget JP, Ravanat JL, Sauvaigo S (March 1999). "Hydroxyl radicals and DNA base damage". *Mutation Research*. **424** (1–2): 9–21. doi:10.1016/S0027-5107(99)00004-4. PMID 10064846.

[82] Beckman KB, Ames BN (August 1997). "Oxidative decay of DNA". *The Journal of Biological Chemistry*. **272** (32): 19633–6. doi:10.1074/jbc.272.32.19633. PMID 9289489.

[83] Valerie K, Povirk LF (September 2003). "Regulation and mechanisms of mammalian double-strand break repair". *Oncogene*. **22** (37): 5792–812. doi:10.1038/sj.onc.1206679. PMID 12947387.

[84] Johnson G (28 December 2010). "Unearthing Prehistoric Tumors, and Debate". *The New York Times*. If we lived long enough, sooner or later we all would get cancer.

[85] Alberts B, Johnson A, Lewis J, et al. (2002). "The Preventable Causes of Cancer". *Molecular biology of the cell* (4th ed.). New York: Garland Science. ISBN 0-8153-4072-9. A certain irreducible background incidence of cancer is to be expected regardless of circumstances: mutations can never be absolutely avoided, because they are an inescapable consequence of fundamental limitations on the accuracy of DNA replication, as discussed in Chapter 5. If a human could live long enough, it is inevitable that at least one of his or her cells would eventually accumulate a set of mutations sufficient for cancer to develop.

[86] Bernstein H, Payne CM, Bernstein C, Garewal H, Dvorak K (2008). "Cancer and aging as consequences of un-repaired DNA damage". In Kimura H, Suzuki A. *New Research on DNA Damage*. New York: Nova Science Publishers. pp. 1–47. ISBN 978-1-60456-581-2.

[87] Hoeijmakers JH (October 2009). "DNA damage, aging, and cancer". *The New England Journal of Medicine*. **361** (15): 1475–85. doi:10.1056/NEJMra0804615. PMID 19812404.

[88] Freitas AA, de Magalhães JP (2011). "A review and appraisal of the DNA damage theory of ageing". *Mutation Research*. **728** (1–2): 12–22. doi:10.1016/j.mrrev.2011.05.001. PMID 21600302.

[89] Ferguson LR, Denny WA (September 1991). "The genetic toxicology of acridines". *Mutation Research*. **258** (2): 123–60. doi:10.1016/0165-1110(91)90006-H. PMID 1881402.

[90] Stephens TD, Bunde CJ, Fillmore BJ (June 2000). "Mechanism of action in thalidomide teratogenesis". *Biochemical Pharmacology*. **59** (12): 1489–99. doi:10.1016/S0006-2952(99)00388-3. PMID 10799645.

[91] Jeffrey AM (1985). "DNA modification by chemical carcinogens". *Pharmacology & Therapeutics*. **28** (2): 237–72. doi:10.1016/0163-7258(85)90013-0. PMID 3936066.

[92] Braña MF, Cacho M, Gradillas A, de Pascual-Teresa B, Ramos A (November 2001). "Intercalators as anticancer drugs". *Current Pharmaceutical Design*. **7** (17): 1745–80. doi:10.2174/1381612013397113. PMID 11562309.

[93] Venter JC, Adams MD, Myers EW, Li PW, Mural RJ, Sutton GG, et al. (February 2001). "The sequence of the human genome". *Science*. **291** (5507): 1304–51. Bibcode:2001Sci...291.1304V. doi:10.1126/science.1058040. PMID 11181995.

[94] Thanbichler M, Wang SC, Shapiro L (October 2005). "The bacterial nucleoid: a highly organized and dynamic structure". *Journal of Cellular Biochemistry*. **96** (3): 506–21. doi:10.1002/jcb.20519. PMID 15988757.

[95] Wolfsberg TG, McEntyre J, Schuler GD (February 2001). "Guide to the draft human genome". *Nature*. **409** (6822): 824–6. Bibcode:2001Natur.409..824W. doi:10.1038/35057000. PMID 11236998.

[96] Gregory TR (January 2005). "The C-value enigma in plants and animals: a review of parallels and an appeal for partnership". *Annals of Botany*. **95** (1): 133–46. doi:10.1093/aob/mci009. PMID 15596463.

[97] Birney E, Stamatoyannopoulos JA, Dutta A, Guigó R, Gingeras TR, Margulies EH, et al. (June 2007). "Identification and analysis of functional elements in 1% of the human genome by the ENCODE pilot project". *Nature*. **447** (7146): 799–816. Bibcode:2007Natur.447..799B. doi:10.1038/nature05874. PMC 2212820. PMID 17571346.

[98] Created from PDB 1MSW

[99] Pidoux AL, Allshire RC (March 2005). "The role of heterochromatin in centromere function". *Philosophical Transactions of the Royal Society of London. Series B, Biological Sciences*. **360** (1455): 569–79. doi:10.1098/rstb.2004.1611. PMC 1569473. PMID 15905142.

[100] Harrison PM, Hegyi H, Balasubramanian S, Luscombe NM, Bertone P, Echols N, Johnson T, Gerstein M (February 2002). "Molecular fossils in the human genome: identification and analysis of the pseudogenes in chromosomes 21 and 22". *Genome Research*. **12** (2): 272–80. doi:10.1101/gr.207102. PMC 155275. PMID 11827946.

[101] Harrison PM, Gerstein M (May 2002). "Studying genomes through the aeons: protein families, pseudogenes and proteome evolution". *Journal of Molecular Biology*. **318** (5): 1155–74. doi:10.1016/S0022-2836(02)00109-2. PMID 12083509.

[102] Albà M (2001). "Replicative DNA polymerases". *Genome Biology*. **2** (1): REVIEWS3002. doi:10.1186/gb-2001-2-1-reviews3002. PMC 150442. PMID 11178285.

[103] Tani K, Nasu M (2010). "Roles of Extracellular DNA in Bacterial Ecosystems". In Kikuchi Y, Rykova EY. *Extracellular Nucleic Acids*. Springer. pp. 25–38. ISBN 978-3-642-12616-1.

[104] Vlassov VV, Laktionov PP, Rykova EY (July 2007). "Extracellular nucleic acids". *BioEssays*. **29** (7): 654–67. doi:10.1002/bies.20604. PMID 17563084.

[105] Finkel SE, Kolter R (November 2001). "DNA as a nutrient: novel role for bacterial competence gene homologs". *Journal of Bacteriology*. **183** (21): 6288–93. doi:10.1128/JB.183.21.6288-6293.2001. PMC 100116. PMID 11591672.

[106] Mulcahy H, Charron-Mazenod L, Lewenza S (November 2008). "Extracellular DNA chelates cations and induces antibiotic resistance in Pseudomonas aeruginosa biofilms". *PLoS Pathogens*. **4** (11): e1000213. doi:10.1371/journal.ppat.1000213. PMC 2581603. PMID 19023416.

[107] Berne C, Kysela DT, Brun YV (August 2010). "A bacterial extracellular DNA inhibits settling of motile progeny cells within a biofilm". *Molecular Microbiology*. **77** (4): 815–29. doi:10.1111/j.1365-2958.2010.07267.x. PMC 2962764. PMID 20598083.

[108] Whitchurch CB, Tolker-Nielsen T, Ragas PC, Mattick JS (February 2002). "Extracellular DNA required for bacterial biofilm formation". *Science*. **295** (5559): 1487. doi:10.1126/science.295.5559.1487. PMID 11859186.

[109] Hu W, Li L, Sharma S, Wang J, McHardy I, Lux R, Yang Z, He X, Gimzewski JK, Li Y, Shi W (2012). "DNA builds and strengthens the extracellular matrix in Myxococcus xanthus biofilms by interacting with exopolysaccharides". *PLoS One*. **7** (12): e51905. doi:10.1371/journal.pone.0051905. PMC 3530553. PMID 23300576.

[110] Hui L, Bianchi DW (February 2013). "Recent advances in the prenatal interrogation of the human fetal genome". *Trends in Genetics*. **29** (2): 84–91.

doi:10.1016/j.tig.2012.10.013. PMC 4378900ⓒ. PMID 23158400.

[111] Sandman K, Pereira SL, Reeve JN (December 1998). "Diversity of prokaryotic chromosomal proteins and the origin of the nucleosome". *Cellular and Molecular Life Sciences.* **54** (12): 1350–64. doi:10.1007/s000180050259. PMID 9893710.

[112] Dame RT (May 2005). "The role of nucleoid-associated proteins in the organization and compaction of bacterial chromatin". *Molecular Microbiology.* **56** (4): 858–70. doi:10.1111/j.1365-2958.2005.04598.x. PMID 15853876.

[113] Luger K, Mäder AW, Richmond RK, Sargent DF, Richmond TJ (September 1997). "Crystal structure of the nucleosome core particle at 2.8 A resolution". *Nature.* **389** (6648): 251–60. Bibcode:1997Natur.389..251L. doi:10.1038/38444. PMID 9305837.

[114] Jenuwein T, Allis CD (August 2001). "Translating the histone code". *Science.* **293** (5532): 1074–80. doi:10.1126/science.1063127. PMID 11498575.

[115] Ito T (2003). "Nucleosome assembly and remodeling". *Current Topics in Microbiology and Immunology.* Current Topics in Microbiology and Immunology. **274**: 1–22. doi:10.1007/978-3-642-55747-7_1. ISBN 978-3-540-44208-0. PMID 12596902.

[116] Thomas JO (August 2001). "HMG1 and 2: architectural DNA-binding proteins". *Biochemical Society Transactions.* **29** (Pt 4): 395–401. doi:10.1042/BST0290395. PMID 11497996.

[117] Grosschedl R, Giese K, Pagel J (March 1994). "HMG domain proteins: architectural elements in the assembly of nucleoprotein structures". *Trends in Genetics.* **10** (3): 94–100. doi:10.1016/0168-9525(94)90232-1. PMID 8178371.

[118] Iftode C, Daniely Y, Borowiec JA (1999). "Replication protein A (RPA): the eukaryotic SSB". *Critical Reviews in Biochemistry and Molecular Biology.* **34** (3): 141–80. doi:10.1080/10409239991209255. PMID 10473346.

[119] Created from PDB 1LMB

[120] Myers LC, Kornberg RD (2000). "Mediator of transcriptional regulation". *Annual Review of Biochemistry.* **69**: 729–49. doi:10.1146/annurev.biochem.69.1.729. PMID 10966474.

[121] Spiegelman BM, Heinrich R (October 2004). "Biological control through regulated transcriptional coactivators". *Cell.* **119** (2): 157–67. doi:10.1016/j.cell.2004.09.037. PMID 15479634.

[122] Li Z, Van Calcar S, Qu C, Cavenee WK, Zhang MQ, Ren B (July 2003). "A global transcriptional regulatory role for c-Myc in Burkitt's lymphoma cells". *Proceedings of the National Academy of Sciences of the United States of America.* **100** (14): 8164–9. Bibcode:2003PNAS..100.8164L. doi:10.1073/pnas.1332764100. PMC 166200ⓒ. PMID 12808131.

[123] Created from PDB 1RVA

[124] Bickle TA, Krüger DH (June 1993). "Biology of DNA restriction". *Microbiological Reviews.* **57** (2): 434–50. PMC 372918ⓒ. PMID 8336674.

[125] Doherty AJ, Suh SW (November 2000). "Structural and mechanistic conservation in DNA ligases". *Nucleic Acids Research.* **28** (21): 4051–8. doi:10.1093/nar/28.21.4051. PMC 113121ⓒ. PMID 11058099.

[126] Schoeffler AJ, Berger JM (December 2005). "Recent advances in understanding structure-function relationships in the type II topoisomerase mechanism". *Biochemical Society Transactions.* **33** (Pt 6): 1465–70. doi:10.1042/BST20051465. PMID 16246147.

[127] Tuteja N, Tuteja R (May 2004). "Unraveling DNA helicases. Motif, structure, mechanism and function". *European Journal of Biochemistry.* **271** (10): 1849–63. doi:10.1111/j.1432-1033.2004.04094.x. PMID 15128295.

[128] Joyce CM, Steitz TA (November 1995). "Polymerase structures and function: variations on a theme?". *Journal of Bacteriology.* **177** (22): 6321–9. doi:10.1128/jb.177.22.6321-6329.1995. PMC 177480ⓒ. PMID 7592405.

[129] Hubscher U, Maga G, Spadari S (2002). "Eukaryotic DNA polymerases". *Annual Review of Biochemistry.* **71**: 133–63. doi:10.1146/annurev.biochem.71.090501.150041. PMID 12045093.

[130] Johnson A, O'Donnell M (2005). "Cellular DNA replicases: components and dynamics at the replication fork". *Annual Review of Biochemistry.* **74**: 283–315. doi:10.1146/annurev.biochem.73.011303.073859. PMID 15952889.

[131] Tarrago-Litvak L, Andréola ML, Nevinsky GA, Sarih-Cottin L, Litvak S (May 1994). "The reverse transcriptase of HIV-1: from enzymology to therapeutic intervention". *FASEB Journal.* **8** (8): 497–503. PMID 7514143.

[132] Martinez E (December 2002). "Multi-protein complexes in eukaryotic gene transcription". *Plant Molecular Biology.* **50** (6): 925–47. doi:10.1023/A:1021258713850. PMID 12516863.

[133] Created from PDB 1M6G

[134] Cremer T, Cremer C (April 2001). "Chromosome territories, nuclear architecture and gene regulation in mammalian cells". *Nature Reviews. Genetics.* **2** (4): 292–301. doi:10.1038/35066075. PMID 11283701.

[135] Pál C, Papp B, Lercher MJ (May 2006). "An integrated view of protein evolution". *Nature Reviews. Genetics.* **7** (5): 337–48. doi:10.1038/nrg1838. PMID 16619049.

[136] O'Driscoll M, Jeggo PA (January 2006). "The role of double-strand break repair - insights from human genetics". *Nature Reviews. Genetics.* **7** (1): 45–54. doi:10.1038/nrg1746. PMID 16369571.

[137] Vispé S, Defais M (October 1997). "Mammalian Rad51 protein: a RecA homologue with pleiotropic functions". *Biochimie.* **79** (9–10): 587–92. doi:10.1016/S0300-9084(97)82007-X. PMID 9466696.

[138] Neale MJ, Keeney S (July 2006). "Clarifying the mechanics of DNA strand exchange in meiotic recombination". *Nature.* **442** (7099): 153–8. Bibcode:2006Natur.442..153N. doi:10.1038/nature04885. PMID 16838012.

[139] Dickman MJ, Ingleston SM, Sedelnikova SE, Rafferty JB, Lloyd RG, Grasby JA, Hornby DP (November 2002). "The RuvABC resolvasome". *European Journal of Biochemistry.* **269** (22): 5492–501. doi:10.1046/j.1432-1033.2002.03250.x. PMID 12423347.

[140] Joyce GF (July 2002). "The antiquity of RNA-based evolution". *Nature.* **418** (6894): 214–21. Bibcode:2002Natur.418..214J. doi:10.1038/418214a. PMID 12110897.

[141] Orgel LE (2004). "Prebiotic chemistry and the origin of the RNA world". *Critical Reviews in Biochemistry and Molecular Biology.* **39** (2): 99–123. doi:10.1080/10409230490460765. PMID 15217990.

[142] Davenport RJ (May 2001). "Ribozymes. Making copies in the RNA world". *Science.* **292** (5520): 1278. doi:10.1126/science.292.5520.1278a. PMID 11360970.

[143] Szathmáry E (April 1992). "What is the optimum size for the genetic alphabet?". *Proceedings of the National Academy of Sciences of the United States of America.* **89** (7): 2614–8. Bibcode:1992PNAS...89.2614S. doi:10.1073/pnas.89.7.2614. PMC 48712. PMID 1372984.

[144] Lindahl T (April 1993). "Instability and decay of the primary structure of DNA". *Nature.* **362** (6422): 709–15. Bibcode:1993Natur.362..709L. doi:10.1038/362709a0. PMID 8469282.

[145] Vreeland RH, Rosenzweig WD, Powers DW (October 2000). "Isolation of a 250 million-year-old halotolerant bacterium from a primary salt crystal". *Nature.* **407** (6806): 897–900. doi:10.1038/35038060. PMID 11057666.

[146] Hebsgaard MB, Phillips MJ, Willerslev E (May 2005). "Geologically ancient DNA: fact or artefact?". *Trends in Microbiology.* **13** (5): 212–20. doi:10.1016/j.tim.2005.03.010. PMID 15866038.

[147] Nickle DC, Learn GH, Rain MW, Mullins JI, Mittler JE (January 2002). "Curiously modern DNA for a "250 million-year-old" bacterium". *Journal of Molecular Evolution.* **54** (1): 134–7. doi:10.1007/s00239-001-0025-x. PMID 11734907.

[148] Callahan MP, Smith KE, Cleaves HJ, Ruzicka J, Stern JC, Glavin DP, House CH, Dworkin JP (August 2011). "Carbonaceous meteorites contain a wide range of extraterrestrial nucleobases". *Proceedings of the National Academy of Sciences of the United States of America.* **108** (34): 13995–8. Bibcode:2011PNAS..10813995C. doi:10.1073/pnas.1106493108. PMC 3161613. PMID 21836052.

[149] Steigerwald J (8 August 2011). "NASA Researchers: DNA Building Blocks Can Be Made in Space". NASA. Retrieved 10 August 2011.

[150] ScienceDaily Staff (9 August 2011). "DNA Building Blocks Can Be Made in Space, NASA Evidence Suggests". ScienceDaily. Retrieved 9 August 2011.

[151] Marlaire R (3 March 2015). "NASA Ames Reproduces the Building Blocks of Life in Laboratory". *NASA.* Retrieved 5 March 2015.

[152] Goff SP, Berg P (December 1976). "Construction of hybrid viruses containing SV40 and lambda phage DNA segments and their propagation in cultured monkey cells". *Cell.* **9** (4 PT 2): 695–705. doi:10.1016/0092-8674(76)90133-1. PMID 189942.

[153] Houdebine LM (2007). "Transgenic animal models in biomedical research". *Methods in Molecular Biology.* **360**: 163–202. doi:10.1385/1-59745-165-7:163. ISBN 1-59745-165-7. PMID 17172731.

[154] Daniell H, Dhingra A (April 2002). "Multigene engineering: dawn of an exciting new era in biotechnology". *Current Opinion in Biotechnology.* **13** (2): 136–41. doi:10.1016/S0958-1669(02)00297-5. PMC 3481857. PMID 11950565.

[155] Job D (November 2002). "Plant biotechnology in agriculture". *Biochimie.* **84** (11): 1105–10. doi:10.1016/S0300-9084(02)00013-5. PMID 12595138.

[156] Collins A, Morton NE (June 1994). "Likelihood ratios for DNA identification". *Proceedings of the National Academy of Sciences of the United States of America.* **91** (13): 6007–11. Bibcode:1994PNAS...91.6007C. doi:10.1073/pnas.91.13.6007. PMC 44126. PMID 8016106.

[157] Weir BS, Triggs CM, Starling L, Stowell LI, Walsh KA, Buckleton J (March 1997). "Interpreting DNA mixtures". *Journal of Forensic Sciences.* **42** (2): 213–22. PMID 9068179.

[158] Jeffreys AJ, Wilson V, Thein SL (1985). "Individual-specific 'fingerprints' of human DNA". *Nature.* **316** (6023): 76–9. Bibcode:1985Natur.316...76J. doi:10.1038/316076a0. PMID 2989708.

[159] Colin Pitchfork — first murder conviction on DNA evidence also clears the prime suspect Forensic Science Service Accessed 23 December 2006

[160] "DNA Identification in Mass Fatality Incidents". National Institute of Justice. September 2006.

[161] "Paternity Blood Tests That Work Early in a Pregnancy" New York Times June 20, 2012

[162] Breaker RR, Joyce GF (December 1994). "A DNA enzyme that cleaves RNA". *Chemistry & Biology*. **1** (4): 223–9. doi:10.1016/1074-5521(94)90014-0. PMID 9383394.

[163] Chandra M, Sachdeva A, Silverman SK (October 2009). "DNA-catalyzed sequence-specific hydrolysis of DNA". *Nature Chemical Biology*. **5** (10): 718–20. doi:10.1038/nchembio.201. PMC 2746877⊘. PMID 19684594.

[164] Carmi N, Shultz LA, Breaker RR (December 1996). "In vitro selection of self-cleaving DNAs". *Chemistry & Biology*. **3** (12): 1039–46. doi:10.1016/S1074-5521(96)90170-2. PMID 9000012.

[165] Torabi SF, Wu P, McGhee CE, Chen L, Hwang K, Zheng N, Cheng J, Lu Y (May 2015). "In vitro selection of a sodium-specific DNAzyme and its application in intracellular sensing". *Proceedings of the National Academy of Sciences of the United States of America*. **112** (19): 5903–8. doi:10.1073/pnas.1420361112. PMC 4434688⊘. PMID 25918425.

[166] Baldi P, Brunak S (2001). *Bioinformatics: The Machine Learning Approach*. MIT Press. ISBN 978-0-262-02506-5. OCLC 45951728.

[167] Gusfield, Dan. *Algorithms on Strings, Trees, and Sequences: Computer Science and Computational Biology*. Cambridge University Press, 15 January 1997. ISBN 978-0-521-58519-4.

[168] Sjölander K (January 2004). "Phylogenomic inference of protein molecular function: advances and challenges". *Bioinformatics*. **20** (2): 170–9. doi:10.1093/bioinformatics/bth021. PMID 14734307.

[169] Mount DM (2004). *Bioinformatics: Sequence and Genome Analysis* (2 ed.). Cold Spring Harbor, NY: Cold Spring Harbor Laboratory Press. ISBN 0-87969-712-1. OCLC 55106399.

[170] Rothemund PW (March 2006). "Folding DNA to create nanoscale shapes and patterns". *Nature*. **440** (7082): 297–302. Bibcode:2006Natur.440..297R. doi:10.1038/nature04586. PMID 16541064.

[171] Andersen ES, Dong M, Nielsen MM, Jahn K, Subramani R, Mamdouh W, Golas MM, Sander B, Stark H, Oliveira CL, Pedersen JS, Birkedal V, Besenbacher F, Gothelf KV, Kjems J (May 2009). "Self-assembly of a nanoscale DNA box with a controllable lid". *Nature*. **459** (7243): 73–6. Bibcode:2009Natur.459...73A. doi:10.1038/nature07971. PMID 19424153.

[172] Ishitsuka Y, Ha T (May 2009). "DNA nanotechnology: a nanomachine goes live". *Nature Nanotechnology*. **4** (5): 281–2. Bibcode:2009NatNa...4..281I. doi:10.1038/nnano.2009.101. PMID 19421208.

[173] Aldaye FA, Palmer AL, Sleiman HF (September 2008). "Assembling materials with DNA as the guide". *Science*. **321** (5897): 1795–9. Bibcode:2008Sci...321.1795A. doi:10.1126/science.1154533. PMID 18818351.

[174] Wray GA (2002). "Dating branches on the tree of life using DNA". *Genome Biology*. **3** (1): REVIEWS0001. doi:10.1046/j.1525-142X.1999.99010.x. PMC 150454⊘. PMID 11806830.

[175] *Lost Tribes of Israel*, Nova, PBS airdate: 22 February 2000. Transcript available from PBS.org. Retrieved 4 March 2006.

[176] Kleiman, Yaakov. "The Cohanim/DNA Connection: The fascinating story of how DNA studies confirm an ancient biblical tradition". *aish.com* (13 January 2000). Retrieved 4 March 2006.

[177] Goldman N, Bertone P, Chen S, Dessimoz C, LeProust EM, Sipos B, Birney E (February 2013). "Towards practical, high-capacity, low-maintenance information storage in synthesized DNA". *Nature*. **494** (7435): 77–80. Bibcode:2013Natur.494...77G. doi:10.1038/nature11875. PMC 3672958⊘. PMID 23354052.

[178] Naik, Gautam (24 January 2013). "Storing Digital Data in DNA". *Wall Street Journal*. Retrieved 24 January 2013.

[179] Comment by Dandekar, T., Lopez, D., Schaack, D. (2013) http://www.nature.com/nature/journal/v494/n7435/abs/nature11875.html#comment-57415

[180] Emerging Technology Final, Dandekar T., Lopez, D., Programmable bacterial membranes with active DNA storage: presentation for the University of Würzburg for the Royal Society for Chemistry, London, 2016-06-29

[181] Patent "Molecular highly integrated data storage via active control DNA", DE102013004584 A1, https://www.google.com/patents/DE102013004584A1"

[182] Miescher, Friedrich (1871) "Ueber die chemische Zusammensetzung der Eiterzellen" (On the chemical composition of pus cells), *Medicinisch-chemische Untersuchungen*. **4** : 441–460. From p. 456: *"Ich habe mich daher später mit meinen Versuchen an die ganzen Kerne gehalten, die Trennung der Körper, die ich einstweilen ohne weiteres Präjudiz als lösliches und unlösliches Nuclein bezeichnen will, einem günstigeren Material überlassend."* (Therefore, in my experiments I subsequently limited myself to the whole nucleus, leaving to a more favorable material the separation of the substances, that for the present, without further prejudice, I will designate as soluble and insoluble nuclear material ("Nuclein").)

[183] Dahm R (January 2008). "Discovering DNA: Friedrich Miescher and the early years of nucleic acid research". *Human Genetics*. **122** (6): 565–81. doi:10.1007/s00439-007-0433-0. PMID 17901982.

[184] See:

- Albrect Kossel (1879) "Ueber Nuclein der Hefe" (On nuclein in yeast) *Zeitschrift für physiologische Chemie*. **3** : 284-291.

- Albrect Kossel (1880) "Ueber Nuclein der Hefe II" (On nuclein in yeast. Part 2) *Zeitschrift für physiologische Chemie*, **4** : 290-295.

- Albrect Kossel (1881) "Ueber die Verbreitung des Hypoxanthins im Thier- und Pflanzenreich" (On the distribution of hypoxanthins in the animal and plant kingdoms) *Zeitschrift für physiologische Chemie*, **5** : 267-271.

- Albrect Kossel, *Untersuchungen über die Nucleine und ihre Spaltungsprodukte* [Investigations into nuclein and its cleavage products] (Strassburg, Germany: K.J. Trübne, 1881), 19 pages.

- Albrect Kossel (1882) "Ueber Xanthin und Hypoxanthin" (On xanthin and hypoxanthin), *Zeitschrift für physiologische Chemie*, **6** : 422-431.

- Albrect Kossel (1883) "Zur Chemie des Zellkerns" (On the chemistry of the cell nucleus), *Zeitschrift für physiologische Chemie*, **7** : 7-22.

- Albrect Kossel (1886) "Weitere Beiträge zur Chemie des Zellkerns" (Further contributions to the chemistry of the cell nucleus), *Zeitschrift für Physiologische Chemie*, **10** : 248-264. Available on-line at: Max Planck Institute for the History of Science, Berlin, Germany. On p. 264, Kossel remarked presciently: *"Der Erforschung der quantitativen Verhältnisse der vier stickstoffreichen Basen, der Abhängigkeit ihrer Menge von den physiologischen Zuständen der Zelle, verspricht wichtige Aufschlüsse über die elementaren physiologisch-chemischen Vorgänge."* (The study of the quantitative relations of the four nitrogenous bases — [and] of the dependence of their quantity on the physiological states of the cell — promises important insights into the fundamental physiological-chemical processes.)

[185] Jones ME (September 1953). "Albrecht Kossel, a biographical sketch". *The Yale Journal of Biology and Medicine*. National Center for Biotechnology Information. **26** (1): 80–97. PMC 2599350. PMID 13103145.

[186] Levene P (1 December 1919). "The structure of yeast nucleic acid". *J Biol Chem*. **40** (2): 415–24.

[187] See:

- W. T. Astbury and Florence O. Bell (1938) "Some recent developments in the X-ray study of proteins and related structures," *Cold Spring Harbor Symposia on Quantitative Biology*, **6** : 109-121. Available on-line at: University of Leeds.

- Astbury, W. T., (1947) "X-ray studies of nucleic acids." *Symposia of the Society for Experimental Biology*, **1** : 66-76. Available on-line at: Oregon State University.

[188] Koltsov proposed that a cell's genetic information was encoded in a long chain of amino acids. See:

- Н. К. Кольцов, "Физико-химические основы морфологии" (The physical-chemical basis of morphology) -- speech given at the 3rd All-Union Meeting of Zoologist, Anatomists, and Histologists at Leningrad, U.S.S.R., December 12, 1927.

- Reprinted in: *Успехи экспериментальной биологии* (Advances in Experimental Biology), series B, 7 (1) : ?-? (1928).

- Reprinted in German as: Nikolaj K. Koltzoff (1928) "Physikalisch-chemische Grundlagen der Morphologie" (The physical-chemical basis of morphology), *Biologisches Zentralblatt*, **48** (6) : 345-369.

- In 1934, Koltsov contended that the proteins that contain a cell's genetic information replicate. See: N. K. Koltzoff (October 5, 1934) "The structure of the chromosomes in the salivary glands of Drosophila," *Science*, **80** (2075) : 312-313. From page 313: "I think that the size of the chromosomes in the salivary glands [of Drosophila] is determined through the multiplication of *genonemes*. By this term I designate the axial thread of the chromosome, in which the geneticists locate the linear combination of genes; ... In the normal chromosome there is usually only one genoneme; before cell-division this genoneme has become divided into two strands."

[189] Soyfer VN (September 2001). "The consequences of political dictatorship for Russian science". *Nature Reviews. Genetics*. **2** (9): 723–9. doi:10.1038/35088598. PMID 11533721.

[190] Griffith F (January 1928). "The Significance of Pneumococcal Types". *The Journal of Hygiene*. **27** (2): 113–59. doi:10.1017/S0022172400031879. PMC 2167760. PMID 20474956.

[191] Lorenz MG, Wackernagel W (September 1994). "Bacterial gene transfer by natural genetic transformation in the environment". *Microbiological Reviews*. **58** (3): 563–602. PMC 372978. PMID 7968924.

[192] Avery OT, Macleod CM, McCarty M (February 1944). "Studies on the Chemical Nature of the Substance Inducing Transformation of Pneumococcal Types: Induction of Transformation by a Desoxyribonucleic Acid Fraction Isolated from Pneumococcus Type III". *The Journal of Experimental Medicine*. **79** (2): 137–58. doi:10.1084/jem.79.2.137. PMC 2135445. PMID 19871359.

[193] Hershey AD, Chase M (May 1952). "Independent functions of viral protein and nucleic acid in growth of bacteriophage". *The Journal of General Physiology*. **36** (1): 39–56. doi:10.1085/jgp.36.1.39. PMC 2147348. PMID 12981234.

[194] The B-DNA X-ray pattern on the right of this linked image was obtained by Rosalind Franklin and Raymond Gosling in May 1952 at high hydration levels of DNA and it has been labeled as "Photo 51"

[195] Nature Archives Double Helix of DNA: 50 Years

[196] "Original X-ray diffraction image". Osulibrary.oregonstate.edu. Retrieved 6 February 2011.

[197] The Nobel Prize in Physiology or Medicine 1962 Nobelprize.org Accessed 22 December 06

[198] Maddox B (January 2003). "The double helix and the 'wronged heroine'" (PDF). *Nature*. **421** (6921): 407–8. Bibcode:2003Natur.421..407M. doi:10.1038/nature01399. PMID 12540909.

[199] Crick, F.H.C. On degenerate templates and the adaptor hypothesis (PDF). genome.wellcome.ac.uk (Lecture, 1955). Retrieved 22 December 2006.

[200] Meselson M, Stahl FW (July 1958). "THE REPLICATION OF DNA IN ESCHERICHIA COLI". *Proceedings of the National Academy of Sciences of the United States of America*. **44** (7): 671–82. Bibcode:1958PNAS...44..671M. doi:10.1073/pnas.44.7.671. PMC 528642⊚. PMID 16590258.

[201] The Nobel Prize in Physiology or Medicine 1968 Nobelprize.org Accessed 22 December 06

## 16.11   Further reading

- Berry A, Watson J (2003). *DNA: the secret of life*. New York: Alfred A. Knopf. ISBN 0-375-41546-7.

- Calladine CR, Drew HR, Luisi BF, Travers AA (2003). *Understanding DNA: the molecule & how it works*. Amsterdam: Elsevier Academic Press. ISBN 0-12-155089-3.

- Carina D, Clayton J (2003). *50 years of DNA*. Basingstoke: Palgrave Macmillan. ISBN 1-4039-1479-6.

- Judson HF (1979). *The Eighth Day of Creation: Makers of the Revolution in Biology* (2nd ed.). Cold Spring Harbor Laboratory Press. ISBN 0-671-22540-5.

- Olby RC (1994). *The path to the double helix: the discovery of DNA*. New York: Dover Publications. ISBN 0-486-68117-3., first published in October 1974 by MacMillan, with foreword by Francis Crick; the definitive DNA textbook, revised in 1994 with a 9-page postscript

- Micklas D (2003). *DNA Science: A First Course*. Cold Spring Harbor Press. ISBN 978-0-87969-636-8.

- Ridley M (2006). *Francis Crick: discoverer of the genetic code*. Ashland, OH: Eminent Lives, Atlas Books. ISBN 0-06-082333-X.

- Olby RC (2009). *Francis Crick: A Biography*. Plainview, N.Y: Cold Spring Harbor Laboratory Press. ISBN 0-87969-798-9.

- Rosenfeld I (2010). *DNA: A Graphic Guide to the Molecule that Shook the World*. Columbia University Press. ISBN 978-0-231-14271-7.

- Schultz M, Cannon Z (2009). *The Stuff of Life: A Graphic Guide to Genetics and DNA*. Hill and Wang. ISBN 0-8090-8947-5.

- Stent GS, Watson J (1980). *The Double Helix: A Personal Account of the Discovery of the Structure of DNA*. New York: Norton. ISBN 0-393-95075-1.

- Watson, James (2004). *DNA: The Secret of Life*. Random House. ISBN 978-0-09-945184-6.

- Wilkins M (2003). *The third man of the double helix the autobiography of Maurice Wilkins*. Cambridge, England: University Press. ISBN 0-19-860665-6.

## 16.12   External links

- DNA at DMOZ

- DNA binding site prediction on protein

- DNA the Double Helix Game From the official Nobel Prize web site

- DNA under electron microscope

- Dolan DNA Learning Center

- Double Helix: 50 years of DNA, *Nature*

- *Proteopedia DNA*

- *Proteopedia Forms_of_DNA*

- ENCODE threads explorer ENCODE home page. Nature (journal)

- Double Helix 1953–2003 National Centre for Biotechnology Education

- Genetic Education Modules for Teachers—*DNA from the Beginning* Study Guide

- PDB Molecule of the Month *DNA*

- Rosalind Franklin's contributions to the study of DNA

- U.S. National DNA Day—watch videos and participate in real-time chat with top scientists

- Clue to chemistry of heredity found The New York Times June 1953. First American newspaper coverage of the discovery of the DNA structure

- Olby R (January 2003). "Quiet debut for the double helix". *Nature*. **421** (6921): 402–5. Bibcode:2003Natur.421..402O. doi:10.1038/nature01397. PMID 12540907.

- DNA from the Beginning Another DNA Learning Center site on DNA, genes, and heredity from Mendel to the human genome project.

- The Register of Francis Crick Personal Papers 1938 – 2007 at Mandeville Special Collections Library, University of California, San Diego

- Seven-page, handwritten letter that Crick sent to his 12-year-old son Michael in 1953 describing the structure of DNA. See Crick's medal goes under the hammer. Nature, 5 April 2013.

- 3D map of DNA reveals hidden loops that allow genes to work together (11 December 2014), *Science (Daily News)*

# Chapter 17

# DNA replication

In molecular biology, **DNA replication** is the biological process of producing two identical replicas of DNA from one original DNA molecule. This process occurs in all living organisms and is the basis for biological inheritance. DNA is made up of a double helix of two complementary strands. During replication, these strands are separated. Each strand of the original DNA molecule then serves as a template for the production of its counterpart, a process referred to as semiconservative replication. Cellular proofreading and error-checking mechanisms ensure near perfect fidelity for DNA replication.[1][2]

In a cell, DNA replication begins at specific locations, or origins of replication, in the genome.[3] Unwinding of DNA at the origin and synthesis of new strands results in replication forks growing bi-directionally from the origin. A number of proteins are associated with the replication fork to help in the initiation and continuation of DNA synthesis. Most prominently, DNA polymerase synthesizes the new strands by adding nucleotides that complement each (template) strand. DNA replication occurs during the S-stage of interphase.

DNA replication can also be performed *in vitro* (artificially, outside a cell). DNA polymerases isolated from cells and artificial DNA primers can be used to initiate DNA synthesis at known sequences in a template DNA molecule. The polymerase chain reaction (PCR), a common laboratory technique, cyclically applies such artificial synthesis to amplify a specific target DNA fragment from a pool of DNA.

## 17.1  DNA structures

DNA usually exists as a double-stranded structure, with both strands coiled together to form the characteristic double-helix. Each single strand of DNA is a chain of four types of nucleotides. Nucleotides in DNA contain a deoxyribose sugar, a phosphate, and a nucleobase. The four types of nucleotide correspond to the four nucleobases

adenine, cytosine, guanine, and thymine, commonly abbreviated as A, C, G and T. Adenine and guanine are purine bases, while cytosine and thymine are pyrimidines. These nucleotides form phosphodiester bonds, creating the phosphate-deoxyribose backbone of the DNA double helix with the nuclei bases pointing inward (i.e., toward the opposing strand). Nucleotides (bases) are matched between strands through hydrogen bonds to form base pairs. Adenine pairs with thymine (two hydrogen bonds), and guanine pairs with cytosine (stronger: three hydrogen bonds).

DNA strands have a directionality, and the different ends of a single strand are called the "3' (three-prime) end" and the "5' (five-prime) end". By convention, if the base sequence of a single strand of DNA is given, the left end of the sequence is the 5' end, while the right end of the sequence is the 3' end. The strands of the double helix are anti-parallel with one being 5' to 3', and the opposite strand 3' to 5'. These terms refer to the carbon atom in deoxyribose to which the next phosphate in the chain attaches. Directionality has consequences in DNA synthesis, because DNA polymerase can synthesize DNA in only one direction by adding nucleotides to the 3' end of a DNA strand.

The pairing of complementary bases in DNA (through hydrogen bonding) means that the information contained within each strand is redundant. Phosphodiester (intra-strand) bonds are stronger than hydrogen (inter-strand) bonds. This allows the strands to be separated from one another. The nucleotides on a single strand can therefore be used to reconstruct nucleotides on a newly synthesized partner strand.[4]

## 17.2  DNA polymerase

Main article: DNA polymerase

DNA polymerases are a family of enzymes that carry out all forms of DNA replication.[6] DNA polymerases in general cannot initiate synthesis of new strands, but can only extend an existing DNA or RNA strand paired with a tem-

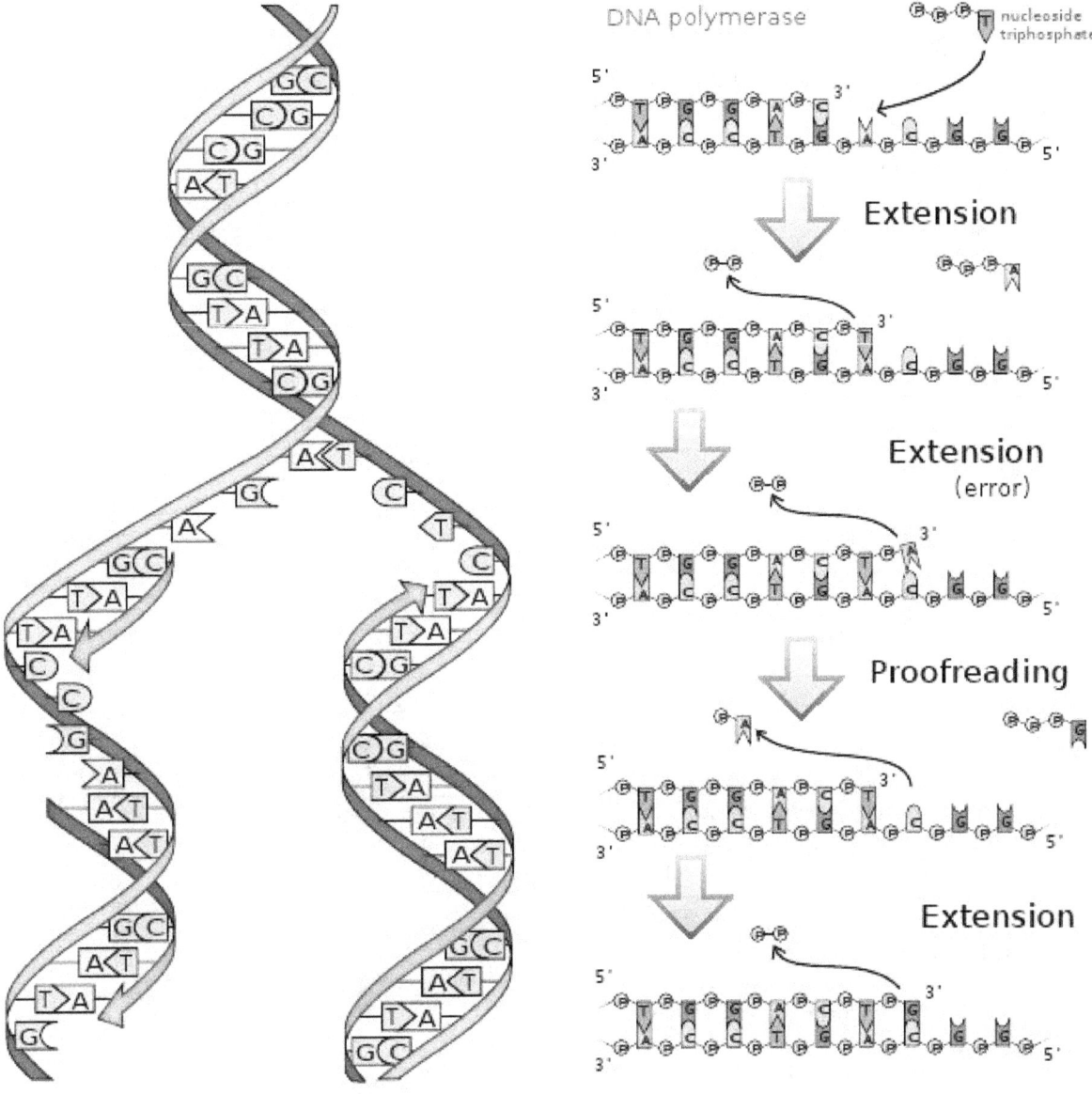

*DNA replication: The double helix is un'zipped' and unwound, then each separated strand (turquoise) acts as a template for replicating a new partner strand (green). Nucleotides (bases) are matched to synthesize the new partner strands into two new double helices.*

*DNA polymerases adds nucleotides to the 3' end of a strand of DNA.[5] If a mismatch is accidentally incorporated, the polymerase is inhibited from further extension. Proofreading removes the mismatched nucleotide and extension continues.*

plate strand. To begin synthesis, a short fragment of RNA, called a primer, must be created and paired with the template DNA strand.

DNA polymerase adds a new strand of DNA by extending the 3' end of an existing nucleotide chain, adding new nucleotides matched to the template strand one at a time via the creation of phosphodiester bonds. The energy for this process of DNA polymerization comes from hydrolysis of the high-energy phosphate (phosphoanhydride) bonds between the three phosphates attached to each unincor-

porated base. Free bases with their attached phosphate groups are called nucleotides; in particular, bases with three attached phosphate groups are called nucleoside triphosphates. When a nucleotide is being added to a growing DNA strand, the formation of a phosphodiester bond between the proximal phosphate of the nucleotide to the growing chain is accompanied by hydrolysis of a high-energy phosphate bond with release of the two distal phosphates as a pyrophosphate. Enzymatic hydrolysis of the resulting pyrophosphate into inorganic phosphate consumes a second high-energy phosphate bond and renders the reaction effectively irreversible.[Note 1]

In general, DNA polymerases are highly accurate, with an intrinsic error rate of less than one mistake for every $10^7$ nucleotides added.[7] In addition, some DNA polymerases also have proofreading ability; they can remove nucleotides from the end of a growing strand in order to correct mismatched bases. Finally, post-replication mismatch repair mechanisms monitor the DNA for errors, being capable of distinguishing mismatches in the newly synthesized DNA strand from the original strand sequence. Together, these three discrimination steps enable replication fidelity of less than one mistake for every $10^9$ nucleotides added.[7]

The rate of DNA replication in a living cell was first measured as the rate of phage T4 DNA elongation in phage-infected E. coli.[8] During the period of exponential DNA increase at 37 °C, the rate was 749 nucleotides per second. The mutation rate per base pair per replication during phage T4 DNA synthesis is 1.7 per $10^8$.[9]

# 17.3 Replication process

Main articles: Prokaryotic DNA replication and Eukaryotic DNA replication

DNA replication, like all biological polymerization processes, proceeds in three enzymatically catalyzed and coordinated steps: initiation, elongation and termination.

## 17.3.1 Initiation

For a cell to divide, it must first replicate its DNA.[10] This process is initiated at particular points in the DNA, known as "origins", which are targeted by initiator proteins.[3] In E. coli this protein is DnaA; in yeast, this is the origin recognition complex.[11] Sequences used by initiator proteins tend to be "AT-rich" (rich in adenine and thymine bases), because A-T base pairs have two hydrogen bonds (rather than the three formed in a C-G pair) and thus are easier to strand separate.[12] Once the origin has been located, these initiators recruit other proteins and form the pre-replication complex, which unzips the double-stranded DNA.

## 17.3.2 Elongation

DNA polymerase has 5'−3' activity. All known DNA replication systems require a free 3' hydroxyl group before synthesis can be initiated (note: the DNA template is read in 3' to 5' direction whereas a new strand is synthesized in the 5' to 3' direction—this is often confused). Four distinct mechanisms for DNA synthesis are recognized:

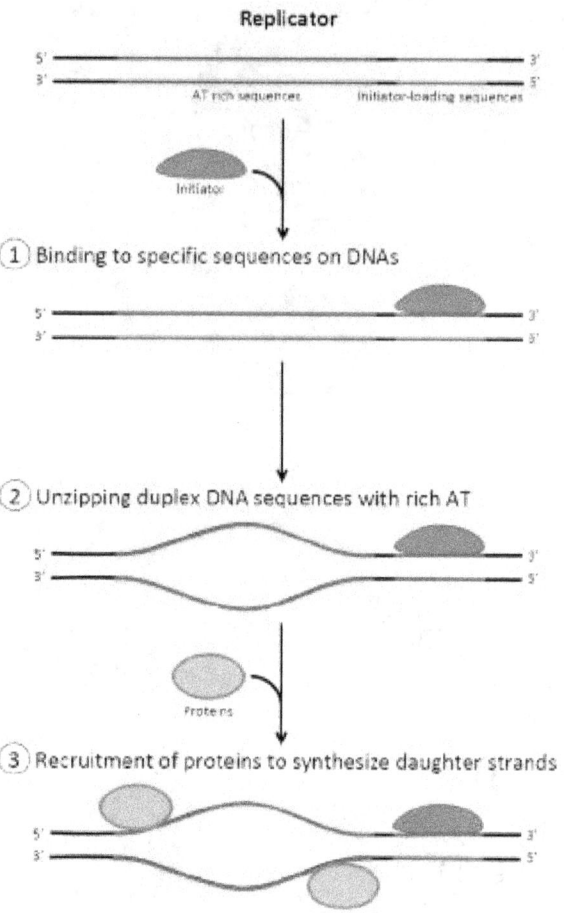

*Role of initiators for initiation of DNA replication.*

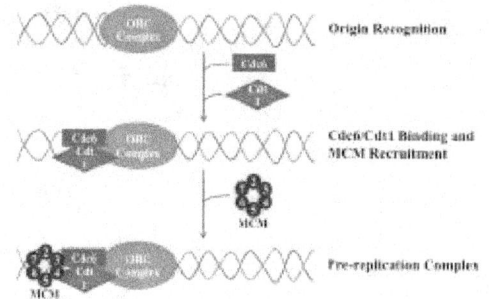

*Formation of pre-replication complex.*

1. All cellular life forms and many DNA viruses, phages and plasmids use a primase to synthesize a short RNA primer with a free 3' OH group which is subsequently elongated by a DNA polymerase.

2. The retroelements (including retroviruses) employ a transfer RNA that primes DNA replication by providing a free 3' OH that is used for elongation by the reverse transcriptase.

3. In the adenoviruses and the φ29 family of bacteriophages, the 3' OH group is provided by the side chain of an amino acid of the genome attached protein (the terminal protein) to which nucleotides are added by the DNA polymerase to form a new strand.

4. In the single stranded DNA viruses — a group that includes the circoviruses, the geminiviruses, the parvoviruses and others — and also the many phages and plasmids that use the rolling circle replication (RCR) mechanism, the RCR endonuclease creates a nick in the genome strand (single stranded viruses) or one of the DNA strands (plasmids). The 5' end of the nicked strand is transferred to a tyrosine residue on the nuclease and the free 3' OH group is then used by the DNA polymerase to synthesize the new strand.

The first is the best known of these mechanisms and is used by the cellular organisms. In this mechanism, once the two strands are separated, primase adds RNA primers to the template strands. The leading strand receives one RNA primer while the lagging strand receives several. The leading strand is continuously extended from the primer by a DNA polymerase with high processivity, while the lagging strand is extended discontinuously from each primer forming Okazaki fragments. RNase removes the primer RNA fragments, and a low processivity DNA polymerase distinct from the replicative polymerase enters to fill the gaps. When this is complete, a single nick on the leading strand and several nicks on the lagging strand can be found. Ligase works to fill these nicks in, thus completing the newly replicated DNA molecule.

The primase used in this process differs significantly between bacteria and archaea/eukaryotes. Bacteria use a primase belonging to the DnaG protein superfamily which contains a catalytic domain of the TOPRIM fold type.[13] The TOPRIM fold contains an α/β core with four conserved strands in a Rossmann-like topology. This structure is also found in the catalytic domains of topoisomerase Ia, topoisomerase II, the OLD-family nucleases and DNA repair proteins related to the RecR protein.

The primase used by archaea and eukaryotes, in contrast, contains a highly derived version of the RNA recognition motif (RRM). This primase is structurally similar to many viral RNA-dependent RNA polymerases, reverse transcriptases, cyclic nucleotide generating cyclases and DNA polymerases of the A/B/Y families that are involved in DNA replication and repair. In eukaryotic replication, the primase forms a complex with Pol α.[14]

Multiple DNA polymerases take on different roles in the DNA replication process. In E. coli, DNA Pol III is the polymerase enzyme primarily responsible for DNA replica-

tion. It assembles into a replication complex at the replication fork that exhibits extremely high processivity, remaining intact for the entire replication cycle. In contrast, DNA Pol I is the enzyme responsible for replacing RNA primers with DNA. DNA Pol I has a 5' to 3' exonuclease activity in addition to its polymerase activity, and uses its exonuclease activity to degrade the RNA primers ahead of it as it extends the DNA strand behind it, in a process called nick translation. Pol I is much less processive than Pol III because its primary function in DNA replication is to create many short DNA regions rather than a few very long regions.

In eukaryotes, the low-processivity enzyme, Pol α, helps to initiate replication because it forms a complex with primase.[15] In eukaryotes, leading strand synthesis is thought to be conducted by Pol ε; however, this view has recently been challenged, suggesting a role for Pol δ.[16] Primer removal is completed Pol δ[17] while repair of DNA during replication is completed by Pol ε.

As DNA synthesis continues, the original DNA strands continue to unwind on each side of the bubble, forming a replication fork with two prongs. In bacteria, which have a single origin of replication on their circular chromosome, this process creates a "theta structure" (resembling the Greek letter theta: θ). In contrast, eukaryotes have longer linear chromosomes and initiate replication at multiple origins within these.[18]

### 17.3.3 Replication fork

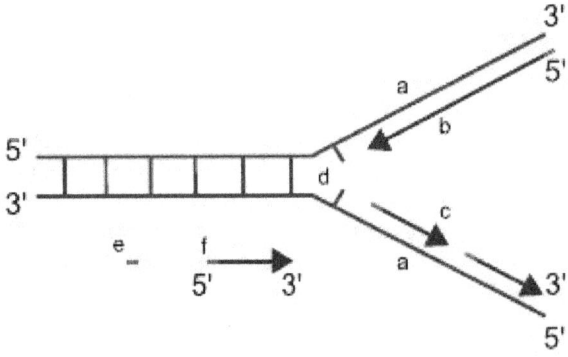

*Scheme of the replication fork.*
*a: template, b: leading strand, c: lagging strand, d: replication*
*fork, e: primer, f: Okazaki fragments*

The replication fork is a structure that forms within the nucleus during DNA replication. It is created by helicases, which break the hydrogen bonds holding the two DNA strands together. The resulting structure has two branching "prongs", each one made up of a single strand of DNA. These two strands serve as the template for the leading and lagging strands, which will be created as DNA polymerase

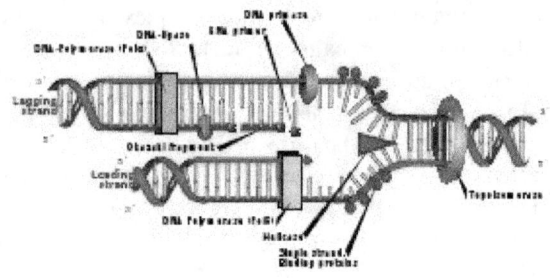

*Many enzymes are involved in the DNA replication fork.*

matches complementary nucleotides to the templates; the templates may be properly referred to as the leading strand template and the lagging strand template.

**DNA is always synthesized in the 5' to 3' direction.** Since the leading and lagging strand templates are oriented in opposite directions at the replication fork, a major issue is how to achieve synthesis of nascent (new) lagging strand DNA, whose direction of synthesis is opposite to the direction of the growing replication fork.

### Leading strand

The leading strand is the strand of nascent DNA which is being synthesized in the same direction as the growing replication fork. A polymerase "reads" the leading strand *template* and adds complementary nucleotides to the nascent leading strand on a continuous basis.

### Lagging strand

The lagging strand is the strand of nascent DNA whose direction of synthesis is opposite to the direction of the growing replication fork. Because of its orientation, replication of the lagging strand is more complicated as compared to that of the leading strand. As a consequence, the DNA polymerase on this strand is seen to "lag behind" the other strand.

The lagging strand is synthesized in short, separated segments. On the lagging strand *template*, a primase "reads" the template DNA and initiates synthesis of a short complementary RNA primer. A DNA polymerase extends the primed segments, forming Okazaki fragments. The RNA primers are then removed and replaced with DNA, and the fragments of DNA are joined together by DNA ligase.

### Dynamics at the replication fork

As helicase unwinds DNA at the replication fork, the DNA ahead is forced to rotate. This process results in a build-

*The assembled human DNA clamp, a trimer of the protein PCNA.*

up of twists in the DNA ahead.[19] This build-up forms a torsional resistance that would eventually halt the progress of the replication fork. Topoisomerases are enzymes that temporarily break the strands of DNA, relieving the tension caused by unwinding the two strands of the DNA helix; topoisomerases (including DNA gyrase) achieve this by adding negative supercoils to the DNA helix.[20]

Bare single-stranded DNA tends to fold back on itself forming secondary structures; these structures can interfere with the movement of DNA polymerase. To prevent this, single-strand binding proteins bind to the DNA until a second strand is synthesized, preventing secondary structure formation.[21]

Clamp proteins form a sliding clamp around DNA, helping the DNA polymerase maintain contact with its template, thereby assisting with processivity. The inner face of the clamp enables DNA to be threaded through it. Once the polymerase reaches the end of the template or detects double-stranded DNA, the sliding clamp undergoes a conformational change that releases the DNA polymerase. Clamp-loading proteins are used to initially load the clamp, recognizing the junction between template and RNA primers.[2]:274-5

### 17.3.4 DNA replication proteins

At the replication fork, many replication enzymes assemble on the DNA into a complex molecular machine called the replisome. The following is a list of major DNA replication enzymes that participate in the replisome:[22]

### 17.3.5  Replication machinery

**Replication machineries** consist of factors involved in DNA replication and appearing on template ssDNAs. Replication machineries include primosotors are replication enzymes: DNA polymerase, DNA helicases, DNA clamps and DNA topoisomerases, and replication proteins; e.g. single-stranded DNA binding proteins (SSB). In the replication machineries these components coordinate. In most of the bacteria, all of the factors involved in DNA replication are located on replication forks and the complexes stay on the forks during DNA replication. These replication machineries are called **replisomes** or **DNA replicase systems**. These terms are generic terms for proteins located on replication forks. In eukaryotic and some bacterial cells the replisomes are not formed.

Since replication machineries do not move relatively to template DNAs such as factories, they are called a **replication factory**.[24] In an alternative figure, DNA factories are similar to projectors and DNAs are like as cinematic films passing constantly into the projectors. In the replication factory model, after both DNA helicases for leading strands and lagging strands are loaded on the template DNAs, the helicases run along the DNAs into each other. The helicases remain associated for the remainder of replication process. Peter Meister et al. observed directly replication sites in budding yeast by monitoring green fluorescent protein(GFP)-tagged DNA polymerases α. They detected DNA replication of pairs of the tagged loci spaced apart symmetrically from a replication origin and found that the distance between the pairs decreased markedly by time.[25] This finding suggests that the mechanism of DNA replication goes with DNA factories. That is, couples of replication factories are loaded on replication origins and the factories associated with each other. Also, template DNAs move into the factories, which bring extrusion of the template ssDNAs and nascent DNAs. Meister's finding is the first direct evidence of replication factory model. Subsequent research has shown that DNA helicases form dimers in many eukaryotic cells and bacterial replication machineries stay in single intranuclear location during DNA synthesis.[24]

The replication factories perform disentanglement of sister chromatids. The disentanglement is essential for distributing the chromatids into daughter cells after DNA replication. Because sister chromatids after DNA replication hold each other by **Cohesin** rings, there is the only chance for the disentanglement in DNA replication. Fixing of replication machineries as replication factories can improve the success rate of DNA replication. If replication forks move freely in chromosomes, catenation of nuclei is aggravated and impedes mitotic segregation.[25]

### 17.3.6  Termination

Eukaryotes initiate DNA replication at multiple points in the chromosome, so replication forks meet and terminate at many points in the chromosome: these are not known to be regulated in any particular way. Because eukaryotes have linear chromosomes, DNA replication is unable to reach the very end of the chromosomes, but ends at the telomere region of repetitive DNA close to the ends. This shortens the telomere of the daughter DNA strand. Shortening of the telomeres is a normal process in somatic cells. As a result, cells can only divide a certain number of times before the DNA loss prevents further division. (This is known as the Hayflick limit.) Within the germ cell line, which passes DNA to the next generation, telomerase extends the repetitive sequences of the telomere region to prevent degradation. Telomerase can become mistakenly active in somatic cells, sometimes leading to cancer formation. Increased telomerase activity is one of the hallmarks of cancer.

Termination requires that the progress of the DNA replication fork must stop or be blocked. Termination at a specific locus, when it occurs, involves the interaction between two components: (1) a termination site sequence in the DNA, and (2) a protein which binds to this sequence to physically stop DNA replication. In various bacterial species, this is named the DNA replication terminus site-binding protein, or Ter protein.

Because bacteria have circular chromosomes, termination of replication occurs when the two replication forks meet each other on the opposite end of the parental chromosome. *E. coli* regulates this process through the use of termination sequences that, when bound by the Tus protein, enable only one direction of replication fork to pass through. As a result, the replication forks are constrained to always meet within the termination region of the chromosome.[26]

## 17.4  Regulation

### 17.4.1  Eukaryotes

Within eukaryotes, DNA replication is controlled within the context of the cell cycle. As the cell grows and divides, it progresses through stages in the cell cycle; DNA replication takes place during the S phase (synthesis phase). The progress of the eukaryotic cell through the cycle is controlled by cell cycle checkpoints. Progression through checkpoints is controlled through complex interactions between various proteins, including cyclins and cyclin-dependent kinases.[27] Unlike bacteria, eukaryotic DNA

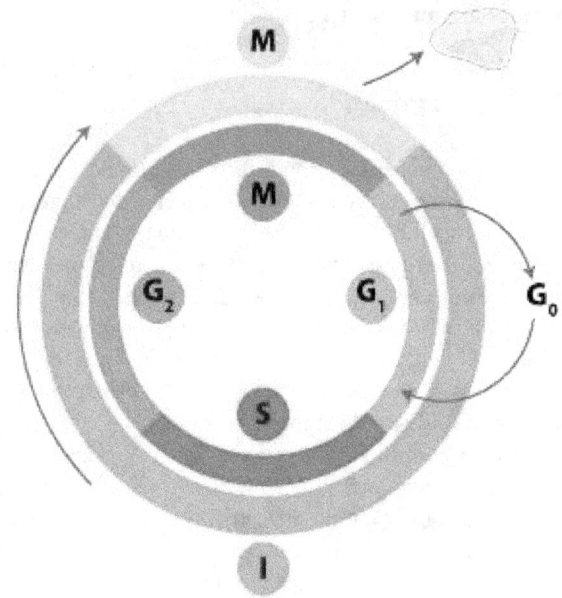

*The cell cycle of eukaryotic cells.*

replicates in the confines of the nucleus.[28]

The G1/S checkpoint (or restriction checkpoint) regulates whether eukaryotic cells enter the process of DNA replication and subsequent division. Cells that do not proceed through this checkpoint remain in the G0 stage and do not replicate their DNA.

Replication of chloroplast and mitochondrial genomes occurs independently of the cell cycle, through the process of D-loop replication.

### Replication focus

In vertebrate cells, replication sites concentrate into positions called **replication foci**.[25] Replication sites can be detected by immunostaining daughter strands and replication enzymes and monitoring GFP-tagged replication factors. By these methods it is found that replication foci of varying size and positions appear in S phase of cell division and their number per nucleus is far smaller than the number of genomic replication forks.

P. Heun et al.(2001) tracked GFP-tagged replication foci in budding yeast cells and revealed that replication origins move constantly in G1 and S phase and the dynamics decreased significantly in S phase.[25] Traditionally, replication sites were fixed on spatial structure of chromosomes by nuclear matrix or lamins. The Heun's results denied the traditional concepts, budding yeasts don't have lamins, and support that replication origins self-assemble and form replication foci.

By firing of replication origins, controlled spatially and temporally, the formation of replication foci is regulated. D. A. Jackson et al.(1998) revealed that neighboring origins fire simultaneously in mammalian cells.[25] Spatial juxtaposition of replication sites brings **clustering** of replication forks. The clustering do **rescue of stalled replication forks** and favors normal progress of replication forks. Progress of replication forks is inhibited by many factors; collision with proteins or with complexes binding strongly on DNA, deficiency of dNTPs, nicks on template DNAs and so on. If replication forks stall and the remaining sequences from the stalled forks are not replicated, the daughter strands have nick obtained un-replicated sites. The un-replicated sites on one parent's strand hold the other strand together but not daughter strands. Therefore, the resulting sister chromatids cannot separate from each other and cannot divide into 2 daughter cells. When neighboring origins fire and a fork from one origin is stalled, fork from other origin access on an opposite direction of the stalled fork and duplicate the un-replicated sites. As other mechanism of the rescue there is application of **dormant replication origins** that excess origins don't fire in normal DNA replication.

### 17.4.2 Bacteria

Most bacteria do not go through a well-defined cell cycle but instead continuously copy their DNA; during rapid growth, this can result in the concurrent occurrence of multiple rounds of replication.[29] In *E. coli*, the best-characterized bacteria, DNA replication is regulated through several mechanisms, including: the hemimethylation and sequestering of the origin sequence, the ratio of adenosine triphosphate (ATP) to adenosine diphosphate (ADP), and the levels of protein DnaA. All these control the binding of initiator proteins to the origin sequences.

Because *E. coli* methylates GATC DNA sequences, DNA synthesis results in hemimethylated sequences. This hemimethylated DNA is recognized by the protein SeqA, which binds and sequesters the origin sequence; in addition, DnaA (required for initiation of replication) binds less well to hemimethylated DNA. As a result, newly replicated origins are prevented from immediately initiating another round of DNA replication.[30]

ATP builds up when the cell is in a rich medium, triggering DNA replication once the cell has reached a specific size. ATP competes with ADP to bind to DnaA, and the DnaA-ATP complex is able to initiate replication. A certain number of DnaA proteins are also required for DNA replication — each time the origin is copied, the number of binding sites for DnaA doubles, requiring the synthesis of more DnaA to enable another initiation of replication.

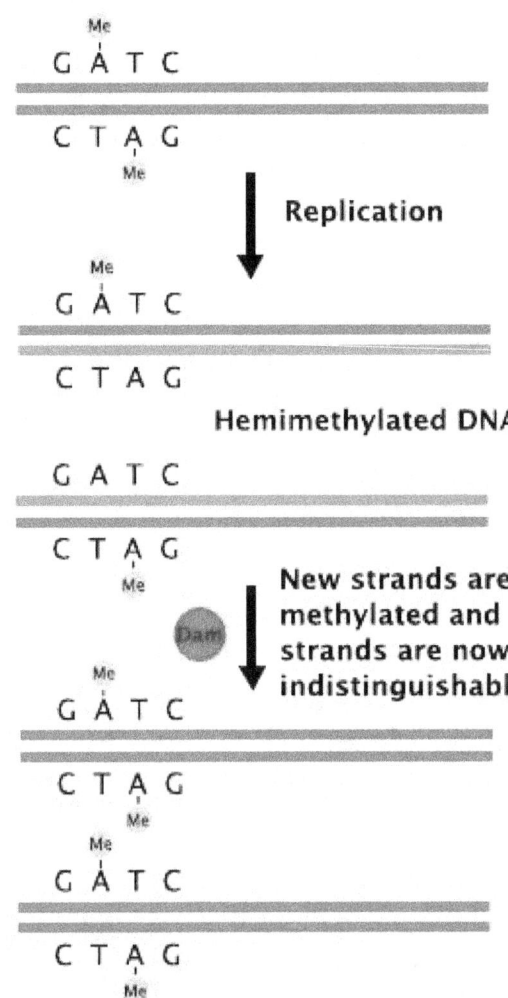

*Dam methylates adenine of GATC sites after replication.*

# 17.5   Polymerase chain reaction

Main article: Polymerase chain reaction

Researchers commonly replicate DNA *in vitro* using the polymerase chain reaction (PCR). PCR uses a pair of primers to span a target region in template DNA, and then polymerizes partner strands in each direction from these primers using a thermostable DNA polymerase. Repeating this process through multiple cycles amplifies the targeted DNA region. At the start of each cycle, the mixture of template and primers is heated, separating the newly synthesized molecule and template. Then, as the mixture cools, both of these become templates for annealing of new primers, and the polymerase extends from these. As a result, the number of copies of the target region doubles each round, increasing exponentially.[31]

# 17.6   Notes

[1] The energetics of this process may also help explain the directionality of synthesis—if DNA were synthesized in the 3' to 5' direction, the energy for the process would come from the 5' end of the growing strand rather than from free nucleotides. The problem is that if the high energy triphosphates were on the growing strand and not on the free nucleotides, proof-reading by removing a mismatched terminal nucleotide would be problematic: Once a nucleotide is added, the triphosphate is lost and a single phosphate remains on the backbone between the new nucleotide and the rest of the strand. If the added nucleotide were mismatched, removal would result in a DNA strand terminated by a monophosphate at the end of the "growing strand" rather than a high energy triphosphate. So strand would be stuck and wouldn't be able to grow anymore. In actuality, the high energy triphosphates hydrolyzed at each step originate from the free nucleotides, not the polymerized strand, so this issue does not exist.

# 17.7   References

[1] Imperfect DNA replication results in mutations. Berg JM, Tymoczko JL, Stryer L, Clarke ND (2002). "Chapter 27: DNA Replication, Recombination, and Repair". *Biochemistry*. W.H. Freeman and Company. ISBN 0-7167-3051-0. External link in |chapter= (help)

[2] Alberts B, Johnson A, Lewis J, Raff M, Roberts K, Walter P (2002). "5DNA Replication, Repair, and Recombination". *Molecular Biology of the Cell*. Garland Science. ISBN 0-8153-3218-1. External link in |chapter= (help)

[3] Berg JM, Tymoczko JL, Stryer L, Clarke ND (2002). "Chapter 27, Section 4: DNA Replication of Both Strands Proceeds Rapidly from Specific Start Sites". *Biochemistry*. W.H. Freeman and Company. ISBN 0-7167-3051-0. External link in |chapter= (help)

[4] Alberts, B., et al., *Molecular Biology of the Cell*, Garland Science, 4th ed., 2002, pp. 238–240 ISBN 0-8153-3218-1

[5] Allison, Lizabeth A. *Fundamental Molecular Biology*. Blackwell Publishing. 2007. p.112 ISBN 978-1-4051-0379-4

[6] Berg JM, Tymoczko JL, Stryer L, Clarke ND (2002). *Biochemistry*. W.H. Freeman and Company. ISBN 0-7167-3051-0. Chapter 27, Section 2: DNA Polymerases Require a Template and a Primer

[7] McCulloch, Scott D; Kunkel, Thomas A (January 2008). "The fidelity of DNA synthesis by eukaryotic replicative and translesion synthesis polymerases". *Cell Research*. **18** (1): 148–161. doi:10.1038/cr.2008.4. PMC 3639319. PMID 18166979.

[8] McCarthy, David; Minner, Charles; Bernstein, Harris; Bernstein, Carol (October 1976). "DNA elongation rates and growing point distributions of wild-type phage T4 and a DNA-delay amber mutant". *Journal of Molecular Biology.* **106** (4): 963–981. doi:10.1016/0022-2836(76)90346-6. PMID 789903.

[9] Drake JW (1970) *The Molecular Basis of Mutation.* Holden-Day, San Francisco ISBN 0816224501 ISBN 978-0816224500

[10] Alberts B, Johnson A, Lewis J, Raff M, Roberts K, Walter P (2002). *Molecular Biology of the Cell.* Garland Science. ISBN 0-8153-3218-1. Chapter 5: DNA Replication Mechanisms

[11] Weigel C, Schmidt A, Rückert B, Lurz R, Messer W (November 1997). "DnaA protein binding to individual DnaA boxes in the Escherichia coli replication origin, oriC". *The EMBO Journal.* **16** (21): 6574–83. doi:10.1093/emboj/16.21.6574. PMC 1170261 . PMID 9351837.

[12] Lodish H, Berk A, Zipursky LS, Matsudaira P, Baltimore D, Darnell J (2000). *Molecular Cell Biology.* W. H. Freeman and Company. ISBN 0-7167-3136-3.12.1. General Features of Chromosomal Replication: Three Common Features of Replication Origins

[13] Aravind, L.; Leipe, D. D.; Koonin, E. V. (1998). "Toprim—a conserved catalytic domain in type IA and II topoisomerases, DnaG-type primases, OLD family nucleases and RecR proteins". *Nucleic Acids Research.* **26** (18): 4205–4213. doi:10.1093/nar/26.18.4205. PMC 147817 . PMID 9722641.

[14] Frick, David; Richardson, Charles (July 2001). "DNA Primases". *Annual Review of Biochemistry.* **70**: 39–80. doi:10.1146/annurev.biochem.70.1.39. PMID 11395402.

[15] Barry, Elizabeth R.; Bell, Stephen D. (8 December 2006). "DNA Replication in the Archaea". *Microbiology and Molecular Biology Reviews.* **70** (4): 876–887. doi:10.1128/MMBR.00029-06. PMC 1698513 . PMID 17158702.

[16] Stillman, Bruce (July 2015). "Reconsidering DNA Polymerases at the Replication Fork in Eukaryotes". *Molecular Cell.* **59** (2): 139–141. doi:10.1016/j.molcel.2015.07.004. PMC 4636199 . PMID 26186286.

[17] Distinguishing the pathways of primer removal during Eukaryotic Okazaki fragment maturation Contributor Author Rossi, Marie Louise. Date Accessioned: 2009-02-23T17:05:09Z. Date Available: 2009-02-23T17:05:09Z. Date Issued: 2009-02-23T17:05:09Z. Identifier Uri: http://hdl.handle.net/1802/6537. Description: Dr. Robert A. Bambara, Faculty Advisor. Thesis (PhD) – School of Medicine and Dentistry, University of Rochester. UR only until January 2010. UR only until January 2010.

[18] Huberman, Joel A.; Riggs, Arthur D. (March 1968). "On the mechanism of DNA replication in mammalian chromosomes". *Journal of Molecular Biology.* **32** (2): 327–341. doi:10.1016/0022-2836(68)90013-2. PMID 5689363.

[19] Alberts B, Johnson A, Lewis J, Raff M, Roberts K, Walter P (2002). *Molecular Biology of the Cell.* Garland Science. ISBN 0-8153-3218-1. DNA Replication Mechanisms: DNA Topoisomerases Prevent DNA Tangling During Replication

[20] Reece, Richard J.; Maxwell, Anthony; Wang, James C. (26 September 2008). "DNA Gyrase: Structure and Function". *Critical Reviews in Biochemistry and Molecular Biology.* **26** (3-4): 335–375. doi:10.3109/10409239109114072. PMID 1657531.

[21] Alberts B, Johnson A, Lewis J, Raff M, Roberts K, Walter P (2002). *Molecular Biology of the Cell.* Garland Science. ISBN 0-8153-3218-1. DNA Replication Mechanisms: Special Proteins Help to Open Up the DNA Double Helix in Front of the Replication Fork

[22] Griffiths A.J.F.; Wessler S.R.; Lewontin R.C.; Carroll S.B. (2008). *Introduction to Genetic Analysis.* W. H. Freeman and Company. ISBN 0-7167-6887-9.[Chapter 7: DNA: Structure and Replication. pg 283–290]

[23] "Will the Hayflick limit keep us from living forever?". *Howstuffworks.* Retrieved January 20, 2015.

[24] James D. Watson et al. (2008), "Molecular Biology of the gene", Pearson Education: 237

[25] Peter Meister, Angela Taddei, Susan M. Gasser(June 2006). "In and out of the Replication Factory", *Cell* **125** (7): 1233–1235

[26] TA Brown (2002). *Genomes.* BIOS Scientific Publishers. ISBN 1-85996-228-9.13.2.3. Termination of replication

[27] Alberts B, Johnson A, Lewis J, Raff M, Roberts K, Walter P (2002). *Molecular Biology of the Cell.* Garland Science. ISBN 0-8153-3218-1. Intracellular Control of Cell-Cycle Events: S-Phase Cyclin-Cdk Complexes (S-Cdks) Initiate DNA Replication Once Per Cycle

[28] Brown, TA (2002). "13". *Genomes* (2nd ed.). Oxford: Wiley-Liss.

[29] Tobiason DM, Seifert HS (2006). "The Obligate Human Pathogen, Neisseria gonorrhoeae, Is Polyploid". *PLoS Biology.* **4** (6): e185. doi:10.1371/journal.pbio.0040185. PMC 1470461 . PMID 16719561.

[30] Slater, Steven; Wold, Sture; Lu, Min; Boye, Erik; Skarstad, Kirsten; Kleckner, Nancy (September 1995). "E. coli SeqA protein binds oriC in two different methyl-modulated reactions appropriate to its roles in DNA replication initiation and origin sequestration". *Cell.* **82** (6): 927–936. doi:10.1016/0092-8674(95)90272-4. PMID 7553853.

[31] Saiki, Randall; Gelfand, David H.; Stoffel, Susanne; Scharf, Stephen J.; Higuchi, Russell; Horn, Glenn T.; Mullis, Kary B.; Erlich, Henry A. (29 January 1988). "Primer-directed enzymatic amplification of DNA with a thermostable DNA polymerase". *Science*. **239** (4839): 487–491. doi:10.1126/science.2448875. PMID 2448875. Retrieved 7 April 2016.

# Chapter 18

# DNA damage theory of aging

The **DNA damage theory of aging** proposes that aging is a consequence of unrepaired accumulation of naturally occurring DNA damages. Damage in this context is a DNA alteration that has an abnormal structure. Although both mitochondrial and nuclear DNA damage can contribute to aging, nuclear DNA is the main subject of this analysis. Nuclear DNA damage can contribute to aging either indirectly (by increasing apoptosis or cellular senescence) or directly (by increasing cell dysfunction).[1][2][3]

In humans and other mammals, DNA damage occurs frequently and DNA repair processes have evolved to compensate. In estimates made for mice, on average approximately 1,500 to 7,000 DNA lesions occur per hour in each mouse cell, or about 36,000 to 160,000 per cell per day.[4] In any cell some DNA damage may remain despite the action of repair processes. The accumulation of unrepaired DNA damage is more prevalent in certain types of cells, particularly in non-replicating or slowly replicating cells, such as cells in the brain, skeletal and cardiac muscle.

## 18.1   DNA damage and mutation

Further information: DNA repair, DNA damage (naturally occurring), and Mutation

To understand the DNA damage theory of aging it is important to distinguish between DNA damage and mutation, the two major types of errors that occur in DNA. Damage and mutation are fundamentally different. DNA damage is any physical abnormality in the DNA, such as single and double strand breaks, 8-hydroxydeoxyguanosine residues and polycyclic aromatic hydrocarbon adducts. DNA damage can be recognized by enzymes, and thus can be correctly repaired using the complementary undamaged sequence in a homologous chromosome if it is available for copying. If a cell retains DNA damage, transcription of a gene can be prevented and thus translation into a protein will also be blocked. Replication may also be blocked and/or the cell

may die. Descriptions of reduced function, characteristic of aging and associated with accumulation of DNA damage, are given later in this article.

In contrast to DNA damage, a mutation is a change in the base sequence of the DNA. A mutation cannot be recognized by enzymes once the base change is present in both DNA strands, and thus a mutation cannot be repaired. At the cellular level, mutations can cause alterations in protein function and regulation. Mutations are replicated when the cell replicates. In a population of cells, mutant cells will increase or decrease in frequency according to the effects of the mutation on the ability of the cell to survive and reproduce. Although distinctly different from each other, DNA damages and mutations are related because DNA damages often cause errors of DNA synthesis during replication or repair and these errors are a major source of mutation.

Given these properties of DNA damage and mutation, it can be seen that DNA damages are a special problem in non-dividing or slowly dividing cells, where unrepaired damages will tend to accumulate over time. On the other hand, in rapidly dividing cells, unrepaired DNA damages that do not kill the cell by blocking replication will tend to cause replication errors and thus mutation. The great majority of mutations that are not neutral in their effect are deleterious to a cell's survival. Thus, in a population of cells comprising a tissue with replicating cells, mutant cells will tend to be lost. However, infrequent mutations that provide a survival advantage will tend to clonally expand at the expense of neighboring cells in the tissue. This advantage to the cell is disadvantageous to the whole organism, because such mutant cells can give rise to cancer. Thus DNA damages in frequently dividing cells, because they give rise to mutations, are a prominent cause of cancer. In contrast, DNA damages in infrequently dividing cells are likely a prominent cause of aging.

The first person to suggest that DNA damage, as distinct from mutation, is the primary cause of aging was Alexander in 1967.[5] By the early 1980s there was significant experimental support for this idea in the literature.[6] By the early

149

1990s experimental support for this idea was substantial, and furthermore it had become increasingly evident that oxidative DNA damage, in particular, is a major cause of aging.[7][8][9][10][11]

In a series of articles from 1970 to 1977, PV Narasimh Acharya, Phd.    (1924–1993) theorized and presented evidence that cells undergo "irreparable DNA damage," whereby DNA crosslinks occur when both normal cellular repair processes fail and cellular apoptosis does not occur. Specifically, Acharya noted that double-strand breaks and a "cross-linkage joining both strands at the same point is irreparable because neither strand can then serve as a template for repair. The cell will die in the next mitosis or in some rare instances, mutate."[12][13][14][15][16]

## 18.2   Age-associated   accumulation of DNA damage and decline in gene expression

Further information: DNA damage (naturally occurring)

In tissues composed of non- or infrequently replicating cells, DNA damage can accumulate with age and lead either to loss of cells, or, in surviving cells, loss of gene expression. Accumulated DNA damage is usually measured directly. Numerous studies of this type have indicated that oxidative damage to DNA is particularly important.[17] The loss of expression of specific genes can be detected at both the mRNA level and protein level.

### 18.2.1   Brain

Further information: Aging brain

The adult brain is composed in large part of terminally differentiated non-dividing neurons. Many of the conspicuous features of aging reflect a decline in neuronal function. Accumulation of DNA damage with age in the mammalian brain has been reported during the period 1971 to 2008 in at least 29 studies.[18] This DNA damage includes the oxidized nucleoside 8-oxo-2'-deoxyguanosine (8-oxo-dG), single- and double-strand breaks, DNA-protein crosslinks and malondialdehyde adducts (reviewed in Bernstein et al.[18]).    Increasing DNA damage with age has been reported in the brains of the mouse, rat, gerbil, rabbit, dog, and human.

Rutten et al.[19] showed that single-strand breaks accumulate in the mouse brain with age.    Young 4-day-old rats have about 3,000 single-strand breaks and 156 double-

strand breaks per neuron, whereas in rats older than 2 years the level of damage increases to about 7,400 single-strand breaks and 600 double-strand breaks per neuron.[20] Sen et al.[21] showed that DNA damages which block the polymerase chain reaction in rat brain accumulate with age. Swain and Rao observed marked increases in several types of DNA damages in aging rat brain, including single-strand breaks, double-strand breaks and modified bases (8-OHdG and uracil).[22] Wolf et al.[23] also showed that the oxidative DNA damage 8-OHdG accumulates in rat brain with age. Similarly, it was shown that as humans age from 48–97 years, 8-OHdG accumulates in the brain.[24]

Lu et al.[25] studied the transcriptional profiles of the human frontal cortex of individuals ranging from 26 to 106 years of age. This led to the identification of a set of genes whose expression was altered after age 40. These genes play central roles in synaptic plasticity, vesicular transport and mitochondrial function.    In the brain, promoters of genes with reduced expression have markedly increased DNA damage.[25] In cultured human neurons, these gene promoters are selectively damaged by oxidative stress. Thus Lu et al.[25] concluded that DNA damage may reduce the expression of selectively vulnerable genes involved in learning, memory and neuronal survival, initiating a program of brain aging that starts early in adult life.

### 18.2.2   Muscle

Further information: Muscle

Muscle strength, and stamina for sustained physical effort, decline in function with age in humans and other species. Skeletal muscle is a tissue composed largely of multinucleated myofibers, elements that arise from the fusion of mononucleated myoblasts. Accumulation of DNA damage with age in mammalian muscle has been reported in at least 18 studies since 1971.[18] We will mention here only two of the more recent studies in rodents plus one in humans. Hamilton et al.[26] reported that the oxidative DNA damage 8-OHdG accumulates in heart and skeletal muscle (as well as in brain, kidney and liver) of both mouse and rat with age. In humans, increases in 8-OHdG with age were reported for skeletal muscle.[27] Catalase is an enzyme that removes hydrogen peroxide, a reactive oxygen species, and thus limits oxidative DNA damage. In mice, when catalase expression is increased specifically in mitochondria, oxidative DNA damage (8-OHdG) in skeletal muscle is decreased and lifespan is increased by about 20%.[28][29] These findings suggest that mitochondria are a significant source of the oxidative damages contributing to aging.

Protein synthesis and protein degradation decline with age in skeletal and heart muscle, as would be expected, since

DNA damage blocks gene transcription. In a recent study Piec et al.[30] found numerous changes in protein expression in rat skeletal muscle with age, including lower levels of several proteins related to myosin and actin. Force is generated in striated muscle by the interactions between myosin thick filaments and actin thin filaments.

### 18.2.3 Liver

Further information: Liver

Liver hepatocytes do not ordinarily divide and appear to be terminally differentiated, but they retain the ability to proliferate when injured. With age, the mass of the liver decreases, blood flow is reduced, metabolism is impaired, and alterations in microcirculation occur. At least 21 studies have reported an increase in DNA damage with age in liver.[18] For instance, Helbock et al.[31] estimated that the steady state level of oxidative DNA base alterations increased from 24,000 per cell in the liver of young rats to 66,000 per cell in the liver of old rats.

### 18.2.4 Kidney

Further information: Kidney

In kidney, changes with age include reduction in both renal blood flow and glomerular filtration rate, and impairment in the ability to concentrate urine and to conserve sodium and water. DNA damages, particularly oxidative DNA damages, increase with age (at least 8 studies).[18] For instance Hashimoto et al.[32] showed that 8-OHdG accumulates in rat kidney DNA with age.

### 18.2.5 Long-lived stem cells

Further information: Stem cell and Stem cell theory of aging

Tissue-specific stem cells produce differentiated cells through a series of increasingly more committed progenitor intermediates. In hematopoiesis (blood cell formation), the process begins with long-term hematopoietic stem cells that self-renew and also produce progeny cells that upon further replication go through a series of stages leading to differentiated cells without self-renewal capacity. In mice, deficiencies in DNA repair appear to limit the capacity of hematopoietic stem cells to proliferate and self-renew with age.[33] Sharpless and Depinho reviewed evidence that hematopoietic stem cells, as well as stem cells in other tissues, undergo intrinsic aging.[34] They speculated that stem cells grow old, in part, as a result of DNA damage. DNA damage may trigger signalling pathways, such as apoptosis, that contribute to depletion of stem cell stocks. This has been observed in several cases of accelerated aging and may occur in normal aging too.[35]

A key aspect of hair loss with age is the aging of the hair follicle.[36] Ordinarily, hair follicle renewal is maintained by the stem cells associated with each follicle. Aging of the hair follicle appears to be due to the DNA damage that accumulates in renewing stem cells during aging.[37]

## 18.3 Mutation theories of aging

Further information: Evolution of ageing

A popular idea, that has failed to gain significant experimental support, is the idea that mutation, as distinct from DNA damage, is the primary cause of aging. As discussed above, mutations tend to arise in frequently replicating cells as a result of errors of DNA synthesis when template DNA is damaged, and can give rise to cancer. However, in mice there is no increase in mutation in the brain with aging.[38][39][40] Mice defective in a gene (Pms2) that ordinarily corrects base mispairs in DNA have about a 100-fold elevated mutation frequency in all tissues, but do not appear to age more rapidly.[41] On the other hand, mice defective in one particular DNA repair pathway show clear premature aging, but do not have elevated mutation.[42]

One variation of the idea that mutation is the basis of aging, that has received much attention, is that mutations specifically in mitochondrial DNA are the cause of aging. Several studies have shown that mutations accumulate in mitochondrial DNA in infrequently replicating cells with age. DNA polymerase gamma is the enzyme that replicates mitochondrial DNA. A mouse mutant with a defect in this DNA polymerase is only able to replicate its mitochondrial DNA inaccurately, so that the mutation rate is 500-fold higher than in normal mice. Yet these mice showed no obvious features of rapidly accelerated aging.[43] The probable explanation for the apparent lack of effect of the additional mutations in mitochondrial DNA is that, within a typical cell, there are large numbers of mitochondria and each mitochondrion can have multiple copies of mitochondrial DNA. Since most mutations are recessive, any particular deleterious mutation would not be expected to have a pronounced effect when many copies of the correct DNA sequence are present in the same and in other mitochondria in the cell. Overall, the observations discussed in this section indicate that mutations are not the primary cause of aging.

## 18.4   Dietary restriction

Further information: Calorie restriction

In rodents, caloric restriction slows aging and extends lifespan. At least 4 studies have shown that caloric restriction reduces 8-OHdG damages in various organs of rodents. One of these studies showed that caloric restriction reduced accumulation of 8-OHdG with age in rat brain, heart and skeletal muscle, and in mouse brain, heart, kidney and liver.[26] More recently, Wolf et al.[23] showed that dietary restriction reduced accumulation of 8-OHdG with age in rat brain, heart, skeletal muscle, and liver. Thus reduction of oxidative DNA damage is associated with a slower rate of aging and increased lifespan.

## 18.5   Inherited defects that cause premature aging

Further information: DNA repair-deficiency disorder

If DNA damage is the underlying cause of aging, it would be expected that humans with inherited defects in the ability to repair DNA damages should age at a faster pace than persons without such a defect. Numerous examples of rare inherited conditions with DNA repair defects are known. Several of these show multiple striking features of premature aging, and others have fewer such features. Perhaps the most striking premature aging conditions are Werner syndrome (mean lifespan 47 years), Huchinson-Gilford Progeria (mean lifespan 13 years), and Cockayne syndrome (mean lifespan 13 years).

Werner syndrome is due to an inherited defect in an enzyme (a helicase and exonuclease) that acts in base excision repair of DNA (e.g. see Harrigan et al.[44]).

Hutchinson-Guilford Progeria is due to a defect in Lamin A protein which forms a scaffolding within the cell nucleus to organize chromatin and is needed for repair of double-strand breaks in DNA.[45] A-type lamins promote genetic stability by maintaining levels of proteins that have key roles in the DNA repair processes of non-homologous end joining and homologous recombination.[46] Mouse cells deficient for maturation of prelamin A show increased DNA damage and chromosome aberrations and are more sensitive to DNA damaging agents.[47]

Cockayne Syndrome is due to a defect in a protein necessary for the repair process, transcription coupled nucleotide excision repair, which can remove damages, particularly oxidative DNA damages, that block transcription.[48]

In addition to these three conditions, several other human syndromes, that also have defective DNA repair, show several features of premature aging. These include ataxia telangiectasia, Nijmegen breakage syndrome, some subgroups of xeroderma pigmentosum, trichothiodystrophy, Fanconi anemia, Bloom syndrome and Rothmund-Thomson syndrome.

In addition to human inherited syndromes, experimental mouse models with genetic defects in DNA repair show features of premature aging and reduced lifespan.(e.g. refs.[49][50][51]) In particular, mutant mice defective in Ku70, or Ku80, or double mutant mice deficient in both Ku70 and Ku80 exhibit early aging.[52] The mean lifespans of the three mutant mouse strains were similar to each other, at about 37 weeks, compared to 108 weeks for the wild-type control. Six specific signs of aging were examined, and the three mutant mice were found to display the same aging signs as the control mice, but at a much earlier age. Cancer incidence was not increased in the mutant mice. Ku70 and Ku80 form the heterodimer Ku protein essential for the non-homologous end joining (NHEJ) pathway of DNA repair, active in repairing DNA double-strand breaks. This suggests an important role of NHEJ in longevity assurance.

## 18.6   Lifespan in different mammalian species

Further information: Maximum life span

Studies comparing DNA repair capacity in different mammalian species have shown that repair capacity correlates with lifespan. The initial study of this type, by Hart and Setlow,[53] showed that the ability of skin fibroblasts of seven mammalian species to perform DNA repair after exposure to a DNA damaging agent correlated with lifespan of the species. The species studied were shrew, mouse, rat, hamster, cow, elephant and human. This initial study stimulated many additional studies involving a wide variety of mammalian species, and the correlation between repair capacity and lifespan generally held up. In one of the more recent studies, Burkle et al.[54] studied the level of a particular enzyme, Poly ADP ribose polymerase, which is involved in repair of single-strand breaks in DNA. They found that the lifespan of 13 mammalian species correlated with the activity of this enzyme.

The DNA repair transcriptomes of the liver of humans, naked mole-rats and mice were compared.[55] The maximum lifespans of humans, naked mole-rat, and mouse are respectively ~120, 30 and 3 years. The longer-lived species, humans and naked mole rats expressed DNA repair genes, including core genes in several DNA repair pathways, at a higher level than did mice. In addition, several DNA repair pathways in humans and naked mole-rats were up-regulated compared with mouse. These findings suggest that increased DNA repair facilitates greater longevity.

## 18.7 Centenarians

Lymphoblastoid cell lines established from blood samples of humans who lived past 100 years (centenarians) have significantly higher activity of the DNA repair protein PARP (poly ADP ribose polymerase) than cell lines from younger (20 to 70 years old) individuals.[56] The lymphocytic cells of centenarians have characteristics typical of cells from young people, both in their capability of priming the mechanism of repair after $H_2O_2$ sublethal oxidative DNA damage and in their PARP capacity.[57]

## 18.8 Menopause

As women age, they experience a decline in reproductive performance leading to menopause. This decline is tied to a decline in the number of ovarian follicles. Although about 1 million oocytes are present at birth in the human ovary, only about 500 (about 0.05%) of these ovulate, and the rest are lost. The decline in ovarian reserve appears to occur at a constantly increasing rate with age,[58] and leads to nearly complete exhaustion of the reserve by about age 51. As ovarian reserve and fertility decline with age, there is also a parallel increase in pregnancy failure and meiotic errors resulting in chromosomally abnormal conceptions.

Titus et al.[59] have proposed an explanation for the decline in ovarian reserve with age. They showed that as women age, double-strand breaks accumulate in the DNA of their primordial follicles. Primordial follicles are immature primary oocytes surrounded by a single layer of granulosa cells. An enzyme system is present in oocytes that normally accurately repairs DNA double-strand breaks. This repair system is referred to as homologous recombinational repair, and it is especially active during meiosis. Titus et al.[59] also showed that expression of four key DNA repair genes that are necessary for homologous recombinational repair (*BRCA1, MRE11, Rad51* and *ATM*) decline in oocytes with age. This age-related decline in ability to repair double-strand damages can account for the accumulation of these damages, which then likely contributes to the decline in ovarian reserve.

Women with an inherited mutation in the DNA repair gene *BRCA1* undergo menopause prematurely,[60] suggesting that naturally occurring DNA damages in oocytes are repaired less efficiently in these women, and this inefficiency leads to early reproductive failure. Genomic data from about 70,000 women were analyzed to identify protein-coding variation associated with age at natural menopause.[61] Pathway analyses identified a major association with DNA damage response genes, particularly those expressed during meiosis and including a common coding variant in the *BRCA1* gene.

## 18.9 Conclusions

Numerous studies have shown that DNA damage accumulates in brain, muscle, liver, kidney, long-lived stem cells, and in oocytes. These accumulated DNA damages are the likely cause of the decline in gene expression and loss of functional capacity observed with increasing age. On the other hand, accumulation of mutations, as distinct from DNA damages, is not a plausible candidate as the primary cause of aging. A calorie-restricted diet in mammals improves lifespan, and this improvement is associated with a decrease in oxidative DNA damage. Several inherited genetic defects in ability to repair DNA damage give rise to premature aging suggesting a causal relationship between DNA damage and aging. In comparisons of different mammalian species that differ in lifespan, DNA repair capacity is found to correlate with lifespan. Also, centenarians tend to have elevated levels of DNA repair. Menopause appears to be a consequence of DNA damage accumulation in oocytes. The principal source of the DNA damages leading to normal aging appears to be reactive oxygen species, produced as byproducts of normal cellular metabolism.

## 18.10 See also

- Aging brain
- Biological immortality
- DNA damage (naturally occurring)
- DNA repair
- Life expectancy
- Longevity
- Maximum life span
- Rejuvenation (aging)
- Senescence
- Telomere

# 18.11    References

[1] Best,BP (2009). "Nuclear DNA damage as a direct cause of aging" (PDF). *Rejuvenation Research*. **12** (3): 199–208. doi:10.1089/rej.2009.0847. PMID 19594328.

[2] Freitas AA, de Magalhàes JP (2011).    "A review and appraisal of the DNA damage theory of aging". *Mutation Research (journal)*. **728** (1-2): 12–22. doi:10.1016/j.mrrev.2011.05.001. PMID 21600302.

[3] Burhans WC1, Weinberger M (2007). "DNA replication stress, genome instability and aging". *Nucleic Acids Research*. **35** (22): 7545–7556. doi:10.1093/nar/gkm1059. PMC 2190710. PMID 18055498.

[4] Vilenchik, MM; Knudson, AG (May 2000). "Inverse radiation dose-rate effects on somatic and germ-line mutations and DNA damage rates". *Proc Natl Acad Sci U S A*. **97** (10): 5381–6. doi:10.1073/pnas.090099497. PMID 10792040.

[5] "The role of DNA lesions in the processes leading to aging in mice.". *Symp Soc Exp Biol*. **21**: 29–50. 1967. PMID 4860956.

[6] Gensler, HL; Bernstein, H (Sep 1981). "DNA damage as the primary cause of aging". *Q Rev Biol*. **56** (3): 279–303. doi:10.1086/412317. PMID 7031747.

[7] Bernstein C, Bernstein H. (1991) Aging, Sex, and DNA Repair. Academic Press, San Diego. ISBN 978-0120928606 partly available at https://books.google.com/books?id=BaXYYUXy71cC&pg=PA3&lpg=PA3&dq=Aging,+Sex,+and+DNA+Repair&source=bl&ots=9E6VrR17tJ&sig=kqUROJfBM6EZZeIrkuEFygsVVpo&hl=en&sa=X&ei=z8BqUpi7D4KQiALC54Ew&ved=0CFUQ6AEwBg#v=onepage&q=Aging%2C%20Sex%2C%20and%20DNA%20Repair&f=false

[8] Ames, BN; Gold, LS (1991). "Endogenous mutagens and the causes of aging and cancer". *Mutation Research/Fundamental and Molecular Mechanisms of Mutagenesis*. **250** (1-2): 3–16. doi:10.1016/0027-5107(91)90157-j. PMID 1944345.

[9] Holmes, GE; Bernstein, C; Bernstein, H (1992). "Oxidative and other DNA damages as the basis of aging: a review". *Mutat Res*. **275** (3-6): 305–315. doi:10.1016/0921-8734(92)90034-M. PMID 1383772.

[10] Rao, KS; Loeb, LA (September 1992). "DNA damage and repair in brain: relationship to aging". *Mutation Research/DNAging*. **275** (3-6): 317–29. doi:10.1016/0921-8734(92)90035-N. PMID 1383773.

[11] Ames, BN; Shigenaga, MK; Hagen, TM (September 1993). "Oxidants, antioxidants, and the degenerative diseases of aging". *Proceedings of the National Academy of Sciences*. **90** (17): 7915–22. doi:10.1073/pnas.90.17.7915. PMC 47258. PMID 8367443.

[12] Acharya PV (1972). "The isolation and partial characterization of age-correlated oligo-deoxyribo-ribonucleotides with covalently linked aspartyl-glutamyl polypeptides". *Johns Hopkins Med. J. Suppl.* (1): 254–60. PMID 5055816.

[13] Acharya, PV; Ashman, SM; Bjorksten, J; The isolation and partial characterization of age-correlated oligo-deoxyribo-ribo nucleo peptides. Finska Kemists Medd. 81 No. 3 (1972) Suomen Kemists. Tied. Chemical Abstacts, Vol 78, No. 19, May 14, 1973. Abs. N. 122001 g.

[14] Acharya, PVN. Isolation and Partial Characterization of Age-Correlated Oligo-nucleotides with Covalently Bound Peptides. 14th Nordic Congress, Umea, Sweden, June 19, 1971.

[15] Acharya, PVN. DNA-damage: The Cause of Aging. Ninth International Congress of Biochemistry: Stockholm. July 1–7, 1973 (Abs.3 m 12).

[16] Acharya, PVN (1977). "Irreparable DNA-damage by Industrial Pollutants in Pre-mature Aging, Chemical Carcinogenesis and Cardiac Hypertrophy: Experiments and Theory". *Israel Journal of Medical Sciences*. **13**: 441.

[17] Sinha, Jitendra Kumar; Ghosh, Shampa; Swain, Umakanta; Giridharan, Nappan Veethil; Raghunath, Manchala (2014). "Increased macromolecular damage due to oxidative stress in the neocortex and hippocampus of WNIN/Ob, a novel rat model of premature aging". *Neuroscience*. **269**: 256–64. doi:10.1016/j.neuroscience.2014.03.040. PMID 24709042.

[18] Bernstein H, Payne CM, Bernstein C, Garewal H, Dvorak K (2008). Cancer and aging as consequences of un-repaired DNA damage. In: New Research on DNA Damages (Editors: Honoka Kimura and Aoi Suzuki) Nova Science Publishers, Inc., New York, Chapter 1, pp. 1–47. open access, but read only https://www.novapublishers.com/catalog/product_info.php?products_id=43247 ISBN 1604565810 ISBN 978-1604565812

[19] Rutten, BP; Schmitz, C; Gerlach, OH; Oyen, HM; de Mesquita, EB; Steinbusch, HW; Korr, H (Jan 2007). "The aging brain:    accumulation of DNA damage or neuron loss?". *Neurobiol Aging*. **28** (1): 91–8. doi:10.1016/j.neurobiolaging.2005.10.019.       PMID 16338029.

[20] Mandavilli BS, Rao KS (1996). "Accumulation of DNA damage in aging neurons occurs through a mechanism other than apoptosis". *J. Neurochem*. **67** (4): 1559–65. doi:10.1046/j.1471-4159.1996.67041559.x. PMID 8858940.

[21] Sen, T; Jana, S; Sreetama, S; Chatterjee, U; Chakrabarti, S (Mar 2007). "Gene-specific oxidative lesions in aged rat brain detected by polymerase chain reaction inhibition assay". *Free Radic Res*. **41** (3): 288–94. doi:10.1080/10715760601083722. PMID 17364957.

[22] Swain, U; Subba Rao, K (Aug 2011). "Study of DNA damage via the comet assay and base excision repair activities in rat brain neurons and astrocytes during aging". *Mech Ageing Dev.* **132** (8-9): 374–81. doi:10.1016/j.mad.2011.04.012. PMID 21600238.

[23] Wolf, FI; Fasanella, S; Tedesco, B; Cavallini, G; Donati, A; Bergamini, E; Cittadini, A (Mar 2005). "Peripheral lymphocyte 8-OHdG levels correlate with age-associated increase of tissue oxidative DNA damage in Sprague-Dawley rats. Protective effects of caloric restriction". *Exp Gerontol.* **40** (3): 181–8. doi:10.1016/j.exger.2004.11.002. PMID 15763395.

[24] Mecocci, P; MacGarvey, U; Kaufman, AE; Koontz, D; Shoffner, JM; Wallace, DC; Beal, MF (Oct 1993). "Oxidative damage to mitochondrial DNA shows marked age-dependent increases in human brain". *Ann Neurol.* **34** (4): 609–16. doi:10.1002/ana.410340416. PMID 8215249.

[25] Lu, T; Pan, Y; Kao, SY; Li, C; Kohane, I; Chan, J; Yankner, BA (Jun 2004). "Gene regulation and DNA damage in the ageing human brain". *Nature.* **429** (6994): 883–91. doi:10.1038/nature02661. PMID 15190254.

[26] Hamilton, ML; Van Remmen, H; Drake, JA; Yang, H; Guo, ZM; Kewitt, K; Walter, CA; Richardson, A (Aug 2001). "Does oxidative damage to DNA increase with age?". *Proc Natl Acad Sci U S A.* **98** (18): 10469–74. doi:10.1073/pnas.171202698. PMC 56984⊙. PMID 11517304.

[27] "Age-dependent increases in oxidative damage to DNA, lipids, and proteins in human skeletal muscle.". *Free Radic Biol Med.* **26** (3-4): 303–8. Feb 1999. doi:10.1016/s0891-5849(98)00208-1. PMID 9895220.

[28] Schriner, SE; Linford, NJ; Martin, GM; Treuting, P; Ogburn, CE; Emond, M; Coskun, PE; Ladiges, W; Wolf, N; Van Remmen, H; Wallace, DC; Rabinovitch, PS (Jun 2005). "Extension of murine life span by overexpression of catalase targeted to mitochondria". *Science.* **308** (5730): 1909–11. doi:10.1126/science.1106653. PMID 15879174.

[29] Linford, NJ; Schriner, SE; Rabinovitch, PS (Mar 2006). "Oxidative damage and aging: spotlight on mitochondria". *Cancer Res.* **66** (5): 2497–9. doi:10.1158/0008-5472.CAN-05-3163. PMID 16510562.

[30] Piec, I; Listrat, A; Alliot, J; Chambon, C; Taylor, RG; Bechet, D (Jul 2005). "Differential proteome analysis of aging in rat skeletal muscle". *FASEB J.* **19** (9): 1143–5. doi:10.1096/fj.04-3084fje. PMID 15831715.

[31] Helbock, HJ; Beckman, KB; Shigenaga, MK; et al. (January 1998). "DNA oxidation matters: the HPLC-electrochemical detection assay of 8-oxo-deoxyguanosine and 8-oxo-guanine". *Proc. Natl. Acad. Sci. U.S.A.* **95** (1): 288–93. doi:10.1073/pnas.95.1.288. PMC 18204⊙. PMID 9419368.

[32] "DNA damage measured by comet assay and 8-OH-dG formation related to blood chemical analyses in aged rats.". *J Toxicol Sci.* **32** (3): 249–59. Aug 2007. doi:10.2131/jts.32.249. PMID 17785942.

[33] Rossi, DJ; Bryder, D; Seita, J; Nussenzweig, A; Hoeijmakers, J; Weissman, IL (Jun 2007). "Deficiencies in DNA damage repair limit the function of haematopoietic stem cells with age". *Nature.* **447** (7145): 725–9. doi:10.1038/nature05862. PMID 17554309.

[34] Sharpless, NE; DePinho, RA (Sep 2007). "How stem cells age and why this makes us grow old". *Nat Rev Mol Cell Biol.* **8** (9): 703–13. doi:10.1038/nrm2241. PMID 17717515.

[35] Freitas AA1, de Magalhàes JP. A review and appraisal of the DNA damage theory of ageing. Mutat Res. 2011 Jul-Oct;728(1-2):12-22. doi: 10.1016/j.mrrev.2011.05.001. PMID 21600302

[36] Lei M, Chuong CM (2016). "STEM CELLS. Aging, alopecia, and stem cells". *Science.* **351** (6273): 559–60. doi:10.1126/science.aaf1635. PMID 26912687.

[37] Matsumura H, Mohri Y, Binh NT, Morinaga H, Fukuda M, Ito M, Kurata S, Hoeijmakers J, Nishimura EK (2016). "Hair follicle aging is driven by transepidermal elimination of stem cells via COL17A1 proteolysis". *Science.* **351** (6273): aad4395. doi:10.1126/science.aad4395. PMID 26912707.

[38] Dollé, ME; Giese, H; Hopkins, CL; Martus, HJ; Hausdorff, JM; Vijg, J (Dec 1997). "Rapid accumulation of genome rearrangements in liver but not in brain of old mice". *Nat Genet.* **17** (4): 431–4. doi:10.1038/ng1297-431. PMID 9398844.

[39] Stuart, GR; Oda, Y; de Boer, JG; Glickman, BW (March 2000). "Mutation frequency and specificity with age in liver, bladder and brain of lacI transgenic mice". *Genetics.* **154** (3): 1291–300. PMC 1460990⊙. PMID 10757770.

[40] Hill, KA; Halangoda, A; Heinmoeller, PW; Gonzalez, K; Chitaphan, C; Longmate, J; Scaringe, WA; Wang, JC; Sommer, SS (Jun 2005). "Tissue-specific time courses of spontaneous mutation frequency and deviations in mutation pattern are observed in middle to late adulthood in Big Blue mice". *Environ Mol Mutagen.* **45** (5): 442–54. doi:10.1002/em.20119. PMID 15690342.

[41] Narayanan, L; Fritzell, JA; Baker, SM; Liskay, RM; Glazer, PM (Apr 1997). "Elevated levels of mutation in multiple tissues of mice deficient in the DNA mismatch repair gene Pms2.". *Proceedings of the National Academy of Sciences.* **94** (7): 3122–7. doi:10.1073/pnas.94.7.3122. PMC 20332⊙. PMID 9096356.

[42] Dollé, ME; Busuttil, RA; Garcia, AM; Wijnhoven, S; van Drunen, E; Niedernhofer, LJ; van der Horst, G; Hoeijmakers, JH; van Steeg, H; Vijg, J (Apr 2006). "Increased genomic instability is not a prerequisite for shortened lifes-

pan in DNA repair deficient mice". *Mutat Res.* **596** (1-2): 22–35. doi:10.1016/j.mrfmmm.2005.11.008. PMID 16472827.

[43] Vermulst, M; Bielas, JH; Kujoth, GC; Ladiges, WC; Rabinovitch, PS; Prolla, TA; Loeb, LA (Apr 2007). "Mitochondrial point mutations do not limit the natural lifespan of mice". *Nat Genet.* **39** (4): 540–3. doi:10.1038/ng1988. PMID 17334366.

[44] Harrigan, JA; Wilson, DM; Prasad, R; Opresko, PL; Beck, G; May, A; Wilson, SH; Bohr, VA (Jan 2006). "The Werner syndrome protein operates in base excision repair and cooperates with DNA polymerase beta". *Nucleic Acids Res.* **34** (2): 745–54. doi:10.1093/nar/gkj475. PMID 16449207.

[45] Liu, Y; Wang, Y; Rusinol, AE; Sinensky, MS; Liu, J; Shell, SM; Zou, Y (Feb 2008). "Involvement of xeroderma pigmentosum group A (XPA) in progeria arising from defective maturation of prelamin A". *FASEB J.* **22** (2): 603–11. doi:10.1096/fj.07-8598com. PMID 17848622.

[46] Redwood AB, Perkins SM, Vanderwaal RP, Feng Z, Biehl KJ, Gonzalez-Suarez I, Morgado-Palacin L, Shi W, Sage J, Roti-Roti JL, Stewart CL, Zhang J, Gonzalo S (2011). "A dual role for A-type lamins in DNA double-strand break repair". *Cell Cycle.* **10** (15): 2549–60. doi:10.4161/cc.10.15.16531. PMC 3180193. PMID 21701264.

[47] Liu B, Wang J, Chan KM, Tjia WM, Deng W, Guan X, Huang JD, Li KM, Chau PY, Chen DJ, Pei D, Pendas AM, Cadiñanos J, López-Otín C, Tse HF, Hutchison C, Chen J, Cao Y, Cheah KS, Tryggvason K, Zhou Z (2005). "Genomic instability in laminopathy-based premature aging". *Nat. Med.* **11** (7): 780–5. doi:10.1038/nm1266. PMID 15980864.

[48] D'Errico, M; Parlanti, E; Teson, M; Degan, P; Lemma, T; Calcagnile, A; Iavarone, I; Jaruga, P; Ropolo, M; Pedrini, AM; Orioli, D; Frosina, G; Zambruno, G; Dizdaroglu, M; Stefanini, M; Dogliotti, E (Jun 2007). "The role of CSA in the response to oxidative DNA damage in human cells". *Oncogene.* **26** (30): 4336–43. doi:10.1038/sj.onc.1210232. PMID 17297471.

[49] Vogel H, Lim DS, Karsenty G, Finegold M, Hasty P (1999). "Deletion of Ku86 causes early onset of senescence in mice". *Proc. Natl. Acad. Sci. U.S.A.* **96** (19): 10770–5. doi:10.1073/pnas.96.19.10770. PMC 17958. PMID 10485901.

[50] Niedernhofer, LJ; Garinis, GA; Raams, A; Lalai, AS; Robinson, AR; Appeldoorn, E; Odijk, H; Oostendorp, R; Ahmad, A; van Leeuwen, W; Theil, AF; Vermeulen, W; van der Horst, GT; Meinecke, P; Kleijer, WJ; Vijg, J; Jaspers, NG; Hoeijmakers, JH (Dec 2006). "A new progeroid syndrome reveals that genotoxic stress suppresses the somatotroph axis". *Nature.* **444** (7122): 1038–43. doi:10.1038/nature05456. PMID 17183314.

[51] Mostoslavsky, R; Chua, KF; Lombard, DB; Pang, WW; Fischer, MR; Gellon, L; Liu, P; Mostoslavsky, G; Franco, S; Murphy, MM; Mills, KD; Patel, P; Hsu, JT; Hong, AL; Ford, E; Cheng, HL; Kennedy, C; Nunez, N; Bronson, R; Frendewey, D; Auerbach, W; Valenzuela, D; Karow, M; Hottiger, MO; Hursting, S; Barrett, JC; Guarente, L; Mulligan, R; Demple, B; Yancopoulos, GD; Alt, FW (Jan 2006). "Genomic instability and aging-like phenotype in the absence of mammalian SIRT6". *Cell.* **124** (2): 315–29. doi:10.1016/j.cell.2005.11.044. PMID 16439206.

[52] Li H, Vogel H, Holcomb VB, Gu Y, Hasty P (2007). "Deletion of Ku70, Ku80, or both causes early aging without substantially increased cancer". *Mol. Cell. Biol.* **27** (23): 8205–14. doi:10.1128/MCB.00785-07. PMC 2169178. PMID 17875923.

[53] "Correlation between deoxyribonucleic acid excision-repair and life-span in a number of mammalian species.". *Proceedings of the National Academy of Sciences.* **71** (6): 2169–73. Jun 1974. doi:10.1073/pnas.71.6.2169. PMID 4526202.

[54] Bürkle, A; Brabeck, C; Diefenbach, J; Beneke, S (May 2005). "The emerging role of poly(ADP-ribose) polymerase-1 in longevity". *Int J Biochem Cell Biol.* **37** (5): 1043–53. doi:10.1016/j.biocel.2004.10.006. PMID 15743677.

[55] MacRae SL, Croken MM, Calder RB, Aliper A, Milholland B, White RR, Zhavoronkov A, Gladyshev VN, Seluanov A, Gorbunova V, Zhang ZD, Vijg J (2015). "DNA repair in species with extreme lifespan differences". *Aging (Albany NY).* **7** (12): 1171–84. doi:10.18632/aging.100866. PMC 4712340. PMID 26729707.

[56] Muiras ML, Müller M, Schächter F, Bürkle A (1998). "Increased poly(ADP-ribose) polymerase activity in lymphoblastoid cell lines from centenarians". *J. Mol. Med.* **76** (5): 346–54. doi:10.1007/s001090050226. PMID 9587069.

[57] Chevanne M, Calia C, Zampieri M, Cecchinelli B, Caldini R, Monti D, Bucci L, Franceschi C, Caiafa P (2007). "Oxidative DNA damage repair and parp 1 and parp 2 expression in Epstein-Barr virus-immortalized B lymphocyte cells from young subjects, old subjects, and centenarians". *Rejuvenation Res.* **10** (2): 191–204. doi:10.1089/rej.2006.0514. PMID 17518695.

[58] Hansen KR, Knowlton NS, Thyer AC, Charleston JS, Soules MR, Klein NA (2008). "A new model of reproductive aging: the decline in ovarian non-growing follicle number from birth to menopause". *Hum. Reprod.* **23** (3): 699–708. doi:10.1093/humrep/dem408. PMID 18192670.

[59] Titus S, Li F, Stobezki R, Akula K, Unsal E, Jeong K, Dickler M, Robson M, Moy F, Goswami S, Oktay K (2013). "Impairment of BRCA1-related DNA double-strand break repair leads to ovarian aging in mice and humans". *Sci Transl Med.* **5** (172): 172ra21. doi:10.1126/scitranslmed.3004925. PMID 23408054.

[60] Rzepka-Górska I, Tarnowski B, Chudecka-Głaz A, Górski B, Zielińska D, Tołoczko-Grabarek A (2006). "Premature menopause in patients with BRCA1 gene mutation". *Breast Cancer Res. Treat.* **100** (1): 59–63. doi:10.1007/s10549-006-9220-1. PMID 16773440.

[61] Day FR, Ruth KS, Thompson DJ, et al. (2015). "Large-scale genomic analyses link reproductive aging to hypothalamic signaling, breast cancer susceptibility and BRCA1-mediated DNA repair". *Nat. Genet.* **47** (11): 1294–303. doi:10.1038/ng.3412. PMC 4661791. PMID 26414677.

# Chapter 19

# DNA damage (naturally occurring)

**DNA damage** is an alteration in the chemical structure of DNA, such as a break in a strand of DNA, a base missing from the backbone of DNA, or a chemically changed base as 8-OHdG. Damage to DNA that occurs naturally can result from metabolic or hydrolytic processes. Metabolism releases compounds that damage DNA including reactive oxygen species, reactive nitrogen species, reactive carbonyl species, lipid peroxidation products and alkylating agents, among others, while hydrolysis cleaves chemical bonds in DNA.[1] Naturally occurring oxidative DNA damages arise at least 10,000 times per cell per day in humans and 50,000 times or more per cell per day in rats,[2] as documented below.

DNA damage is distinctly different from mutation, although both are types of error in DNA. DNA damage is an abnormal chemical structure in DNA, while a mutation is a change in the sequence of standard base pairs.

DNA damage and mutation have different biological consequences. While most DNA damages can undergo DNA repair, such repair is not 100% efficient. Un-repaired DNA damages accumulate in non-replicating cells, such as cells in the brains or muscles of adult mammals and can cause aging.[3][4][5] (Also see DNA damage theory of aging.) In replicating cells, such as cells lining the colon, errors occur upon replication of past damages in the template strand of DNA or during repair of DNA damages. These errors can give rise to mutations or epigenetic alterations.[6] Both of these types of alteration can be replicated and passed on to subsequent cell generations. These alterations can change gene function or regulation of gene expression and possibly contribute to progression to cancer.

## 19.1 A major problem for life

One indication that DNA damages are a major problem for life is that DNA repair processes, to cope with DNA damages, have been found in all cellular organisms in which DNA repair has been investigated. For example, in bacteria, a regulatory network aimed at repairing DNA damages (called the SOS response in *Escherichia coli*) has been found in many bacterial species. *E. coli* RecA, a key enzyme in the SOS response pathway, is the defining member of a ubiquitous class of DNA strand-exchange proteins that are essential for homologous recombination, a pathway that maintains genomic integrity by repairing broken DNA.[7] Genes homologous to *RecA* and to other central genes in the SOS response pathway are found in almost all the bacterial genomes sequenced to date, covering a large number of phyla, suggesting both an ancient origin and a widespread occurrence of recombinational repair of DNA damage.[8] Eukaryotic recombinases that are homologues of RecA are also widespread in eukaryotic organisms. For example, in fission yeast and humans, RecA homologues promote duplex-duplex DNA-strand exchange needed for repair of many types of DNA lesions.[9][10]

Another indication that DNA damages are a major problem for life is that cells make large investments in DNA repair processes. As pointed out by Hoeijmakers,[4] repairing just one double-strand break could require more than 10,000 ATP molecules, as used in signaling the presence of the damage, the generation of repair foci, and the formation (in humans) of the RAD51 nucleofilament (an intermediate in homologous recombinational repair). (RAD51 is a homologue of bacterial RecA.)

## 19.2 Frequencies

The list below shows some frequencies with which new naturally occurring DNA damages arise per day, due to endogenous cellular processes.

- Oxidative damages
    - Humans, per cell per day
        - 10,000[11]
          11,500[12]

2,800[13] specific damages 8-oxoGua, 8-oxodG plus 5-HMUra

2,800[14] specific damages 8-oxoGua, 8-oxodG plus 5-HMUra

- Rats, per cell per day
  - 74,000[12]
    86,000[15]
    100,000[11]

- Mice, per cell per day
  - 34,000[13] specific damages 8-oxoGua, 8-oxodG plus 5-HMUra
    47,000[16] specific damages oxo8dG in mouse liver
    28,000[14] specific damages 8-oxoGua, 8-oxodG, 5-HMUra

- Depurinations
  - Mammalian cells, per cell per day
    - 2,000 to 10,000[17][18]
      9,000[19]
      12,000[20]
      13,920[21]

- Depyrimidinations
  - Mammalian cells, per cell per day
    - 600[20]
      696[21]

- Single-strand breaks
  - Mammalian cells, per cell per day
    - 55,200[21]

- Double-strand breaks
  - Human cells, per cell cycle
    - 10[22]
      50[23]

- O6-methylguanines
  - Mammalian cells, per cell per day
    - 3,120[21]

- Cytosine deamination
  - Mammalian cells, per cell per day
    - 192[21]

Another important endogenous DNA damage is M1dG, short for (3-(2'-deoxy-beta-D-erythro-pentofuranosyl)-pyrimido[1,2-a]-purin-10(3H)-one). The excretion in urine (likely reflecting rate of occurrence) of M1dG may

be as much as 1,000-fold lower than that of 8-oxodG.[24] However, a more important measure may be the steady-state level in DNA, reflecting both rate of occurrence and rate of DNA repair. The steady-state level of M1dG is higher than that of 8-oxodG.[25] This points out that some DNA damages produced at a low rate may be difficult to repair and remain in DNA at a high steady-state level. Both M1dG[26] and 8-oxodG[27] are mutagenic.

## 19.3 Steady-state levels

Steady-state levels of DNA damages represent the balance between formation and repair. More than 100 types of oxidative DNA damage have been characterized, and 8-oxodG constitutes about 5% of the steady state oxidative damages in DNA.[28] Helbock et al.[29] estimated that there were 24,000 steady state oxidative DNA adducts per cell in young rats and 66,000 adducts per cell in old rats. This reflects the accumulation of DNA damage with age. DNA damage accumulation with age is further described in DNA damage theory of aging.

Swenberg et al.[30] measured average amounts of selected steady state endogenous DNA damages in mammalian cells. The seven most common damages they evaluated are shown in Table 1.

Evaluating steady-state damages in specific tissues of the rat, Nakamura and Swenberg[31] indicated that the number of abasic sites varied from about 50,000 per cell in liver, kidney and lung to about 200,000 per cell in the brain.

## 19.4 Consequences

Differentiated somatic cells of adult mammals generally replicate infrequently or not at all. Such cells, including, for example, brain neurons and muscle myocytes, have little or no cell turnover. Non-replicating cells do not generally generate mutations due to DNA damage-induced errors of replication. These non-replicating cells do not commonly give rise to cancer, but they do accumulate DNA damages with time that likely contribute to aging (*see DNA damage theory of aging*). In a non-replicating cell, a single-strand break or other type of damage in the transcribed strand of DNA can block RNA polymerase II catalysed transcription.[32] This would interfere with the synthesis of the protein coded for by the gene in which the blockage occurred.

Brasnjevic et al.[33] summarized the evidence showing that single-strand breaks accumulate with age in the brain (though accumulation differed in different regions of the brain) and that single-strand breaks are the most frequent steady-state DNA damages in the brain. As discussed above, these accumulated single-strand breaks would be expected to block transcription of genes. Consistent with this, as reviewed by Hetman et al.,[134] 182 genes were identified and shown to have reduced transcription in the brains of individuals older than 72 years, compared to transcription in the brains of those less than 43 years old. When 40 particular proteins were evaluated in a muscle of rats, the majority of the proteins showed significant decreases during aging from 18 months (mature rat) to 30 months (aged rat) of age.[135]

Another type of DNA damage, the double strand break, was shown to cause cell death (loss of cells) through apoptosis.[136] This type of DNA damage would not accumulate with age, since once a cell was lost through apoptosis, its double strand damage would be lost with it.

## 19.5   See also

- Ageing

- Aging brain

- AP site

- Direct DNA damage

- DNA

- DNA adduct

- DNA damage theory of aging

- DNA repair

- DNA replication

- Free radical damage to DNA

- Homologous recombination

- Meiosis

- Mutation

- Natural competence

- Origin and function of meiosis

- Reactive oxygen species

## 19.6   References

[1] De Bont R, van Larebeke N. (2004) Endogenous DNA damage in humans: a review of quantitative data. Mutagenesis 19(3):169-185. Review. PMID 15123782

[2] Bernstein C, Prasad AR, Nfonsam V, Bernstein H (2013). "DNA Damage, DNA Repair and Cancer". In Chen C. New Research Directions in DNA Repair. Rijeka, Croatia: In-Tech. doi:10.5772/53919. ISBN 978-953-51-1114-6.

[3] Bernstein H, Payne CM, Bernstein C, Garewal H, Dvorak K (2008). Cancer and aging as consequences of unrepaired DNA damage. In: New Research on DNA Damages (Editors: Honoka Kimura and Aoi Suzuki) Nova Science Publishers, Inc., New York, Chapter 1, pp. 1-47. open access, but read only https://www.novapublishers.com/catalog/product_info.php?products_id=43247 ISBN 978-1604565812

[4] Hoeijmakers JH. (2009) DNA damage, aging, and cancer. N Engl J Med. 361(15):1475-1485. Review. PMID 19812404

[5] Freitas AA, de Magalhães JP. (2011) A review and appraisal of the DNA damage theory of ageing. Mutat Res. 728(1-2):12-22. Review. doi:10.1016/j.mrrev.2011.05.001 PMID 21600302

[6] O'Hagan HM, Mohammad HP, Baylin SB. (2008) Double strand breaks can initiate gene silencing and SIRT1-dependent onset of DNA methylation in an exogenous promoter CpG island. PLoS Genet. 4(8):e1000155. doi:10.1371/journal.pgen.1000155 PMID 18704159

[7] Bell JC, Plank JL, Dombrowski CC, Kowalczykowski SC. (2012) Direct imaging of RecA nucleation and growth on single molecules of SSB-coated ssDNA. Nature 491(7423):274-278. doi:10.1038/nature11598. PMID 23103864

[8] Erill I, Campoy S, Barbé J. (2007) Aeons of distress: an evolutionary perspective on the bacterial SOS response. FEMS Microbiol Rev. 31(6):637-656. Review. doi:10.1111/j.1574-6976.2007.00082.x PMID 17883408

[9] Murayama Y, Kurokawa Y, Mayanagi K, Iwasaki H. (2008) Formation and branch migration of Holliday junctions mediated by eukaryotic recombinases. Nature 451(7181):1018-1021. PMID 18256600

[10] Holthausen JT, Wyman C, Kanaar R. (2010) Regulation of DNA strand exchange in homologous recombination. DNA Repair (Amst) 9(12):1264-1272. PMID 20971042

[11] Ames BN, Shigenaga MK, Hagen TM. (1993) Oxidants, antioxidants, and the degenerative diseases of aging. Proc Natl Acad Sci U S A. 90(17):7915-7922. Review. PMID 8367443

[12] Helbock HJ, Beckman KB, Shigenaga MK, Walter PB, Woodall AA, Yeo HC, Ames BN. (1998) DNA oxidation

matters: the HPLC-electrochemical detection assay of 8-oxo-deoxyguanosine and 8-oxo-guanine. *Proc Natl Acad Sci U S A.* 95(1): 288-293. PMID 9419368

[13] Foksinski M, Rozalski R, Guz J, Ruszkowska B, Sztukowska P, Piwowarski M, Klungland A, Olinski R. (2004) Urinary excretion of DNA repair products correlates with metabolic rates as well as with maximum life spans of different mammalian species. Free Radic Biol Med 37(9) 1449-1454. PMID 15454284

[14] Tudek B, Winczura A, Janik J, Siomek A, Foksinski M, Oliński R. (2010). Involvement of oxidatively damaged DNA and repair in cancer development and aging. Am J Transl Res 2(3):254-284. PMID 20589166

[15] Fraga CG, Shigenaga MK, Park JW, Degan P, Ames BN. Oxidative damage to DNA during aging: 8-hydroxy-2'-deoxyguanosine in rat organ DNA and urine. *Proc Natl Acad Sci U S A* 1990:87(12) 4533-4537. PMID 2352934

[16] Hamilton ML, Guo Z, Fuller CD, Van Remmen H, Ward WF, Austad SN, Troyer DA, Thompson I, Richardson A. (2001). A reliable assessment of 8-oxo-2-deoxyguanosine levels in nuclear and mitochondrial DNA using the sodium iodide method to isolate DNA. *Nucleic Acids Res* 29(10):2117-2126. PMID 11353081

[17] Lindahl T, Nyberg B. (1972) Rate of depurination of native deoxyribonucleic acid. *Biochemistry* 11(19) 3610-3618.doi:10.1038/362709a0 PMID 4626532

[18] Lindahl T. (1993) Instability and decay of the primary structure of DNA. *Nature* 362(6422) 709-715. PMID 8469282

[19] Nakamura J, Walker VE, Upton PB, Chiang SY, Kow YW, Swenberg JA. Highly sensitive apurinic/apyrimidinic site assay can detect spontaneous and chemically induced depurination under physiological conditions. *Cancer Res* 1998:58(2) 222-225. PMID 9443396

[20] Lindahl T. (1977) DNA repair enzymes acting on spontaneous lesions in DNA. In: Nichols WW and Murphy DG (eds.) DNA Repair Processes. Symposia Specialists, Miami p225-240. ISBN 088372099X ISBN 978-0883720998

[21] Tice, R.R., and Setlow, R.B. (1985) DNA repair and replication in aging organisms and cells. In: Finch EE and Schneider EL (eds.) Handbook of the Biology of Aging. Van Nostrand Reinhold, New York. Pages 173-224. ISBN 0442225296 ISBN 978-0442225292

[22] Haber JE. (1999) DNA recombination: the replication connection. *Trends Biochem Sci* 24(7) 271-275. PMID 10390616

[23] Vilenchik MM, Knudson AG. (2003) Endogenous DNA double-strand breaks: production, fidelity of repair, and induction of cancer. *Proc Natl Acad Sci U S A* 100(22) 12871-12876. PMID 14566050

[24] Chan SW, Dedon PC. (2010) The biological and metabolic fates of endogenous DNA damage products. J Nucleic Acids 2010:929047. PMID 21209721

[25] Kadlubar FF, Anderson KE, Häussermann S, Lang NP, Barone GW, Thompson PA, MacLeod SL, Chou MW, Mikhailova M, Plastaras J, Marnett LJ, Nair J, Velic I, Bartsch H. (1998) Comparison of DNA adduct levels associated with oxidative stress in human pancreas. Mutat Res. 405(2):125-33. PMID 9748537

[26] VanderVeen LA, Hashim MF, Shyr Y, Marnett LJ. Induction of frameshift and base pair substitution mutations by the major DNA adduct of the endogenous carcinogen malondialdehyde. (2003) Proc Natl Acad Sci U S A 100(24):14247-14252. PMID 14603032

[27] Tan X, Grollman AP, Shibutani S. (1999) Comparison of the mutagenic properties of 8-oxo-7,8-dihydro-2'-deoxyadenosine and 8-oxo-7,8-dihydro-2'-deoxyguanosine DNA lesions in mammalian cells. *Carcinogenesis* 20(12):2287-2292. PMID 10590221

[28] Hamilton ML, Guo Z, Fuller CD, Van Remmen H, Ward WF, Austad SN, Troyer DA, Thompson I, Richardson A. (2001) A reliable assessment of 8-oxo-2-deoxyguanosine levels in nuclear and mitochondrial DNA using the sodium iodide method to isolate DNA. *Nucleic Acids Res.* 29(10):2117-26. PMID 11353081

[29] Helbock HJ, Beckman KB, Shigenaga MK, Walter PB, Woodall AA, Yeo HC, Ames BN. (1998) DNA oxidation matters: the HPLC-electrochemical detection assay of 8-oxo-deoxyguanosine and 8-oxo-guanine. *Proc Natl Acad Sci U S A* 95(1):288-293. PMID 9419368

[30] Swenberg JA, Lu K, Moeller BC, Gao L, Upton PB, Nakamura J, Starr TB. (2011) Endogenous versus exogenous DNA adducts: their role in carcinogenesis, epidemiology, and risk assessment. Toxicol Sci. 120(Suppl 1):S130-45. PMID 21163908

[31] Nakamura J, Swenberg JA. (1999) Endogenous apurinic/apyrimidinic sites in genomic DNA of mammalian tissues. *Cancer Res.* 59(11):2522-2526. PMID 10363965

[32] Kathe SD, Shen GP, Wallace SS. (2004) Single-stranded breaks in DNA but not oxidative DNA base damages block transcriptional elongation by RNA polymerase II in HeLa cell nuclear extracts. *J Biol Chem.* 279(18):18511-18520. PMID 14978042

[33] Brasnjevic I, Hof PR, Steinbusch HW, Schmitz C. (2008) Accumulation of nuclear DNA damage or neuron loss: molecular basis for a new approach to understanding selective neuronal vulnerability in neurodegenerative diseases. DNA Repair (Amst). 7(7):1087-1097. PMID 18458001

[34] Hetman M, Vashishta A, Rempala G. (2010) Neurotoxic mechanisms of DNA damage: focus on transcriptional inhibition. J Neurochem. 114(6):1537-1549.

doi: 10.1111/j.1471-4159.2010.06859.x.  Review.  PMID 20557419

[35] Piec I, Listrat A, Alliot J, Chambon C, Taylor RG, Bechet D. (2005) Differential proteome analysis of aging in rat skeletal muscle. *FASEB J.* 19(9):1143-5. PMID 15831715

[36] Carnevale J, Palander O, Seifried LA, Dick FA. (2012) DNA damage signals through differentially modified E2F1 molecules to induce apoptosis. *Mol Cell Biol.* 32(5):900-912. PMID 22184068

# Chapter 20

# DNA repair

For the journal, see DNA Repair (journal).

**DNA repair** is a collection of processes by which a cell

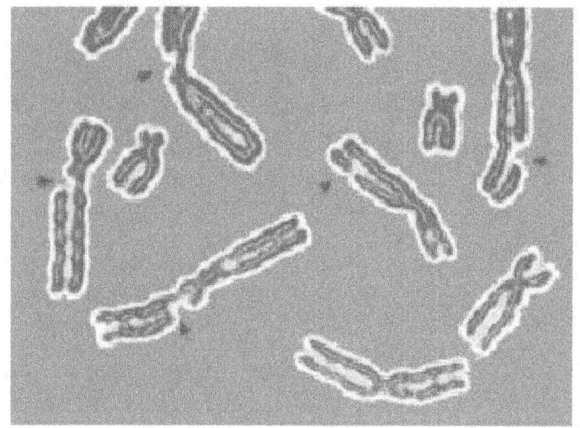

*DNA damage resulting in multiple broken chromosomes*

identifies and corrects damage to the DNA molecules that encode its genome. In human cells, both normal metabolic activities and environmental factors such as radiation can cause DNA damage, resulting in as many as 1 million individual molecular lesions per cell per day.[1] Many of these lesions cause structural damage to the DNA molecule and can alter or eliminate the cell's ability to transcribe the gene that the affected DNA encodes. Other lesions induce potentially harmful mutations in the cell's genome, which affect the survival of its daughter cells after it undergoes mitosis. As a consequence, the DNA repair process is constantly active as it responds to damage in the DNA structure. When normal repair processes fail, and when cellular apoptosis does not occur, irreparable DNA damage may occur, including double-strand breaks and DNA crosslinkages (interstrand crosslinks or ICLs).[2][3] This can eventually lead to malignant tumors, or cancer as per the two hit hypothesis.

The rate of DNA repair is dependent on many factors, including the cell type, the age of the cell, and the extracellular environment. A cell that has accumulated a large amount of DNA damage, or one that no longer effectively repairs damage incurred to its DNA, can enter one of three

possible states:

1. an irreversible state of dormancy, known as senescence

2. cell suicide, also known as apoptosis or programmed cell death

3. unregulated cell division, which can lead to the formation of a tumor that is cancerous

The DNA repair ability of a cell is vital to the integrity of its genome and thus to the normal functionality of that organism. Many genes that were initially shown to influence life span have turned out to be involved in DNA damage repair and protection.[4]

*Paul Modrich*

The 2015 Nobel Prize in Chemistry was awarded to Tomas Lindahl, Paul Modrich, and Aziz Sancar for their work on the molecular mechanisms of DNA repair processes.[5][6] There are two types: nucleotide excision repair and base excision repair.

## 20.1  DNA damage

Further information: DNA damage (naturally occurring) and Free radical damage to DNA

DNA damage, due to environmental factors and normal metabolic processes inside the cell, occurs at a rate of 10,000 to 1,000,000 molecular lesions per cell per day.[1] While this constitutes only 0.000165% of the human genome's approximately 6 billion bases (3 billion base pairs), unrepaired lesions in critical genes (such as tumor suppressor genes) can impede a cell's ability to carry out its function and appreciably increase the likelihood of tumor formation and contribute to tumour heterogeneity.

The vast majority of DNA damage affects the primary structure of the double helix; that is, the bases themselves are chemically modified. These modifications can in turn disrupt the molecules' regular helical structure by introducing non-native chemical bonds or bulky adducts that do not fit in the standard double helix. Unlike proteins and RNA, DNA usually lacks tertiary structure and therefore damage or disturbance does not occur at that level. DNA is, however, supercoiled and wound around "packaging" proteins called histones (in eukaryotes), and both superstructures are vulnerable to the effects of DNA damage.

### 20.1.1 Sources of damage

DNA damage can be subdivided into two main types:

1. endogenous damage such as attack by reactive oxygen species produced from normal metabolic byproducts (spontaneous mutation), especially the process of oxidative deamination

    (a) also includes replication errors

2. exogenous damage caused by external agents such as

    (a) ultraviolet [UV 200-400 nm] radiation from the sun

    (b) other radiation frequencies, including x-rays and gamma rays

    (c) hydrolysis or thermal disruption

    (d) certain plant toxins

    (e) human-made mutagenic chemicals, especially aromatic compounds that act as DNA intercalating agents

    (f) viruses[7]

The replication of damaged DNA before cell division can lead to the incorporation of wrong bases opposite damaged ones. Daughter cells that inherit these wrong bases carry mutations from which the original DNA sequence is unrecoverable (except in the rare case of a back mutation, for example, through gene conversion).

### 20.1.2 Types of damage

There are several types of damage to DNA due to endogenous cellular processes:

1. *oxidation* of bases [e.g. 8-oxo-7,8-dihydroguanine (8-oxoG)] and generation of DNA strand interruptions from reactive oxygen species.

2. *alkylation* of bases (usually methylation), such as formation of 7-methylguanosine, 1-methyladenine, 6-O-Methylguanine

3. *hydrolysis* of bases, such as deamination, depurination, and depyrimidination.

4. "bulky adduct formation" (i.e., benzo[a]pyrene diol epoxide-dG adduct, aristolactam I-dA adduct)

5. *mismatch* of bases, due to errors in DNA replication, in which the wrong DNA base is stitched into place in a newly forming DNA strand, or a DNA base is skipped over or mistakenly inserted.

6. Monoadduct damage cause by change in single nitrogenous base of DNA

7. Diadduct damage

Damage caused by exogenous agents comes in many forms. Some examples are:

1. *UV-B light* causes crosslinking between adjacent cytosine and thymine bases creating *pyrimidine dimers*. This is called direct DNA damage.

2. *UV-A light* creates mostly free radicals. The damage caused by free radicals is called indirect DNA damage.

3. *Ionizing radiation* such as that created by radioactive decay or in *cosmic rays* causes breaks in DNA strands. Intermediate-level ionizing radiation may induce irreparable DNA damage (leading to replicational and transcriptional errors needed for neoplasia or may trigger viral interactions) leading to pre-mature aging and cancer.

4. *Thermal disruption* at elevated temperature increases the rate of depurination (loss of purine bases from the DNA backbone) and single-strand breaks. For example, hydrolytic depurination is seen in the thermophilic bacteria, which grow in hot springs at 40-80 °C.[8][9] The rate of depurination (300 purine residues per genome per generation) is too high in these species to be repaired by normal repair machinery, hence a possibility of an adaptive response cannot be ruled out.

5. *Industrial chemicals* such as vinyl chloride and hydrogen peroxide, and environmental chemicals such as polycyclic aromatic hydrocarbons found in smoke, soot and tar create a huge diversity of DNA adducts-ethenobases, oxidized bases, alkylated phosphotriesters and crosslinking of DNA, just to name a few.

UV damage, alkylation/methylation, X-ray damage and oxidative damage are examples of induced damage. Spontaneous damage can include the loss of a base, deamination, sugar ring puckering and tautomeric shift.

### 20.1.3 Nuclear versus mitochondrial DNA damage

In human cells, and eukaryotic cells in general, DNA is found in two cellular locations — inside the nucleus and inside the mitochondria. Nuclear DNA (nDNA) exists as chromatin during non-replicative stages of the cell cycle and is condensed into aggregate structures known as chromosomes during cell division. In either state the DNA is highly compacted and wound up around bead-like proteins called histones. Whenever a cell needs to express the genetic information encoded in its nDNA the required chromosomal region is unravelled, genes located therein are expressed, and then the region is condensed back to its resting conformation. Mitochondrial DNA (mtDNA) is located inside mitochondria organelles, exists in multiple copies, and is also tightly associated with a number of proteins to form a complex known as the nucleoid. Inside mitochondria, reactive oxygen species (ROS), or free radicals, byproducts of the constant production of adenosine triphosphate (ATP) via oxidative phosphorylation, create a highly oxidative environment that is known to damage mtDNA. A critical enzyme in counteracting the toxicity of these species is superoxide dismutase, which is present in both the mitochondria and cytoplasm of eukaryotic cells.

### 20.1.4 Senescence and apoptosis

Senescence, an irreversible process in which the cell no longer divides, is a protective response to the shortening of the chromosome ends. The telomeres are long regions of repetitive noncoding DNA that cap chromosomes and undergo partial degradation each time a cell undergoes division (see Hayflick limit).[10] In contrast, quiescence is a reversible state of cellular dormancy that is unrelated to genome damage (see cell cycle). Senescence in cells may serve as a functional alternative to apoptosis in cases where the physical presence of a cell for spatial reasons is required by the organism,[11] which serves as a "last resort" mechanism to prevent a cell with damaged DNA from replicating

inappropriately in the absence of pro-growth cellular signaling. Unregulated cell division can lead to the formation of a tumor (see cancer), which is potentially lethal to an organism. Therefore, the induction of senescence and apoptosis is considered to be part of a strategy of protection against cancer.[12]

### 20.1.5 DNA damage and mutation

It is important to distinguish between DNA damage and mutation, the two major types of error in DNA. DNA damages and mutation are fundamentally different. Damages are physical abnormalities in the DNA, such as single- and double-strand breaks, 8-hydroxydeoxyguanosine residues, and polycyclic aromatic hydrocarbon adducts. DNA damages can be recognized by enzymes, and, thus, they can be correctly repaired if redundant information, such as the undamaged sequence in the complementary DNA strand or in a homologous chromosome, is available for copying. If a cell retains DNA damage, transcription of a gene can be prevented, and, thus, translation into a protein will also be blocked. Replication may also be blocked or the cell may die.

In contrast to DNA damage, a mutation is a change in the base sequence of the DNA. A mutation cannot be recognized by enzymes once the base change is present in both DNA strands, and, thus, a mutation cannot be repaired. At the cellular level, mutations can cause alterations in protein function and regulation. Mutations are replicated when the cell replicates. In a population of cells, mutant cells will increase or decrease in frequency according to the effects of the mutation on the ability of the cell to survive and reproduce. Although distinctly different from each other, DNA damages and mutations are related because DNA damages often cause errors of DNA synthesis during replication or repair; these errors are a major source of mutation.

Given these properties of DNA damage and mutation, it can be seen that DNA damages are a special problem in non-dividing or slowly dividing cells, where unrepaired damages will tend to accumulate over time. On the other hand, in rapidly dividing cells, unrepaired DNA damages that do not kill the cell by blocking replication will tend to cause replication errors and thus mutation. The great majority of mutations that are not neutral in their effect are deleterious to a cell's survival. Thus, in a population of cells composing a tissue with replicating cells, mutant cells will tend to be lost. However, infrequent mutations that provide a survival advantage will tend to clonally expand at the expense of neighboring cells in the tissue. This advantage to the cell is disadvantageous to the whole organism, because such mutant cells can give rise to cancer. Thus, DNA damages in frequently dividing cells, because they give rise to mutations, are a prominent cause of cancer. In contrast, DNA damages in infrequently dividing cells are likely a prominent cause of aging.[13]

## 20.2 DNA repair mechanisms

Cells cannot function if DNA damage corrupts the integrity and accessibility of essential information in the genome (but cells remain superficially functional when non-essential genes are missing or damaged). Depending on the type of damage inflicted on the DNA's double helical structure, a variety of repair strategies have evolved to restore lost information. If possible, cells use the unmodified complementary strand of the DNA or the sister chromatid as a template to recover the original information. Without access to a template, cells use an error-prone recovery mechanism known as translesion synthesis as a last resort.

Damage to DNA alters the spatial configuration of the helix, and such alterations can be detected by the cell. Once damage is localized, specific DNA repair molecules bind at or near the site of damage, inducing other molecules to bind and form a complex that enables the actual repair to take place.

### 20.2.1 Direct reversal

Cells are known to eliminate three types of damage to their DNA by chemically reversing it. These mechanisms do not require a template, since the types of damage they counteract can occur in only one of the four bases. Such direct reversal mechanisms are specific to the type of damage incurred and do not involve breakage of the phosphodiester backbone. The formation of pyrimidine dimers upon irradiation with UV light results in an abnormal covalent bond between adjacent pyrimidine bases. The photoreactivation process directly reverses this damage by the action of the enzyme photolyase, whose activation is obligately dependent on energy absorbed from blue/UV light (300–500 nm wavelength) to promote catalysis.[14] Photolyase, an old enzyme present in bacteria, fungi, and most animals no longer functions in humans,[15] who instead use nucleotide excision repair to repair damage from UV irradiation. Another type of damage, methylation of guanine bases, is directly reversed by the protein methyl guanine methyl transferase (MGMT), the bacterial equivalent of which is called ogt. This is an expensive process because each MGMT molecule can be used only once; that is, the reaction is stoichiometric rather than catalytic.[16] A generalized response to methylating agents in bacteria is known as the adaptive response and confers a level of resistance to alkylating agents upon sustained exposure by upregulation of

alkylation repair enzymes.[17] The third type of DNA damage reversed by cells is certain methylation of the bases cytosine and adenine.

### 20.2.2 Single-strand damage

When only one of the two strands of a double helix has a defect, the other strand can be used as a template to guide the correction of the damaged strand. In order to repair damage to one of the two paired molecules of DNA, there exist a number of excision repair mechanisms that remove the damaged nucleotide and replace it with an undamaged nucleotide complementary to that found in the undamaged DNA strand.[16]

1. Base excision repair (BER) repairs damage to a single nitrogenous base by deploying enzymes called glycosylases.[18] These enzymes remove a single nitrogenous base to create an apurinic or apyrimidinic site (AP site).[18] Enzymes called AP endonucleases nick the damaged DNA backbone at the AP site. DNA polymerase then removes the damaged region using its 5' to 3' exonuclease activity and correctly synthesizes the new strand using the complementary strand as a template.[18]

2. Nucleotide excision repair (NER) repairs damaged DNA which commonly consists of bulky, helix-distorting damage, such as pyrimidine dimerization caused by UV light. Damaged regions are removed in 12-24 nucleotide-long strands in a three-step process which consists of recognition of damage, excision of damaged DNA both upstream and downstream of damage by endonucleases, and resynthesis of removed DNA region.[19] NER is a highly evolutionarily conserved repair mechanism and is used in nearly all eukaryotic and prokaryotic cells.[19] In prokaryotes, NER is mediated by Uvr proteins.[19] In eukaryotes, many more proteins are involved, although the general strategy is the same.[19]

3. Mismatch repair systems are present in essentially all cells to correct errors that are not corrected by proofreading. These systems consist of at least two proteins. One detects the mismatch, and the other recruits an endonuclease that cleaves the newly synthesized DNA strand close to the region of damage. In *E. coli*, the proteins involved are the Mut class proteins. This is followed by removal of damaged region by an exonuclease, resynthesis by DNA polymerase, and nick sealing by DNA ligase.[20]

### 20.2.3    Double-strand breaks

Double-strand breaks, in which both strands in the double helix are severed, are particularly hazardous to the cell because they can lead to genome rearrangements. Three mechanisms exist to repair double-strand breaks (DSBs): non-homologous end joining (NHEJ), microhomology-mediated end joining (MMEJ), and homologous recombination.[16] PVN Acharya noted that double-strand breaks and a "cross-linkage joining both strands at the same point is irreparable because neither strand can then serve as a template for repair. The cell will die in the next mitosis or in some rare instances, mutate."[2][3]

In NHEJ, DNA Ligase IV, a specialized DNA ligase that forms a complex with the cofactor XRCC4, directly joins the two ends.[21] To guide accurate repair, NHEJ relies on short homologous sequences called microhomologies present on the single-stranded tails of the DNA ends to be joined. If these overhangs are compatible, repair is usually accurate.[22][23][24][25] NHEJ can also introduce mutations during repair. Loss of damaged nucleotides at the break site can lead to deletions, and joining of nonmatching termini forms insertions or translocations. NHEJ is especially important before the cell has replicated its DNA, since there is no template available for repair by homologous recombination. There are "backup" NHEJ pathways in higher eukaryotes.[26] Besides its role as a genome caretaker, NHEJ is required for joining hairpin-capped double-strand breaks induced during V(D)J recombination, the process that generates diversity in B-cell and T-cell receptors in the vertebrate immune system.[27]

MMEJ starts with short-range end resection by MRE11 nuclease on either side of a double-strand break to reveal microhomology regions.[28] In further steps,[29] PARP1 is required and may be an early step in MMEJ. There is pairing of microhomology regions followed by recruitment of flap structure-specific endonuclease 1 (FEN1) to remove overhanging flaps. This is followed by recruitment of XRCC1–LIG3 to the site for ligating the DNA ends, leading to an intact DNA.

DNA double strand breaks in mammalian cells are primarily repaired by homologous recombination (HR) and non-homologous end joining (NHEJ).[30] In an *in vitro* system, MMEJ occurred in mammalian cells at the levels of 10–20% of HR when both HR and NHEJ mechanisms were also available.[28] MMEJ is always accompanied by a deletion, so that MMEJ is a mutagenic pathway for DNA repair.[31]

Homologous recombination requires the presence of an identical or nearly identical sequence to be used as a template for repair of the break. The enzymatic machinery responsible for this repair process is nearly identical to the machinery responsible for chromosomal crossover during meiosis. This pathway allows a damaged chromosome to be repaired using a sister chromatid (available in G2 after DNA replication) or a homologous chromosome as a template. DSBs caused by the replication machinery attempting to synthesize across a single-strand break or unrepaired lesion cause collapse of the replication fork and are typically repaired by recombination.

Topoisomerases introduce both single- and double-strand breaks in the course of changing the DNA's state of supercoiling, which is especially common in regions near an open replication fork. Such breaks are not considered DNA damage because they are a natural intermediate in the topoisomerase biochemical mechanism and are immediately repaired by the enzymes that created them.

A team of French researchers bombarded *Deinococcus radiodurans* to study the mechanism of double-strand break DNA repair in that bacterium. At least two copies of the genome, with random DNA breaks, can form DNA fragments through annealing. Partially overlapping fragments are then used for synthesis of homologous regions through a moving D-loop that can continue extension until they find complementary partner strands. In the final step there is crossover by means of RecA-dependent homologous recombination.[32]

### 20.2.4    Translesion synthesis

Translesion synthesis (TLS) is a DNA damage tolerance process that allows the DNA replication machinery to replicate past DNA lesions such as thymine dimers or AP sites.[33] It involves switching out regular DNA polymerases for specialized translesion polymerases (i.e. DNA polymerase IV or V, from the Y Polymerase family), often with larger active sites that can facilitate the insertion of bases opposite damaged nucleotides. The polymerase switching is thought to be mediated by, among other factors, the post-translational modification of the replication processivity factor PCNA. Translesion synthesis polymerases often have low fidelity (high propensity to insert wrong bases) on undamaged templates relative to regular polymerases. However, many are extremely efficient at inserting correct bases opposite specific types of damage. For example, Pol η mediates error-free bypass of lesions induced by UV irradiation, whereas Pol ι introduces mutations at these sites. Pol η is known to add the first adenine across the T^T photodimer using Watson-Crick base pairing and the second adenine will be added in its syn conformation using Hoogsteen base pairing. From a cellular perspective, risking the introduction of point mutations during translesion synthesis may be preferable to resorting to more drastic mechanisms of DNA repair, which may cause gross chromosomal aberrations or cell death. In short, the process involves specialized polymerases either bypassing or repairing lesions at locations of stalled DNA replication. For example, Human DNA polymerase eta can bypass complex DNA lesions like guanine-thymine intra-strand crosslink, G[8,5-Me]T, although can cause targeted and semi-targeted mutations.[34] Paromita Raychaudhury and Ashis Basu[35] studied the toxicity and mutagenesis of the same lesion in *Escherichia coli* by replicating a G[8,5-Me]T-modified plasmid in *E. coli* with specific DNA polymerase knockouts. Viability was very low in a strain lacking pol II, pol IV, and pol V, the three SOS-inducible DNA polymerases, indicating that translesion synthesis is conducted primarily by these specialized DNA polymerases. A bypass platform is provided to these polymerases by Proliferating cell nuclear antigen (PCNA). Under normal circumstances, PCNA bound to polymerases replicates the DNA. At a site of lesion, PCNA is ubiquitinated, or modified, by the RAD6/RAD18 proteins to provide a platform for the specialized polymerases to bypass the lesion and resume DNA replication.[36][37] After translesion synthesis, extension is required. This extension can be carried out by a replicative polymerase if the TLS is error-free, as in the case of Pol η, yet if TLS results in a mismatch, a specialized polymerase is needed to extend it; Pol ζ. Pol ζ is unique in that it can extend terminal mismatches, whereas more processive polymerases cannot. So when a lesion is encountered, the replication fork will stall, PCNA will switch from a processive polymerase to a TLS polymerase such as Pol ι to fix the lesion, then PCNA may switch to Pol ζ to extend the mismatch, and last PCNA will switch to the processive polymerase to continue replication.

# 20.3   Global response to DNA damage

Cells exposed to ionizing radiation, ultraviolet light or chemicals are prone to acquire multiple sites of bulky DNA lesions and double-strand breaks. Moreover, DNA damaging agents can damage other biomolecules such as proteins, carbohydrates, lipids, and RNA. The accumulation of damage, to be specific, double-strand breaks or adducts stalling the replication forks, are among known stimulation signals for a global response to DNA damage.[38] The global response to damage is an act directed toward the cells' own preservation and triggers multiple pathways of macromolecular repair, lesion bypass, tolerance, or apoptosis. The common features of global response are induction of multiple genes, cell cycle arrest, and inhibition of cell division.

## 20.3.1   DNA damage checkpoints

After DNA damage, cell cycle checkpoints are activated. Checkpoint activation pauses the cell cycle and gives the cell time to repair the damage before continuing to divide. DNA damage checkpoints occur at the G1/S and G2/M boundaries. An intra-S checkpoint also exists. Checkpoint activation is controlled by two master kinases, ATM and ATR. ATM responds to DNA double-strand breaks and disruptions in chromatin structure,[39] whereas ATR primarily responds to stalled replication forks. These kinases phosphorylate downstream targets in a signal transduction cascade, eventually leading to cell cycle arrest. A class of checkpoint mediator proteins including BRCA1, MDC1, and 53BP1 has also been identified.[40] These proteins seem to be required for transmitting the checkpoint activation signal to downstream proteins.

**DNA damage checkpoint** is a signal transduction pathway that blocks cell cycle progression in G1, G2 and metaphase and slows down the rate of S phase progression when DNA is damaged. It leads to a pause in cell cycle allowing the cell time to repair the damage before continuing to divide.

Checkpoint Proteins can be separated into four groups: phosphatidylinositol 3-kinase (PI3K)-like protein kinase, proliferating cell nuclear antigen (PCNA)-like group, two serine/threonine(S/T) kinases and their adaptors. Central to all DNA damage induced checkpoints responses is a pair of large protein kinases belonging to the first group of PI3K-like protein kinases-the ATM (Ataxia telangiectasia mutated) and ATR (Ataxia- and Rad-related) kinases, whose sequence and functions have been well conserved in evolution. All DNA damage response requires either ATM or ATR because they have the ability to bind to the chromosomes at the site of DNA damage, together with accessory proteins that are platforms on which DNA damage response components and DNA repair complexes can be assembled.

An important downstream target of ATM and ATR is p53, as it is required for inducing apoptosis following DNA damage.[41] The cyclin-dependent kinase inhibitor p21 is induced by both p53-dependent and p53-independent mechanisms and can arrest the cell cycle at the G1/S and G2/M checkpoints by deactivating cyclin/cyclin-dependent kinase complexes.[42]

## 20.3.2   The prokaryotic SOS response

The SOS response is the changes in gene expression in *Escherichia coli* and other bacteria in response to extensive DNA damage. The prokaryotic SOS system is regulated by two key proteins: LexA and RecA. The LexA homodimer is a transcriptional repressor that binds to operator sequences commonly referred to as SOS boxes. In *Escherichia coli* it is known that LexA regulates transcription of approximately 48 genes including the lexA and recA genes.[43] The SOS response is known to be widespread in the Bacteria domain, but it is mostly absent in some bacterial phyla, like the Spirochetes.[44] The most common cellular signals activating the SOS response are regions of single-stranded DNA (ssDNA), arising from stalled replication forks or double-strand breaks, which are processed by DNA helicase to separate the two DNA strands.[38] In the initiation step, RecA protein binds to ssDNA in an ATP hydrolysis driven reaction creating RecA–ssDNA filaments. RecA–ssDNA filaments activate LexA autoprotease activity, which ultimately leads to cleavage of LexA dimer and subsequent LexA degradation. The loss of LexA repressor induces transcription of the SOS genes and allows for further signal induction, inhibition of cell division and an increase in levels of proteins responsible for damage processing.

In *Escherichia coli*, SOS boxes are 20-nucleotide long sequences near promoters with palindromic structure and a high degree of sequence conservation. In other classes and phyla, the sequence of SOS boxes varies considerably, with different length and composition, but it is always highly conserved and one of the strongest short signals in the genome.[44] The high information content of SOS boxes permits differential binding of LexA to different promoters and allows for timing of the SOS response. The lesion repair genes are induced at the beginning of SOS response. The error-prone translesion polymerases, for example, UmuCD'2 (also called DNA polymerase V), are induced later on as a last resort.[45] Once the DNA damage is repaired or bypassed using polymerases or through recombination, the amount of single-stranded DNA in cells is decreased, lowering the amounts of RecA filaments decreases cleavage activity of LexA homodimer, which then binds to the SOS boxes near promoters and restores normal gene expression.

### 20.3.3  Eukaryotic transcriptional responses to DNA damage

Eukaryotic cells exposed to DNA damaging agents also activate important defensive pathways by inducing multiple proteins involved in DNA repair, cell cycle checkpoint control, protein trafficking and degradation. Such genome wide transcriptional response is very complex and tightly regulated, thus allowing coordinated global response to damage. Exposure of yeast *Saccharomyces cerevisiae* to DNA damaging agents results in overlapping but distinct transcriptional profiles. Similarities to environmental shock response indicates that a general global stress response pathway exist at the level of transcriptional activation. In contrast, different human cell types respond to damage differ-

ently indicating an absence of a common global response. The probable explanation for this difference between yeast and human cells may be in the heterogeneity of mammalian cells. In an animal different types of cells are distributed among different organs that have evolved different sensitivities to DNA damage.[46]

In general global response to DNA damage involves expression of multiple genes responsible for postreplication repair, homologous recombination, nucleotide excision repair, DNA damage checkpoint, global transcriptional activation, genes controlling mRNA decay, and many others. A large amount of damage to a cell leaves it with an important decision: undergo apoptosis and die, or survive at the cost of living with a modified genome. An increase in tolerance to damage can lead to an increased rate of survival that will allow a greater accumulation of mutations. Yeast Rev1 and human polymerase η are members of [Y family translesion DNA polymerases present during global response to DNA damage and are responsible for enhanced mutagenesis during a global response to DNA damage in eukaryotes.[38]

## 20.4  DNA repair and aging

Main article: DNA damage theory of aging

### 20.4.1  Pathological effects of poor DNA repair

Experimental animals with genetic deficiencies in DNA repair often show decreased life span and increased cancer incidence.[13] For example, mice deficient in the dominant NHEJ pathway and in telomere maintenance mechanisms get lymphoma and infections more often, and, as a consequence, have shorter lifespans than wild-type mice.[47] In similar manner, mice deficient in a key repair and transcription protein that unwinds DNA helices have premature onset of aging-related diseases and consequent shortening of lifespan.[48] However, not every DNA repair deficiency creates exactly the predicted effects; mice deficient in the NER pathway exhibited shortened life span without correspondingly higher rates of mutation.[49]

If the rate of DNA damage exceeds the capacity of the cell to repair it, the accumulation of errors can overwhelm the cell and result in early senescence, apoptosis, or cancer. Inherited diseases associated with faulty DNA repair functioning result in premature aging,[13] increased sensitivity to carcinogens, and correspondingly increased cancer risk (see below). On the other hand, organisms with enhanced DNA repair systems, such as *Deinococcus radiodurans*, the most radiation-resistant known organism, exhibit remarkable resistance to the double-strand break-inducing effects of radioactivity, likely due to enhanced efficiency of DNA repair and especially NHEJ.[50]

### 20.4.2   Longevity and caloric restriction

A number of individual genes have been identified as influencing variations in life span within a population of organisms. The effects of these genes is strongly dependent on the environment, in particular, on the organism's diet. Caloric restriction reproducibly results in extended lifespan in a variety of organisms, likely via nutrient sensing pathways and decreased metabolic rate. The molecular mechanisms by which such restriction results in lengthened lifespan are as yet unclear (see[51] for some discussion); however, the behavior of many genes known to be involved in DNA repair is altered under conditions of caloric restriction.

For example, increasing the gene dosage of the gene SIR-2, which regulates DNA packaging in the nematode worm *Caenorhabditis elegans*, can significantly extend lifespan.[52] The mammalian homolog of SIR-2 is known to induce downstream DNA repair factors involved in NHEJ, an activity that is especially promoted under conditions of caloric restriction.[53] Caloric restriction has been closely linked to the rate of base excision repair in the nuclear DNA of rodents,[54] although similar effects have not been observed in mitochondrial DNA.[55]

The *C. elegans* gene AGE-1, an upstream effector of DNA repair pathways, confers dramatically extended life span under free-feeding conditions but leads to a decrease in reproductive fitness under conditions of caloric restriction.[56] This observation supports the pleiotropy theory of the biological origins of aging, which suggests that genes conferring a large survival advantage early in life will be selected for even if they carry a corresponding disadvantage late in life.

## 20.5   Medicine and DNA repair modulation

Main article: DNA repair-deficiency disorder

### 20.5.1   Hereditary DNA repair disorders

Defects in the NER mechanism are responsible for several genetic disorders, including:

- Xeroderma pigmentosum: hypersensitivity to sunlight/UV, resulting in increased skin cancer incidence

and premature aging

- Cockayne syndrome: hypersensitivity to UV and chemical agents
- Trichothiodystrophy: sensitive skin, brittle hair and nails

Mental retardation often accompanies the latter two disorders, suggesting increased vulnerability of developmental neurons.

Other DNA repair disorders include:

- Werner's syndrome: premature aging and retarded growth
- Bloom's syndrome: sunlight hypersensitivity, high incidence of malignancies (especially leukemias).
- Ataxia telangiectasia: sensitivity to ionizing radiation and some chemical agents

All of the above diseases are often called "segmental progerias" ("accelerated aging diseases") because their victims appear elderly and suffer from aging-related diseases at an abnormally young age, while not manifesting all the symptoms of old age.

Other diseases associated with reduced DNA repair function include Fanconi anemia, hereditary breast cancer and hereditary colon cancer.

## 20.6   DNA repair and cancer

Because of inherent limitations in the DNA repair mechanisms, if humans lived long enough, they would all eventually develop cancer.[57][58] There are at least 34 Inherited human DNA repair gene mutations that increase cancer risk. Many of these mutations cause DNA repair to be less effective than normal. In particular, Hereditary non-polyposis colorectal cancer (HNPCC) is strongly associated with specific mutations in the DNA mismatch repair pathway. *BRCA1* and *BRCA2*, two famous genes whose mutations confer a hugely increased risk of breast cancer on carriers, are both associated with a large number of DNA repair pathways, especially NHEJ and homologous recombination.

Cancer therapy procedures such as chemotherapy and radiotherapy work by overwhelming the capacity of the cell to repair DNA damage, resulting in cell death. Cells that are most rapidly dividing — most typically cancer cells — are preferentially affected. The side-effect is that other non-cancerous but rapidly dividing cells such as progenitor cells in the gut, skin, and hematopoietic system are also affected. Modern cancer treatments attempt to localize the DNA damage to cells and tissues only associated with cancer, either by physical means (concentrating the therapeutic agent in the region of the tumor) or by biochemical means (exploiting a feature unique to cancer cells in the body). ...; in the context of therapies targeting DNA damage response genes, the latter approach has been termed 'synthetic lethality'.[59]

Perhaps the most well-known of these 'synthetic lethality' drugs is the poly(ADP-ribose) polymerase 1 (PARP1) inhibitor olaparib, which was approved by the Food and Drug Administration in 2015 for the treatment in women of BRCA-defective ovarian cancer. Tumor cells with partial loss of DNA damage response (specifically, homologous recombination repair) are dependent on another mechanism - single-strand break repair - which is a mechanism consisting, in part, of the PARP1 gene product.[60] Olaparib is combined with chemotherapeutics to inhibit single-strand break repair induced by DNA damage caused by the co-administered chemotherapy. Tumor cells relying on this residual DNA repair mechanism are unable to repair the damage and hence are not able to survive and proliferate, whereas normal cells can repair the damage with the functioning homologous recombination mechanism.

Many other drugs for use against other residual DNA repair mechanisms commonly found in cancer are currently under investigation. However, synthetic lethality therapeutic approaches have been questioned due to emerging evidence of acquired resistance, achieved through rewiring of DNA damage response pathways and reversion of previously-inhibited defects.[61]

### 20.6.1 DNA repair defects in cancer

It has become apparent over the past several years that the DNA damage response acts as a barrier to the malignant transformation of preneoplastic cells.[62] Previous studies have shown an elevated DNA damage response in cell-culture models with oncogene activation[63] and preneoplastic colon adenomas.[64] DNA damage response mechanisms trigger cell-cycle arrest, and attempt to repair DNA lesions or promote cell death/senescence if repair is not possible. Replication stress is observed in preneoplastic cells due to increased proliferation signals from oncogenic mutations. Replication stress is characterized by: increased replication initiation/origin firing; increased transcription and collisions of transcription-replication complexes; nucleotide deficiency; increase in reactive oxygen species (ROS).[65]

Replication stress, along with the selection for inactivating mutations in DNA damage response genes in the evolution of the tumor,[66] leads to downregulation and/or loss of some DNA damage response mechanisms, and hence

loss of DNA repair and/or senescence/programmed cell death. In experimental mouse models, loss of DNA damage response-mediated cell senescence was observed after using a short hairpin RNA (shRNA) to inhibit the double-strand break response kinase ataxia telangiectasia (ATM), leading to increased tumor size and invasiveness.[64] Humans born with inherited defects in DNA repair mechanisms (for example, Li-Fraumeni syndrome) have a higher cancer risk.[67]

The prevalence of DNA damage response mutations differs across cancer types; for example, 30% of breast invasive carcinomas have mutations in genes involved in homologous recombination.[62] In cancer, downregulation is observed across all DNA damage response mechanisms (base excision repair (BER), nucleotide excision repair (NER), DNA mismatch repair (MMR), homologous recombination repair (HR), non-homologous end joining (NHEJ) and translesion DNA synthesis (TLS).[68] As well as mutations to DNA damage repair genes, mutations also arise in the genes responsible for arresting the cell cycle to allow sufficient time for DNA repair to occur, and some genes are involved in both DNA damage repair and cell cycle checkpoint control, for example ATM and checkpoint kinase 2 (CHEK2) - a tumor suppressor that is often absent or downregulated in non-small cell lung cancer.[69]

Table: Genes involved in DNA damage response pathways and frequently mutated in cancer (HR = homologous recombination; NHEJ = non-homologous end joining; SSA = single-strand annealing; FA = fanconi anemia pathway; BER = base excision repair; NER = nucleotide excision repair; MMR = mismatch repair)

### 20.6.2 Epigenetic DNA repair defects in cancer

Classically, cancer has been viewed as a set of diseases that are driven by progressive genetic abnormalities that include mutations in tumour-suppressor genes and oncogenes, and chromosomal aberrations. However, it has become apparent that cancer is also driven by epigenetic alterations.[70]

Epigenetic alterations refer to functionally relevant modifications to the genome that do not involve a change in the nucleotide sequence. Examples of such modifications are changes in DNA methylation (hypermethylation and hypomethylation) and histone modification,[71] changes in chromosomal architecture (caused by inappropriate expression of proteins such as HMGA2 or HMGA1)[72] and changes caused by microRNAs. Each of these epigenetic alterations serves to regulate gene expression without altering the underlying DNA sequence. These changes usually remain through cell divisions, last for multiple cell generations, and can be considered to be epimutations (equivalent to mutations).

While large numbers of epigenetic alterations are found in cancers, the epigenetic alterations in DNA repair genes, causing reduced expression of DNA repair proteins, appear to be particularly important. Such alterations are thought to occur early in progression to cancer and to be a likely cause of the genetic instability characteristic of cancers.[73][74][75][76]

Reduced expression of DNA repair genes causes deficient DNA repair. When DNA repair is deficient DNA damages remain in cells at a higher than usual level and these excess damages cause increased frequencies of mutation or epimutation. Mutation rates increase substantially in cells defective in DNA mismatch repair[77][78] or in homologous recombinational repair (HRR).[79] Chromosomal rearrangements and aneuploidy also increase in HRR defective cells.[80]

Higher levels of DNA damage not only cause increased mutation, but also cause increased epimutation. During repair of DNA double strand breaks, or repair of other DNA damages, incompletely cleared sites of repair can cause epigenetic gene silencing.[81][82]

Deficient expression of DNA repair proteins due to an inherited mutation can cause increased risk of cancer. Individuals with an inherited impairment in any of 34 DNA repair genes (see article DNA repair-deficiency disorder) have an increased risk of cancer, with some defects causing up to a 100% lifetime chance of cancer (e.g. p53 mutations).[83] However, such germline mutations (which cause highly penetrant cancer syndromes) are the cause of only about 1 percent of cancers.[84]

### 20.6.3 Frequencies of epimutations in DNA repair genes

Deficiencies in DNA repair enzymes are occasionally caused by a newly arising somatic mutation in a DNA repair gene, but are much more frequently caused by epigenetic alterations that reduce or silence expression of DNA repair genes. For example, when 113 colorectal cancers were examined in sequence, only four had a missense mutation in the DNA repair gene MGMT, while the majority had reduced MGMT expression due to methylation of the MGMT promoter region (an epigenetic alteration).[85] Five different studies found that between 40% and 90% of colorectal cancers have reduced MGMT expression due to methylation of the MGMT promoter region.[86][87][88][89][90]

Similarly, out of 119 cases of mismatch repair-deficient colorectal cancers that lacked DNA repair gene PMS2 expression, PMS2 was deficient in 6 due to mutations in the PMS2 gene, while in 103 cases PMS2 expression was deficient because its pairing partner MLH1 was repressed due to promoter methylation (PMS2 protein is unstable in the absence of MLH1).[91] In the other 10 cases, loss of PMS2 expression was likely due to epigenetic overexpression of the microRNA, miR-155, which down-regulates MLH1.[92]

In further examples (tabulated in Table 4 of this reference[93]), epigenetic defects were found at frequencies of between 13% – 100% for the DNA repair genes BRCA1, WRN, FANCB, FANCF, MGMT, MLH1, MSH2, MSH4, ERCC1, XPF, NEIL1 and ATM. These epigenetic defects occurred in various cancers (e.g. breast, ovarian, colorectal and head and neck). Two or three deficiencies in the expression of ERCC1, XPF or PMS2 occur simultaneously in the majority of the 49 colon cancers evaluated by Facista et al.[94]

The chart in this section shows some frequent DNA damaging agents, examples of DNA lesions they cause, and the pathways that deal with these DNA damages. At least 169 enzymes are either directly employed in DNA repair or influence DNA repair processes.[95] Of these, 83 are directly employed in repairing the 5 types of DNA damages illustrated in the chart.

Some of the more well studied genes central to these repair processes are shown in the chart. The gene designations shown in red, gray or cyan indicate genes frequently epigenetically altered in various types of cancers. Wikipedia articles on each of the genes highlighted by red, gray or cyan describe the epigenetic alteration(s) and the cancer(s) in which these epimutations are found. Two review articles,[93][96] and two broad experimental survey articles[97][98] also document most of these epigenetic DNA repair deficiencies in cancers.

Red-highlighted genes are frequently reduced or silenced by epigenetic mechanisms in various cancers. When these genes have low or absent expression, DNA damages can accumulate. Replication errors past these damages (see translesion synthesis) can lead to increased mutations and, ultimately, cancer. Epigenetic repression of DNA repair genes in **accurate** DNA repair pathways appear to be central to carcinogenesis.

The two gray-highlighted genes *RAD51* and *BRCA2*, are required for homologous recombinational repair. They are sometimes epigenetically over-expressed and sometimes under-expressed in certain cancers. As indicated in the Wikipedia articles on RAD51 and BRCA2, such cancers ordinarily have epigenetic deficiencies in other DNA repair genes. These repair deficiencies would likely cause increased unrepaired DNA damages. The over-expression of *RAD51* and *BRCA2* seen in these cancers may reflect selective pressures for compensatory *RAD51* or *BRCA2* over-expression and increased homologous recombinational repair to at least partially deal with such excess DNA damages. In those cases where *RAD51* or *BRCA2* are under-expressed, this would itself lead to increased unrepaired DNA damages. Replication errors past these damages (see translesion synthesis) could cause increased mutations and cancer, so that under-expression of *RAD51* or *BRCA2* would be carcinogenic in itself.

Cyan-highlighted genes are in the microhomology-mediated end joining (MMEJ) pathway and are up-regulated in cancer. MMEJ is an additional error-prone **inaccurate** repair pathway for double-strand breaks. In MMEJ repair of a double-strand break, an homology of 5-25 complementary base pairs between both paired strands is sufficient to align the strands, but mismatched ends (flaps) are usually present. MMEJ removes the extra nucleotides (flaps) where strands are joined, and then ligates the strands to create an intact DNA double helix. MMEJ almost always involves at least a small deletion, so that it is a mutagenic pathway.[99] FEN1, the flap endonuclease in MMEJ, is epigenetically increased by promoter hypomethylation and is over-expressed in the majority of cancers of the breast,[100] prostate,[101] stomach,[102][103] neuroblastomas,[104] pancreas,[105] and lung.[106] PARP1 is also over-expressed when its promoter region ETS site is epigenetically hypomethylated, and this contributes to progression to endometrial cancer,[107] BRCA-mutated ovarian cancer,[108] and BRCA-mutated serous ovarian cancer.[109] Other genes in the MMEJ pathway are also over-expressed in a number of cancers (see MMEJ for summary), and are also shown in cyan.

## 20.7 DNA repair and evolution

The basic processes of DNA repair are highly conserved among both prokaryotes and eukaryotes and even among bacteriophages (viruses which infect bacteria); however, more complex organisms with more complex genomes have correspondingly more complex repair mechanisms.[110] The ability of a large number of protein structural motifs to catalyze relevant chemical reactions has played a significant role in the elaboration of repair mechanisms during evolution. For an extremely detailed review of hypotheses relating to the evolution of DNA repair, see.[111]

The fossil record indicates that single-cell life began to proliferate on the planet at some point during the Precambrian period, although exactly when recognizably modern life first emerged is unclear. Nucleic acids became the sole and universal means of encoding genetic information, requiring DNA repair mechanisms that in their basic form have been inherited by all extant life forms from their common ancestor. The emergence of Earth's oxygen-rich atmosphere (known as the "oxygen catastrophe") due to photosynthetic organisms, as well as the presence of potentially damaging free radicals in the cell due to oxidative phosphorylation, necessitated the evolution of DNA repair mechanisms that act specifically to counter the types of damage induced by oxidative stress.

### 20.7.1 Rate of evolutionary change

On some occasions, DNA damage is not repaired, or is repaired by an error-prone mechanism that results in a change from the original sequence. When this occurs, mutations may propagate into the genomes of the cell's progeny. Should such an event occur in a germ line cell that will eventually produce a gamete, the mutation has the potential to be passed on to the organism's offspring. The rate of evolution in a particular species (or, in a particular gene) is a function of the rate of mutation. As a consequence, the rate and accuracy of DNA repair mechanisms have an influence over the process of evolutionary change.[112] Since the normal adaptation of populations of organisms to changing circumstances (for instance the adaptation of the beaks of a population of finches to the changing presence of hard seeds or insects) proceeds by gene regulation and the recombination and selection of gene variations – alleles – and not by passing on irreparable DNA damages to the offspring,[113] DNA damage protection and repair does not influence the rate of adaptation by gene regulation and by recombination and selection of alleles. On the other hand, DNA damage repair and protection does influence the rate of accumulation of irreparable, advantageous, code expanding, inheritable mutations, and slows down the evolutionary mechanism for expansion of the genome of organisms with new functionalities. The tension between evolvability and mutation repair and protection needs further investigation.

## 20.8 DNA repair technology

A technology named clustered regularly interspaced short palindromic repeat shortened to CRISPR-Cas9 was discovered in 2012. The new technology allows anyone with molecular biology training to alter the genes of any species with precision.[114]

## 20.9 See also

- Accelerated aging disease
- Aging DNA
- Cell cycle
- DNA damage (naturally occurring)

- DNA damage theory of aging

- DNA replication

- Direct DNA damage

- Gene therapy

- Human mitochondrial genetics

- Indirect DNA damage

- Life extension

- Progeria

- Senescence

- SiDNA

- The scientific journal *DNA Repair* under Mutation Research

## 20.10    References

[1] Lodish H, Berk A, Matsudaira P, Kaiser CA, Krieger M, Scott MP, Zipursky SL, Darnell J. (2004). Molecular Biology of the Cell, p963. WH Freeman: New York, NY. 5th ed.

[2] Acharya, PV (1971). "The isolation and partial characterization of age-correlated oligo-deoxyribo-ribonucleotides with covalently linked aspartyl-glutamyl polypeptides.". *Johns Hopkins medical journal. Supplement* (1): 254–60. PMID 5055816.

[3] Bjorksten, J; Acharya, PV; Ashman, S; Wetlaufer, DB (1971). "Gerogenic fractions in the tritiated rat.". *Journal of the American Geriatrics Society.* **19** (7): 561–74. doi:10.1111/j.1532-5415.1971.tb02577.x. PMID 5106728.

[4] Browner, WS; Kahn, AJ; Ziv, E; Reiner, AP; Oshima, J; Cawthon, RM; Hsueh, WC; Cummings, SR. (2004). "The genetics of human longevity". *Am J Med.* **117** (11): 851–60. doi:10.1016/j.amjmed.2004.06.033. PMID 15589490.

[5] Broad, William J. (7 October 2015). "Nobel Prize in Chemistry Awarded to Tomas Lindahl, Paul Modrich and Aziz Sancar for DNA Studies". *New York Times.* Retrieved 7 October 2015.

[6] Staff (7 October 2015). "THE NOBEL PRIZE IN CHEMISTRY 2015 - DNA repair – providing chemical stability for life" (PDF). *Nobel Prize.* Retrieved 7 October 2015.

[7] Roulston A, Marcellus RC, Branton PE (1999). "Viruses and apoptosis". *Annu. Rev. Microbiol.* **53**: 577–628. doi:10.1146/annurev.micro.53.1.577. PMID 10547702.

[8] Madigan MT, Martino JM (2006). *Brock Biology of Microorganisms* (11th ed.). Pearson. p. 136. ISBN 0-13-196893-9.

[9] Ohta, Toshihiro; Shin-ichi, Tokishita; Mochizuki, Kayo; Kawase, Jun; Sakahira, Masahide; Yamagata, Hideo (2006). "UV Sensitivity and Mutagenesis of the Extremely Thermophilic Eubacterium Thermus thermophilus HB27". *Genes and Environment.* **28** (2): 56–61. doi:10.3123/jemsge.28.56.

[10] Braig, M; Schmitt, CA. (2006). "Oncogene-induced senescence: putting the brakes on tumor development". *Cancer Res.* **66** (6): 2881–2884. doi:10.1158/0008-5472.CAN-05-4006. PMID 16540631.

[11] Lynch, MD. (2006). "How does cellular senescence prevent cancer?". *DNA Cell Biol.* **25** (2): 69–78. doi:10.1089/dna.2006.25.69. PMID 16460230.

[12] Campisi J, d'Adda di Fagagna F (2007). "Cellular senescence: when bad things happen to good cells.". *Rev Mol Cell Biol.* **8** (9): 729–40. doi:10.1038/nrm2233. PMID 17667954.

[13] Best, Benjamin P (2009). "Nuclear DNA damage as a direct cause of aging" (PDF). *Rejuvenation Research.* **12** (3): 199–208. doi:10.1089/rej.2009.0847. PMID 19594328.

[14] Sancar, A. (2003). "Structure and function of DNA photolyase and cryptochrome blue-light photoreceptors". *Chem Rev.* **103** (6): 2203–37. doi:10.1021/cr0204348. PMID 12797829.

[15] Michael Lynch, José Ignacio Lucas-Lledó; Lynch, M. (19 February 2009). "Evolution of Mutation Rates: Phylogenomic Analysis of the Photolyase/Cryptochrome Family". *Molecular Biology and Evolution.* **26** (5): 1143–1153. doi:10.1093/molbev/msp029. PMC 2668831. PMID 19228922.

[16] Watson JD, Baker TA, Bell SP, Gann A, Levine M, Losick R. (2004). *Molecular Biology of the Gene.* Pearson Benjamin Cummings; CSHL Press. 5th ed., chapters 9 and 10

[17] Volkert MR (1988). "Adaptive response of Escherichia coli to alkylation damage". *Environmental and Molecular Mutagenesis.* **11** (2): 241–55. doi:10.1002/em.2850110210. PMID 3278898.

[18] Willey, J; Sherwood, L; Woolverton, C (2014). *Prescott's Microbiology.* New York, New York: McGraw Hill. p. 381. ISBN 978-0-07-3402-40-6.

[19] Reardon, J; Sancar, A (2006). "Purification and Characterization of Escherichia coli and Human Nucleotide Excision Repair Enzyme Systems". *Methods in Enzymology.* **408**: 189–213. doi:10.1016/S0076-6879(06)08012-8. PMID 16793370.

[20] Berg, M; Tymoczko, J; Stryer, L (2012). *Biochemistry 7th edition.* New York: W.H. Freeman and Company. p. 840. ISBN 9781429229364.

[21] Wilson TE, Grawunder U, Lieber MR (July 1997). "Yeast DNA ligase IV mediates non-homologous DNA end joining". *Nature*. **388** (6641): 495–8. doi:10.1038/41365. PMID 9242411.

[22] Moore JK, Haber JE (May 1996). "Cell cycle and genetic requirements of two pathways of nonhomologous end-joining repair of double-strand breaks in Saccharomyces cerevisiae". *Molecular and Cellular Biology*. **16** (5): 2164–73. PMC 231204⊚. PMID 8628283.

[23] Boulton SJ, Jackson SP (September 1996). "Saccharomyces cerevisiae Ku70 potentiates illegitimate DNA double-strand break repair and serves as a barrier to error-prone DNA repair pathways". *The EMBO Journal*. **15** (18): 5093–103. PMC 452249⊚. PMID 8890183.

[24] Wilson TE, Lieber MR (August 1999). "Efficient processing of DNA ends during yeast nonhomologous end joining. Evidence for a DNA polymerase beta (Pol4)-dependent pathway". *The Journal of Biological Chemistry*. **274** (33): 23599–609. doi:10.1074/jbc.274.33.23599. PMID 10438542.

[25] Budman J, Chu G (February 2005). "Processing of DNA for nonhomologous end-joining by cell-free extract". *The EMBO Journal*. **24** (4): 849–60. doi:10.1038/sj.emboj.7600563. PMC 549622⊚. PMID 15692565.

[26] Wang H, Perrault AR, Takeda Y, Qin W, Wang H, Iliakis G (September 2003). "Biochemical evidence for Ku-independent backup pathways of NHEJ". *Nucleic Acids Research*. **31** (18): 5377–88. doi:10.1093/nar/gkg728. PMC 203313⊚. PMID 12954774.

[27] Jung D, Alt FW (January 2004). "Unraveling V(D)J recombination; insights into gene regulation". *Cell*. **116** (2): 299–311. doi:10.1016/S0092-8674(04)00039-X. PMID 14744439.

[28] Truong LN, Li Y, Shi LZ, Hwang PY, He J, Wang H, Razavian N, Berns MW, Wu X (2013). "Microhomology-mediated End Joining and Homologous Recombination share the initial end resection step to repair DNA double-strand breaks in mammalian cells". *Proc. Natl. Acad. Sci. U.S.A.* **110** (19): 7720–5. doi:10.1073/pnas.1213431110. PMC 3651503⊚. PMID 23610439.

[29] Sharma S, Javadekar SM, Pandey M, Srivastava M, Kumari R, Raghavan SC (2015). "Homology and enzymatic requirements of microhomology-dependent alternative end joining". *Cell Death Dis*. **6**: e1697. doi:10.1038/cddis.2015.58. PMC 4385936⊚. PMID 25789972.

[30] Liang L, Deng L, Chen Y, Li GC, Shao C, Tischfield JA (2005). "Modulation of DNA end joining by nuclear proteins". *J. Biol. Chem*. **280** (36): 31442–9. doi:10.1074/jbc.M503776200. PMID 16012167.

[31] Decottignies A (2013). "Alternative end-joining mechanisms: a historical perspective". *Front Genet*. **4**: 48. doi:10.3389/fgene.2013.00048. PMC 3613618⊚. PMID 23565119.

[32] Zahradka K, Slade D, Bailone A, et al. (October 2006). "Reassembly of shattered chromosomes in Deinococcus radiodurans". *Nature*. **443** (7111): 569–73. doi:10.1038/nature05160. PMID 17006450.

[33] Waters LS, Minesinger BK, Wiltrout ME, D'Souza S, Woodruff RV, Walker GC (March 2009). "Eukaryotic Translesion Polymerases and Their Roles and Regulation in DNA Damage Tolerance". *Microbiol. Mol. Biol. Rev.* **73** (1): 134–54. doi:10.1128/MMBR.00034-08. PMC 2650891⊚. PMID 19258535.

[34] LC Colis; P Raychaudhury; AK Basu (2008). "Mutational specificity of gamma-radiation-induced guanine-thymine and thymine-guanine intrastrand cross-links in mammalian cells and translesion synthesis past the guanine-thymine lesion by human DNA polymerase eta". *Biochemistry*. **47** (6): 8070–8079. doi:10.1021/bi800529f. PMID 18616294.

[35] P Raychaudhury; Basu, Ashis K. (2011). "Genetic requirement for mutagenesis of the G[8,5-Me]T cross-link in Escherichia coli: DNA polymerases IV and V compete for error-prone bypass". *Biochemistry*. **50** (12): 2330–2338. doi:10.1021/bi102064z. PMID 21302943.

[36] "Translesion Synthesis". Research.chem.psu.edu. Archived from the original on 2012-03-10. Retrieved 2012-08-14.

[37] Wang, Z (2001). "Translesion synthesis by the UmuC family of DNA polymerases". *Mutat. Res.* **486** (2): 59–70. doi:10.1016/S0921-8777(01)00089-1. PMID 11425512.

[38] Friedberg EC, Walker GC, Siede W, Wood RD, Schultz RA, Ellenberger T. (2006). DNA Repair and Mutagenesis, part 3. ASM Press. 2nd ed.

[39] Bakkenist CJ, Kastan MB (January 2003). "DNA damage activates ATM through intermolecular autophosphorylation and dimer dissociation". *Nature*. **421** (6922): 499–506. doi:10.1038/nature01368. PMID 12556884.

[40] Wei, Qingyi; Lei Li; David Chen (2007). *DNA Repair, Genetic Instability, and Cancer*. World Scientific. ISBN 981-270-014-5.

[41] Schonthal, Axel H. (2004). *Checkpoint Controls and Cancer*. Humana Press. ISBN 1-58829-500-1.

[42] Gartel AL, Tyner AL (June 2002). "The role of the cyclin-dependent kinase inhibitor p21 in apoptosis". *Molecular Cancer Therapeutics*. **1** (8): 639–49. PMID 12479224.

[43] Janion, C. (2001). "Some aspects of the SOS response system-a critical survey". *Acta Biochim Pol*. **48** (3): 599–610. PMID 11833768.

[44] Erill, I; Campoy, S; Barbé, J (2007). "Aeons of distress: an evolutionary perspective on the bacterial SOS response". *FEMS Microbiol Rev.* **31** (6): 637–656. doi:10.1111/j.1574-6976.2007.00082.x. PMID 17883408.

[45] Schlacher, K; Pham, P; Cox, MM; Goodman, MF. (2006). "Roles of DNA Polymerase V and RecA Protein in SOS Damage-Induced Mutation". *Chem. Rev.* **106** (2): 406–419. doi:10.1021/cr0404951. PMID 16464012.

[46] Fry, RC; Begley, TJ; Samson, LD (2004). "Genome-wide responses to DNA-damaging agents". *Annu Rev Microbiol.* **59**: 357–77. doi:10.1146/annurev.micro.59.031805.133658. PMID 16153173.

[47] Espejel, S; Martin, M; Klatt, P; Martin-Caballero, J; Flores, JM; Blasco, MA. (2004). "Shorter telomeres, accelerated ageing and increased lymphoma in DNA-PKcs-deficient mice". *EMBO Rep.* **5** (5): 503–9. doi:10.1038/sj.embor.7400127. PMC 1299048. PMID 15105825.

[48] De Boer, J; Andressoo, JO; De Wit, J; Huijmans, J; Beems, RB; Van Steeg, H; Weeda, G; Van Der Horst, GT; et al. (2002). "Premature aging in mice deficient in DNA repair and transcription". *Science.* **296** (5571): 1276–9. doi:10.1126/science.1070174. PMID 11950998.

[49] Dolle, ME; Busuttil, RA; Garcia, AM; Wijnhoven, S; van Drunen, E; Niedernhofer, LJ; der Horst, G; Hoeijmakers, JH; et al. (2006). "Increased genomic instability is not a prerequisite for shortened life span in DNA repair deficient mice". *Mutation Research.* **596** (1–2): 22–35. doi:10.1016/j.mrfmmm.2005.11.008. PMID 16472827.

[50] Kobayashi, Y; Narumi, I; Satoh, K; Funayama, T; Kikuchi, M; Kitayama, S; Watanabe, H. (2004). "Radiation response mechanisms of the extremely radioresistant bacterium Deinococcus radiodurans". *Biol Sci Space.* **18** (3): 134–5. PMID 15858357.

[51] Spindler, SR. (2005). "Rapid and reversible induction of the longevity, anticancer and genomic effects of caloric restriction". *Mech Ageing Dev.* **126** (9): 960–6. doi:10.1016/j.mad.2005.03.016. PMID 15927235.

[52] Tissenbaum, HA; Guarente, L. (2001). "Increased dosage of a sir-2 gene extends life span in Caenorhabditis elegans". *Nature.* **410** (6825): 227–30. doi:10.1038/35065638. PMID 11242085.

[53] Cohen, HY; Miller, C; Bitterman, KJ; Wall, NR; Hekking, B; Kessler, B; Howitz, KT; Gorospe, M; et al. (2004). "Calorie restriction promotes mammalian cell survival by inducing the SIRT1 deacetylase". *Science.* **305** (5682): 390–2. doi:10.1126/science.1099196. PMID 15205477.

[54] Cabelof, DC; Yanamadala, S; Raffoul, JJ; Guo, Z; Soofi, A; Heydari, AR. (2003). "Caloric restriction promotes genomic stability by induction of base excision repair and reversal of its age-related decline". *DNA Repair (Amst.).* **2** (3):

295–307. doi:10.1016/S1568-7864(02)00219-7. PMID 12547392.

[55] Stuart, JA; Karahalil, B; Hogue, BA; Souza-Pinto, NC; Bohr, VA. (2004). "Mitochondrial and nuclear DNA base excision repair are affected differently by caloric restriction". *FASEB J.* **18** (3): 595–7. doi:10.1096/fj.03-0890fje. PMID 14734635.

[56] Walker, DW; McColl, G; Jenkins, NL; Harris, J; Lithgow, GJ. (2000). "Evolution of lifespan in C. elegans". *Nature.* **405** (6784): 296–7. doi:10.1038/35012693. PMID 10830948.

[57] Johnson, George (28 December 2010). "Unearthing Prehistoric Tumors, and Debate". *The New York Times.* If we lived long enough, sooner or later we all would get cancer.

[58] Alberts, B; Johnson A; Lewis J; et al. (2002). "The Preventable Causes of Cancer". *Molecular biology of the cell* (4th ed.). New York: Garland Science. ISBN 0-8153-4072-9. A certain irreducible background incidence of cancer is to be expected regardless of circumstances: mutations can never be absolutely avoided, because they are an inescapable consequence of fundamental limitations on the accuracy of DNA replication, as discussed in Chapter 5. If a human could live long enough, it is inevitable that at least one of his or her cells would eventually accumulate a set of mutations sufficient for cancer to develop.

[59] Gavande NS, VanderVere-Carozza PS, Hinshaw HD, Jalal SI, Sears CR, Pawelczak KS, Turchi JJ (April 2016). "DNA repair targeted therapy: The past or future of cancer treatment?". *Pharmacol Ther.* **160**: 65–83. doi:10.1016/j.pharmthera.2016.02.003. PMC 4811676. PMID 26896565.

[60] Bryant HE, Schultz N, Thomas HD, Parker KM, Flower D, Lopez E, Kyle S, Meuth M, Curtin NJ, Helleday T (April 2005). "Specific killing of BRCA2-deficient tumours with inhibitors of poly(ADP-ribose) polymerase". *Nature.* **434** (7035): 913–7. doi:10.1038/nature03443. PMID 15829966.

[61] Goldstein M, Kastan MB (2015). "The DNA damage response: implications for tumor responses to radiation and chemotherapy". *Annu Rev Med.* **66**: 129–43. doi:10.1146/annurev-med-081313-121208. PMID 25423595.

[62] Jeggo PA, Pearl LH, Carr AM (Jan 2016). "DNA repair, genome stability and cancer: a historical perspective". *Nat Rev Cancer.* **16** (1): 35–42. doi:10.1038/nrc.2015.4. PMID 26667849.

[63] Bartkova J, Horejsi Z, et al. (April 2005). "DNA damage response as a candidate anti-cancer barrier in early human tumorigenesis". *Nature.* **434** (7035): 864–70. doi:10.1038/nature03482. PMID 15829956.

[64] Bartkova J, Rezaei N, et al. (Nov 2006). "Oncogene-induced senescence is part of the tumorigenesis barrier imposed by DNA damage checkpoints". *Nature*. **444** (7119): 633–7. doi:10.1038/nature05268. PMID 17136093.

[65] Gaillard H, Garcia-Muse T, Aguilera A (May 2015). "Replication stress and cancer". *Nat Rev Cancer*. **15** (5): 276–89. doi:10.1038/nrc3916. PMID 25907220.

[66] Halazonetis TD, Gorgoulis VG, Bartek J (Mar 2008). "An oncogene-induced DNA damage model for cancer development". *Science*. **319**: 1352–5. doi:10.1126/science.1140735. PMID 18323444.

[67] de Boer J, Heoijmakers JH (Mar 2000). "Nucleotide excision repair and human syndromes". *Carcinogenesis*. **21** (3): 453–60. doi:10.1093/carcin/21.3.453. PMID 10688865.

[68] Broustas CG, Lieberman HB (Feb 2014). "DNA damage response genes and the development of cancer metastasis". *Radiat Res*. **181** (2): 111–30. doi:10.1667/RR13515.1. PMC 4064942. PMID 24397478.

[69] Zhang P, Wang J, Gao W, Yuan BZ, Rogers J, Reed E (May 2004). "CHK2 kinase expression is down-regulated due to promoter methylation in non-small cell lung cancer". *Mol Cancer*. **3** (4): 3–14. doi:10.1186/1476-4598-3-14. PMC 419366. PMID 15125777.

[70] Baylin SB; Ohm, Joyce E. (February 2006). "Epigenetic gene silencing in cancer - a mechanism for early oncogenic pathway addiction?". *Nat. Rev. Cancer*. **6** (2): 107–16. doi:10.1038/nrc1799. PMID 16491070.

[71] Kanwal R, Gupta S (April 2012). "Epigenetic modifications in cancer". *Clinical Genetics*. **81** (4): 303–11. doi:10.1111/j.1399-0004.2011.01809.x. PMC 3590802. PMID 22082348.

[72] Baldassarre G, Battista S, Belletti B, et al. (April 2003). "Negative regulation of BRCA1 gene expression by HMGA1 proteins accounts for the reduced BRCA1 protein levels in sporadic breast carcinoma". *Molecular and Cellular Biology*. **23** (7): 2225–38. doi:10.1128/MCB.23.7.2225-2238.2003. PMC 150734. PMID 12640109.

[73] Jacinto FV, Esteller M; Esteller, M. (July 2007). "Mutator pathways unleashed by epigenetic silencing in human cancer". *Mutagenesis*. **22** (4): 247–53. doi:10.1093/mutage/gem009. PMID 17412712.

[74] Lahtz C; Pfeifer, G. P. (February 2011). "Epigenetic changes of DNA repair genes in cancer". *J Mol Cell Biol*. **3** (1): 51–8. doi:10.1093/jmcb/mjq053. PMC 3030973. PMID 21278452. http://jmcb.oxfordjournals.org/content/3/1/51.long

[75] Bernstein C, Nfonsam V, Prasad AR, Bernstein H (March 2013). "Epigenetic field defects in progression to cancer". *World J Gastrointest Oncol*. **5** (3): 43–9. doi:10.4251/wjgo.v5.i3.43. PMC 3648662. PMID 23671730.

[76] Bernstein C, Prasad AR, Nfonsam V, Bernstein H. (2013). "Chapter 16: DNA Damage, DNA Repair and Cancer", *New Research Directions in DNA Repair*, Prof. Clark Chen (Ed.), ISBN 978-953-51-1114-6, InTech.

[77] Narayanan L, Fritzell JA, Baker SM, Liskay RM, Glazer PM (April 1997). "Elevated levels of mutation in multiple tissues of mice deficient in the DNA mismatch repair gene Pms2". *Proceedings of the National Academy of Sciences of the United States of America*. **94** (7): 3122–7. doi:10.1073/pnas.94.7.3122. PMC 20332. PMID 9096356.

[78] Hegan DC, Narayanan L, Jirik FR, Edelmann W, Liskay RM, Glazer PM (December 2006). "Differing patterns of genetic instability in mice deficient in the mismatch repair genes Pms2, Mlh1, Msh2, Msh3 and Msh6". *Carcinogenesis*. **27** (12): 2402–8. doi:10.1093/carcin/bgl079. PMC 2612936. PMID 16728433.

[79] Tutt AN, van Oostrom CT, Ross GM, van Steeg H, Ashworth A (March 2002). "Disruption of Brca2 increases the spontaneous mutation rate in vivo: synergism with ionizing radiation". *EMBO Reports*. **3** (3): 255–60. doi:10.1093/embo-reports/kvf037. PMC 1084010. PMID 11850397.

[80] German J (March 1969). "Bloom's syndrome. I. Genetical and clinical observations in the first twenty-seven patients". *American Journal of Human Genetics*. **21** (2): 196–227. PMC 1706430. PMID 5770175.

[81] O'Hagan HM, Mohammad HP, Baylin SB (2008). "Double strand breaks can initiate gene silencing and SIRT1-dependent onset of DNA methylation in an exogenous promoter CpG island". *PLoS Genetics*. **4** (8): e1000155. doi:10.1371/journal.pgen.1000155. PMC 2491723. PMID 18704159.

[82] Cuozzo C, Porcellini A, Angrisano T, et al. (July 2007). "DNA damage, homology-directed repair, and DNA methylation". *PLoS Genetics*. **3** (7): e110. doi:10.1371/journal.pgen.0030110. PMC 1913100. PMID 17616978.

[83] Malkin D (April 2011). "Li-fraumeni syndrome". *Genes & Cancer*. **2** (4): 475–84. doi:10.1177/1947601911413466. PMC 3135649. PMID 21779515.

[84] Fearon ER (November 1997). "Human cancer syndromes: clues to the origin and nature of cancer". *Science*. **278** (5340): 1043–50. doi:10.1126/science.278.5340.1043. PMID 9353177.

[85] Halford S, Rowan A, Sawyer E, Talbot I, Tomlinson I (June 2005). "O(6)-methylguanine methyltransferase in colorectal cancers: detection of mutations, loss of expression, and weak association with G:C>A:T transitions". *Gut*. **54** (6): 797–802. doi:10.1136/gut.2004.059535. PMC 1774551. PMID 15888787.

[86] Shen L, Kondo Y, Rosner GL, et al. (September 2005). "MGMT promoter methylation and field defect in sporadic colorectal cancer". *Journal of the National Cancer Institute.* **97** (18): 1330–8. doi:10.1093/jnci/dji275. PMID 16174854.

[87] Psofaki V, Kalogera C, Tzambouras N, et al. (July 2010). "Promoter methylation status of hMLH1, MGMT, and CDKN2A/p16 in colorectal adenomas". *World Journal of Gastroenterology.* **16** (28): 3553–60. doi:10.3748/wjg.v16.i28.3553. PMC 2909555☉. PMID 20653064.

[88] Lee KH, Lee JS, Nam JH, et al. (October 2011). "Promoter methylation status of hMLH1, hMSH2, and MGMT genes in colorectal cancer associated with adenoma-carcinoma sequence". *Langenbeck's Archives of Surgery.* **396** (7): 1017–26. doi:10.1007/s00423-011-0812-9. PMID 21706233.

[89] Amatu A, Sartore-Bianchi A, Moutinho C, et al. (April 2013). "Promoter CpG island hypermethylation of the DNA repair enzyme MGMT predicts clinical response to dacarbazine in a phase II study for metastatic colorectal cancer". *Clinical Cancer Research.* **19** (8): 2265–72. doi:10.1158/1078-0432.CCR-12-3518. PMID 23422094.

[90] Mokarram P, Zamani M, Kavousipour S, et al. (May 2013). "Different patterns of DNA methylation of the two distinct O6-methylguanine-DNA methyltransferase (O6-MGMT) promoter regions in colorectal cancer". *Molecular Biology Reports.* **40** (5): 3851–7. doi:10.1007/s11033-012-2465-3. PMID 23271133.

[91] Truninger K, Menigatti M, Luz J, et al. (May 2005). "Immunohistochemical analysis reveals high frequency of PMS2 defects in colorectal cancer". *Gastroenterology.* **128** (5): 1160–71. doi:10.1053/j.gastro.2005.01.056. PMID 15887099.

[92] Valeri N, Gasparini P, Fabbri M, et al. (April 2010). "Modulation of mismatch repair and genomic stability by miR-155". *Proceedings of the National Academy of Sciences of the United States of America.* **107** (15): 6982–7. doi:10.1073/pnas.1002472107. PMC 2872463☉. PMID 20351277.

[93] Carol Bernstein and Harris Bernstein (2015). Epigenetic Reduction of DNA Repair in Progression to Cancer. Advances in DNA Repair, Prof. Clark Chen (Ed.), ISBN 978-953-51-2209-8. InTech. Available from: http://www.intechopen.com/books/advances-in-dna-repair/epigenetic-reduction-of-dna-repair-in-progression-to-cancer

[94] Facista A, Nguyen H, Lewis C, et al. (2012). "Deficient expression of DNA repair enzymes in early progression to sporadic colon cancer". *Genome Integrity.* **3** (1): 3. doi:10.1186/2041-9414-3-3. PMC 3351028☉. PMID 22494821.

[95] Human DNA Repair Genes. 15 April 2014. MD Anderson Cancer Center. University of Texas

[96] Xinrong Chen and Tao Chen (2011). "Chapter 18: Roles of MicroRNA in DNA Damage and Repair", *DNA Repair,* Inna Kruman (ed.), ISBN 978-953-307-697-3, InTech.

[97] Krishnan K, Steptoe AL, Martin HC, et al. (February 2013). "MicroRNA-182-5p targets a network of genes involved in DNA repair". *RNA.* **19** (2): 230–42. doi:10.1261/rna.034926.112. PMC 3543090☉. PMID 23249749.

[98] Chaisaingmongkol J, Popanda O, Warta R, et al. (December 2012). "Epigenetic screen of human DNA repair genes identifies aberrant promoter methylation of NEIL1 in head and neck squamous cell carcinoma". *Oncogene.* **31** (49): 5108–16. doi:10.1038/onc.2011.660. PMID 22286769.

[99] Liang L, Deng L, Chen Y, Li GC, Shao C, Tischfield JA (2005). "Modulation of DNA end joining by nuclear proteins". *J. Biol. Chem.* **280** (36): 31442–9. doi:10.1074/jbc.M503776200. PMID 16012167.

[100] Singh P, Yang M, Dai H, Yu D, Huang Q, Tan W, Kernstine KH, Lin D, Shen B (2008). "Overexpression and hypomethylation of flap endonuclease 1 gene in breast and other cancers". *Mol. Cancer Res.* **6** (11): 1710–7. doi:10.1158/1541-7786.MCR-08-0269. PMC 2948671☉. PMID 19010819.

[101] Lam JS, Seligson DB, Yu H, Li A, Eeva M, Pantuck AJ, Zeng G, Horvath S, Belldegrun AS (2006). "Flap endonuclease 1 is overexpressed in prostate cancer and is associated with a high Gleason score". *BJU Int.* **98** (2): 445–51. doi:10.1111/j.1464-410X.2006.06224.x. PMID 16879693.

[102] Kim JM, Sohn HY, Yoon SY, Oh JH, Yang JO, Kim JH, Song KS, Rho SM, Yoo HS, Yoo HS, Kim YS, Kim JG, Kim NS (2005). "Identification of gastric cancer-related genes using a cDNA microarray containing novel expressed sequence tags expressed in gastric cancer cells". *Clin. Cancer Res.* **11** (2 Pt 1): 473–82. PMID 15701830.

[103] Wang K, Xie C, Chen D (2014). "Flap endonuclease 1 is a promising candidate biomarker in gastric cancer and is involved in cell proliferation and apoptosis". *Int. J. Mol. Med.* **33** (5): 1268–74. doi:10.3892/ijmm.2014.1682. PMID 24590400.

[104] Krause A, Combaret V, Iacono I, Lacroix B, Compagnon C, Bergeron C, Valsesia-Wittmann S, Leissner P, Mougin B, Puisieux A (2005). "Genome-wide analysis of gene expression in neuroblastomas detected by mass screening". *Cancer Lett.* **225** (1): 111–20. doi:10.1016/j.canlet.2004.10.035. PMID 15922863.

[105] Iacobuzio-Donahue CA, Maitra A, Olsen M, Lowe AW, van Heek NT, Rosty C, Walter K, Sato N, Parker A, Ashfaq R, Jaffee E, Ryu B, Jones J, Eshleman JR, Yeo CJ, Cameron JL, Kern SE, Hruban RH, Brown PO, Goggins M (2003). "Exploration of global gene expression patterns in pancreatic adenocarcinoma using cDNA microarrays". *Am. J. Pathol.*

**162** (4): 1151–62. doi:10.1016/S0002-9440(10)63911-9. PMC 1851213☉. PMID 12651607.

[106] Sato M, Girard L, Sekine I, Sunaga N, Ramirez RD, Kamibayashi C, Minna JD (2003). "Increased expression and no mutation of the Flap endonuclease (FEN1) gene in human lung cancer". *Oncogene*. **22** (46): 7243–6. doi:10.1038/sj.onc.1206977. PMID 14562054.

[107] Bi FF, Li D, Yang Q (2013). "Hypomethylation of ETS transcription factor binding sites and upregulation of PARP1 expression in endometrial cancer". *Biomed Res Int*. **2013**: 946268. doi:10.1155/2013/946268. PMC 3666359☉. PMID 23762867.

[108] Li D, Bi FF, Cao JM, Cao C, Li CY, Liu B, Yang Q (2014). "Poly (ADP-ribose) polymerase 1 transcriptional regulation: a novel crosstalk between histone modification H3K9ac and ETS1 motif hypomethylation in BRCA1-mutated ovarian cancer". *Oncotarget*. **5** (1): 291–7. doi:10.18632/oncotarget.1549. PMC 3960209☉. PMID 24448423.

[109] Bi FF, Li D, Yang Q (2013). "Promoter hypomethylation, especially around the E26 transformation-specific motif, and increased expression of poly (ADP-ribose) polymerase 1 in BRCA-mutated serous ovarian cancer". *BMC Cancer*. **13**: 90. doi:10.1186/1471-2407-13-90. PMC 3599366☉. PMID 23442605.

[110] Cromie, GA; Connelly, JC; Leach, DR (2001). "Recombination at double-strand breaks and DNA ends: conserved mechanisms from phage to humans". *Mol Cell*. **8** (6): 1163–74. doi:10.1016/S1097-2765(01)00419-1. PMID 11779493.

[111] O'Brien, PJ. (2006). "Catalytic promiscuity and the divergent evolution of DNA repair enzymes". *Chem Rev*. **106** (2): 720–52. doi:10.1021/cr040481v. PMID 16464022.

[112] Maresca, B; Schwartz, JH (2006). "Sudden origins: a general mechanism of evolution based on stress protein concentration and rapid environmental change". *Anat Rec B New Anat*. **289** (1): 38–46. doi:10.1002/ar.b.20089. PMID 16437551.

[113] DeJong W, Degens H (2011). "The Evolutionary Dynamics of Digital and Nucleotide Codes: A Mutation Protection Perspective" (PDF). *Open Evolution Journal, 5: 1-4*.

[114] "CRISPR gene-editing tool has scientists thrilled — but nervous" CBC news. Author Kelly Crowe. November 30, 2015.

## 20.11 External links

- Roswell Park Cancer Institute DNA Repair Lectures

- A comprehensive list of Human DNA Repair Genes

- 3D structures of some DNA repair enzymes

- Human DNA repair diseases

- DNA repair special interest group

- DNA Repair

- DNA Damage and DNA Repair

- Segmental Progeria

- DNA-damage repair; the good, the bad, and the ugly

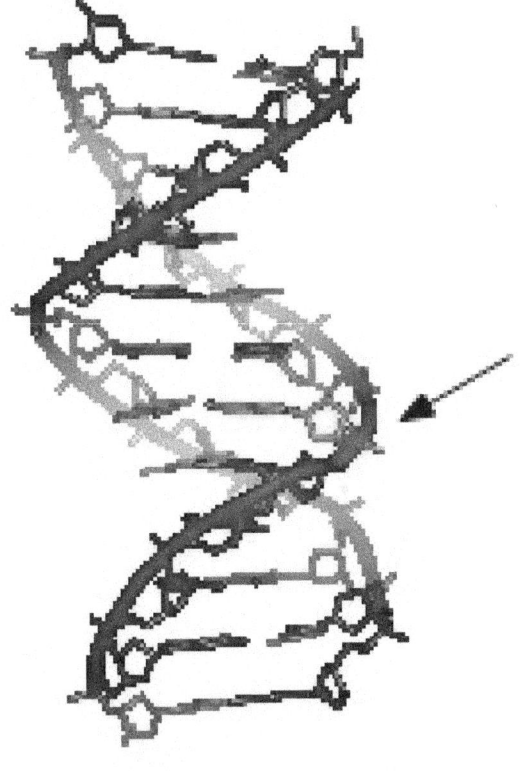

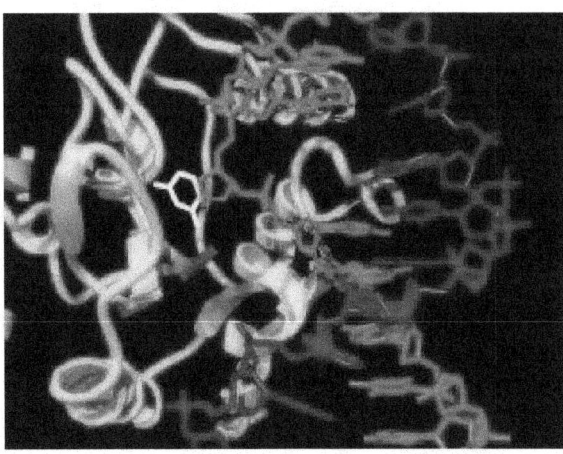

Structure of the base-excision repair enzyme uracil-DNA glycosylase excising a hydrolytically-produced uracil residue from DNA. The uracil residue is shown in yellow.

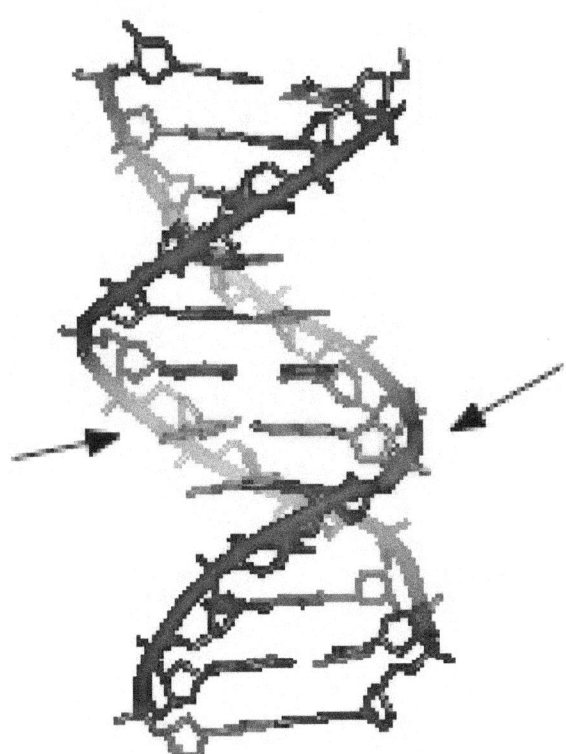

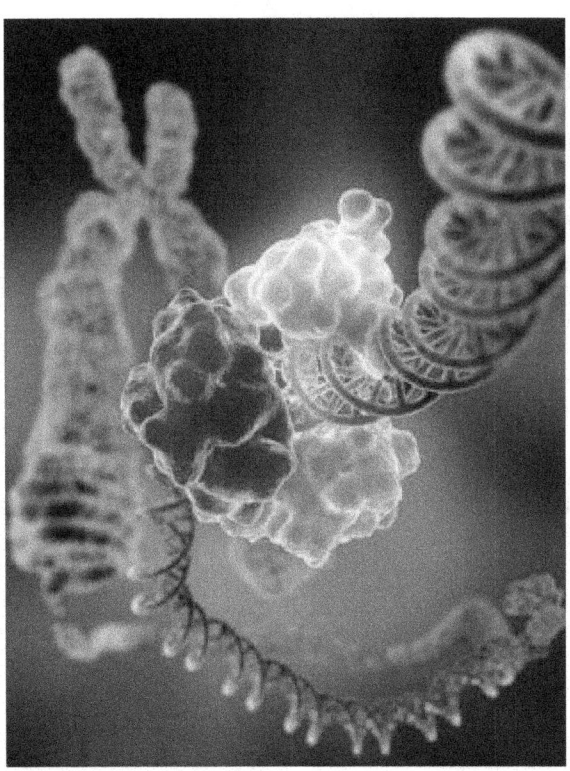

DNA ligase, shown above repairing chromosomal damage, is an enzyme that joins broken nucleotides together by catalyzing the formation of an internucleotide ester bond between the phosphate backbone and the deoxyribose nucleotides.

Single-strand and double-strand DNA damage

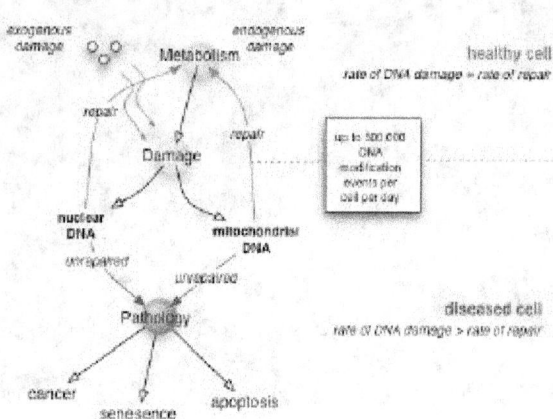

*DNA repair rate is an important determinant of cell pathology*

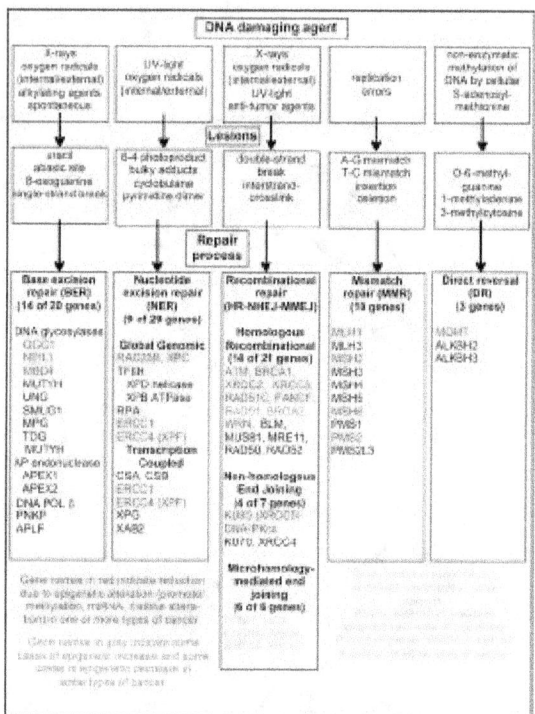

A chart of common DNA damaging agents, examples of lesions they cause in DNA, and pathways used to repair these lesions. Also shown are many of the genes in these pathways, an indication of which genes are epigenetically regulated to have reduced (or increased) expression in various cancers. It also shows genes in the error prone microhomology-mediated end joining pathway with increased expression in various cancers.

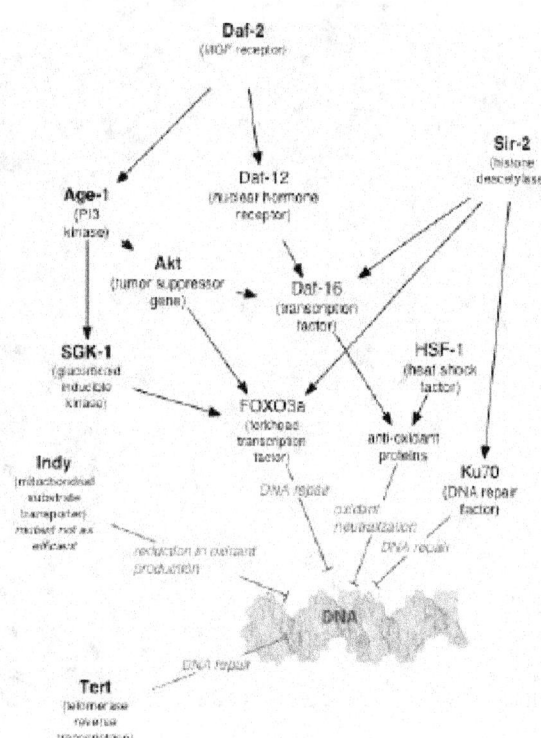

*Most life span influencing genes affect the rate of DNA damage*

# Chapter 21

# Okazaki fragments

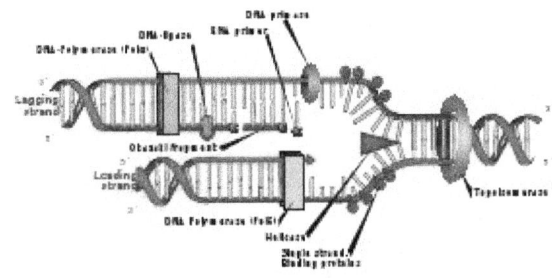

*DNA replication*

**Okazaki fragments** are short, newly synthesized DNA fragments that are formed on the lagging template strand during DNA replication. They are complementary to the lagging template strand, together forming short double-stranded DNA sections. Okazaki fragments are between 1000 and 2000 nucleotides long in prokaryotes (e.g. *Escherichia coli*) and are roughly 100 to 200 nucleotides long in eukaryotes.[1] They are separated by ~§120-nucleotide RNA primers and are unligated until RNA primers are removed, followed by enzyme ligase connecting (ligating) an Okazaki fragment onto the (now continuous) newly synthesized complementary strand.

On the leading strand DNA replication proceeds continuously along the DNA molecule as the parent double-stranded DNA is unwound, but on the lagging strand the new DNA is made in installments, which are later joined together by a DNA ligase enzyme. This is because the enzymes that synthesise the new DNA can only work in one direction along the parent DNA molecule and the two strands are anti-parallel . On the leading strand this route is continuous, but on the lagging strand it is discontinuous.[2]

DNA is synthesised from 5' to 3', so when copying the 3' to 5' strand, replication is continuous. Phosphodiester links form between the 3' to 5' and nucleotides can be added with the aid of the enzyme DNA polymerase for the continuous leading strand. However, in order to synthesise the lagging strand (the 5' to 3' strand) synthesis must occur in small sections (100-200 nucleotides at a time in eukaryotes). These

new stretches of DNA are" called Okazaki fragments and each one requires its own RNA primer.

A series of experiments eventually lead to the discovery of Okazaki fragments. The experiments (see below) were conducted during the 1960s with Reiji Okazaki, Tsuneko Okazaki, Kiwako Sakabe and their colleagues during their research on DNA replication of *Escherichia coli*.[3] In 1966, Kiwako Sakabe and Reiji Okazaki first showed that DNA replication was a discontinuous process involving fragments.[4] The fragments were further investigated by the researchers and their colleagues through their research including the study on bacteriophage DNA replication in *Escherichia coli*.[5][6][7]

## 21.1 Experiments

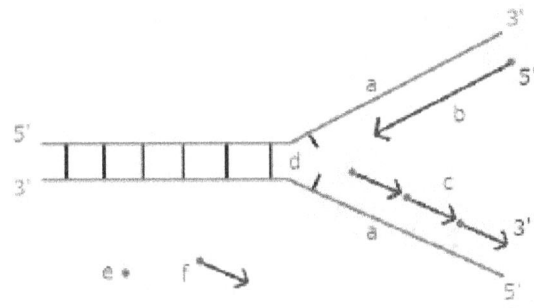

*Synthesis of Okazaki fragments*

The work of Kiwako Sakabe and Reiji Okazaki provided experimental evidence supporting the hypothesis that DNA replication is a discontinuous process. Previously, it was commonly accepted that replication was continuous in both the 3' to 5' and 5' to 3' directions. 3' and 5' are specifically numbered carbons on the deoxyribose ring in nucleic acids, and refer to the orientation or directionality of a strand. In 1967, the Okazakis and their colleagues suggested that there is no found mechanism that showed continuous replication in the 3' to 5' direction, only 5' to 3' using DNA polymerase, a replication enzyme. The team hypothesized that if discontinuous replication was used, short strands of DNA, synthesized at the replicating point, could be attached in the 5' to 3' direction to the older strand.[7]

To distinguish the method of replication used by DNA experimentally, the team pulse-labeled newly replicated areas of *Escherichia coli* chromosomes, denatured, and extracted the DNA. A large amount of radioactive short units meant that the replication method was likely discontinuous. The hypothesis was further supported by the discovery of polynucleotide ligase, an enzyme that links short DNA strands together.[8]

In 1968, Reiji and Tsuneko Okazaki gathered additional evidence of nascent DNA strands. They hypothesized that if discontinuous replication, involving short DNA chains linked together by polynucleotide ligase, is the mechanism used in DNA synthesis, then "newly synthesized short DNA chains would accumulate in the cell under conditions where the function of ligase is temporarily impaired." *E. coli* were infected with bacteriophage T4 that produce temperature-sensitive polynucleotide ligase. The cells infected with the T4 phages accumulated a large amount of short, newly synthesized DNA chains, as predicted in the hypothesis, when exposed to high temperatures. This experiment further supported the Okazakis' hypothesis of discontinuous replication and linkage by polynucleotide ligase. It disproved the notion that short chains were produced during the extraction process as well.[9]

The Okazakis' experiments provided extensive information on the replication process of DNA and the existence of short, newly synthesized DNA chains that later became known as Okazaki fragments.

## 21.2 Process

### 21.2.1 Pathways

There are two pathways that have been proposed to process Okazaki fragments. In the first pathway, only the nuclease FEN1 is involved. FEN1 cleaves the short "flaps" (or short sections of single stranded DNA that "hang off" because their nucleotide bases are prevented from binding to their complementary base pair—despite any base pairing downstream) immediately when they form. While this pathway can process basically all flaps, an issue with this pathway is that some flaps may escape cleavage and thus become long. These flaps then bind to replication protein A (RPA) which inhibits FEN1 cleavage.[10] The second pathway thus becomes involved and is able to utilize both FEN1 and Dna2 nucleases to process the long flaps. Dna2 can cleave the

RPA bound flap as it is able to displace to RPA, while creating a flap to which RPA cannot bind. Then, FEN1 will complete the cleavage of the flap. Dna2 is a key part of this process. Without the Dna2, the RPA bound flaps could not be processed which would ultimately lead to cell instability. The Pif1 helicase is also involved in this pathway as it aids creation of long flaps. Without the Pif1 helicase, the flaps would not become long enough to need cleavage by Dna2. Recently, it has been suggested that an alternative pathway for Okazaki fragment processing exists. This alternative pathway occurs when the Pif1 helicase removes entire Okazaki fragments initiated by fold back flaps.[11]

### 21.2.2 Alternate pathway

Until recently, there were only two known pathways to process Okazaki fragments. However, current investigations have concluded that a new pathway for Okazaki fragmentation and DNA replication exists. This alternate pathway involves the enzymes Pol δ with Pif1 which perform the same flap removal process as Polδ and FEN1.[11]

## 21.3 Enzymes involved in fragment formation

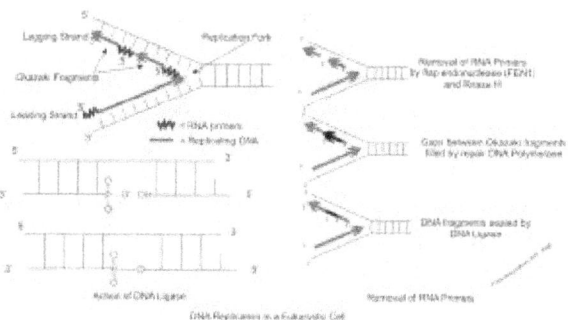

### 21.3.1 Primase

Main article: Primase

Primase adds RNA primers onto the lagging strand, which allows synthesis of Okazaki fragments from 5' to 3'. However, primase creates RNA primers at a much lower rate than that at which DNA polymerase synthesizes DNA on the leading strand. DNA polymerase on the lagging strand also has to be continually recycled to construct Okazaki fragments following RNA primers. This makes the speed of lagging strand synthesis much lower than that of the leading strand. To solve this, primase acts as a temporary stop signal, briefly halting the progression of the replication fork during DNA replication. This molecular process prevents the leading strand from overtaking the lagging strand.[12]

### 21.3.2   DNA polymerase δ

Following creation of RNA primers by primase on the lagging strand, DNA polymerase δ synthesizes Okazaki fragments. DNA polymerase δ also carries out a 3' to 5' exonuclease role, proofreading newly synthesized DNA strands during DNA replication. When the polymerase encounters an erroneous base pair, it removes one of the nucleotides and replaces it with a correct one. A third function of DNA polymerase δ is to supplement FEN1/RAD27 5' Flap Endonuclease activity. This includes preventing and removing the strand displacement of 5' flaps, and creating ligatable nicks at the border of Okazaki fragments.[13][14]

### 21.3.3   DNA ligase I

Main article: DNA ligase

During lagging strand synthesis, DNA ligase I connects the Okazaki fragments, following replacement of the RNA primers with DNA nucleotides by DNA polymerase δ. Okazaki fragments that are not ligated could cause double-strand-breaks, which cleaves the DNA. Since only a small number of double-strand breaks are tolerated, and only a small number can be repaired, enough ligation failures could be lethal to the cell.

Further research implicates the supplementary role of proliferating cell nuclear antigen (PCNA) to DNA ligase I's function of joining Okazaki fragments. When the PCNA binding site on DNA ligase I is inactive, DNA ligase I's ability to connect Okazaki fragments is severely impaired. Thus, a proposed mechanism follows: after a PCNA-DNA polymerase δ complex synthesizes Okazaki fragments, the DNA polymerase δ is released. Then, DNA ligase I binds to the PCNA, which is clamped to the nicks of the lagging strand, and catalyzes the formation of phosphodiester bonds.[14][15][16]

### 21.3.4   Flap endonuclease 1

Main article: Flap structure-specific endonuclease 1

Flap endonuclease 1 (FEN1) is responsible for processing Okazaki fragments. It works with DNA polymerase to remove the RNA primer of an Okazaki fragment and can remove the 5' ribonucleotide and 5' flaps when DNA poly-

merase displaces the strands during lagging strand synthesis. The removal of these flaps involves a process called nick translation and creates a nick for ligation. Thus, FEN1's function is necessary to Okazaki fragment maturation in forming a long continuous DNA strand. Likewise, during DNA base repair, the damaged nucleotide is displaced into a flap and subsequently removed by FEN1.[13][17]

### 21.3.5   Dna2 endonuclease

In the presence of a single stranded DNA-binding protein RPA, the DNA 5' flaps become too long, and the nicks no longer fit as substrate for FEN1. This prevents the FEN1 from removing the 5'-flaps. Thus, Dna2's role is to reduce the 3' end of these fragments, making it possible for FEN1 to cut the flaps, and the Okazaki fragment maturation more efficient.[18]

## 21.4   Biological function

Although synthesis of the lagging strand involves only half the DNA in the nucleus, the complexity associated with processing Okazaki fragments is about twice that required to synthesize the leading strand. Even in small species such as yeast, Okazaki fragment maturation happens approximately a million times during a single round of DNA replication. Processing of Okazaki fragments is therefore very common and crucial for DNA replication and cell proliferation.

During this process, RNA and DNA primers are removed, allowing the Okazaki fragments to attach to the lagging DNA strand. While this process seems quite simple and repetitive, defects in Okazaki fragment maturation can cause DNA strand breakage which can cause varying forms of "chromosome aberrations".[19] Severe defects of Okazaki fragment maturation may halt DNA replication and induce cell death. However, while subtle defects do not affect growth, they do result in future varying forms of genome instabilities. Based on the dangers associated with a failure in the DNA process, Okazaki fragments maintain our evolutionary development.

## 21.5   In prokaryotes and eukaryotes

DNA molecules in eukaryotes differ from the circular molecules of prokaryotes in that they are larger and usually have multiple origins of replication. This means that each eukaryotic chromosome is composed of many replicating units of DNA with multiple origins of replication. In comparison, the prokaryotic E. coli chromosome has only

a single origin of replication. In eukaryotes, these replicating forks, which are numerous all along the DNA, form
"bubbles" in the DNA during replication. The replication
fork forms at a specific point called autonomously replicating sequences (ARS). Eukaryotes have a clamp loader complex and a six-unit clamp called the proliferating cell nuclear antigen.[20] The efficient movement of the replication
fork also relies critically on the rapid placement of sliding
clamps at newly primed sites on the lagging DNA strand by
ATP-dependent clamp loader complexes. This means that
the piecewise generation of Okazaki fragments can keep
up with the continuous synthesis of DNA on the leading
strand. These clamp loader complexes are characteristic
of all eukaryotes and separate some of the minor differences in the synthesis of Okazaki fragments in prokaryotes
and eukaryotes.[21] The lengths of Okazaki fragments in
prokaryotes and eukaryotes are different as well. Prokaryotes have Okazaki fragments that are quite longer than
those of eukaryotes. Eukaryotes typically have Okazaki
fragments that are 100 to 200 nucleotides long, whereas
prokaryotic *E. coli* can be 2,000 nucleotides long. The reason for this discrepancy is unknown.

## 21.6  Uses in technology

### 21.6.1  Medical concepts associated with Okazaki fragments

Although cells undergo multiple steps in order to ensure
there are no mutations in the genetic sequence, sometimes
specific deletions and other genetic changes during Okazaki
fragment maturation go unnoticed. Because Okazaki fragments are the set of nucleotides for the lagging strand, any
alteration including deletions, insertions, or duplications
from the original strand can cause a mutation if it is not
detected and fixed. Other causes of mutations include problems with the proteins that aid in DNA replication. For example, a mutation related to primase affects RNA primer
removal and can make the DNA strand more fragile and susceptible to breaks. Another mutation concerns polymerase
α, which impairs the editing of the Okazaki fragment sequence and incorporation of the protein into the genetic
material. Both alterations can lead to chromosomal aberrations, unintentional genetic rearrangement, and a variety
of cancers later in life.[19]

To test the effects of the protein mutations on living organisms, researchers genetically altered lab mice to be homozygous for another mutation in protein related to DNA replication, flap endonuclease 1, or FEN1. The results varied
based on the specific gene alterations. Homozygous knock-
out mutant mice experienced a "failure of cell proliferation"
and "early embryonic lethality" (27). Mice with mutation

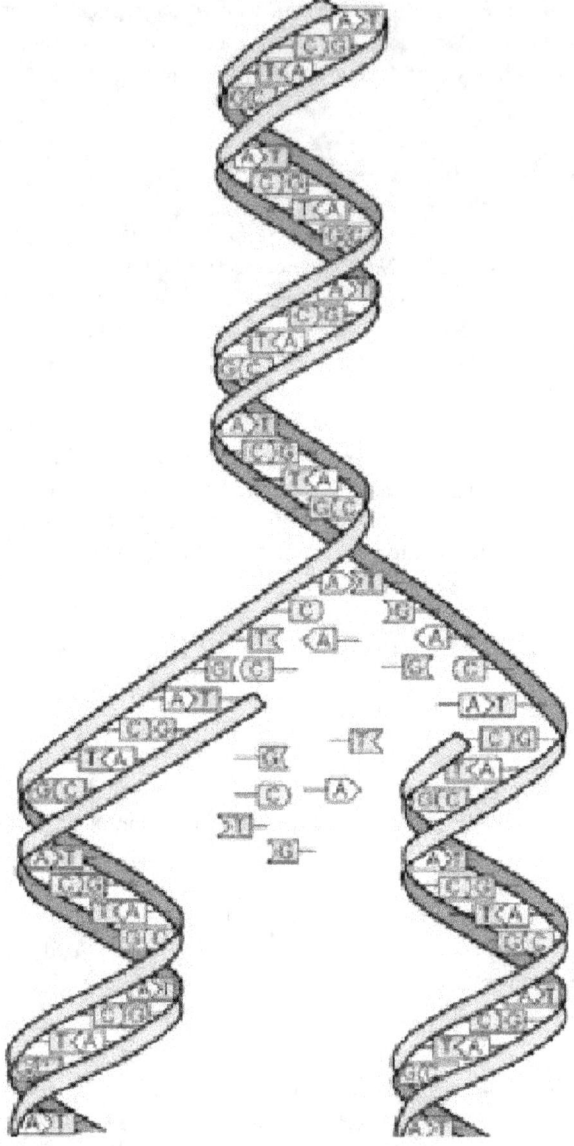

*A slight mutation in the matched nucleotides can lead to chromoso-
mal aberrations and unintentional genetic rearrangement.*

F343A and F344A (also known as FFAA) died directly after birth due to complications including pancytopenia and
pulmonary hypoplasia. This is because the FFAA mutation
keeps FEN1 from interacting with PCNA (proliferating cell
nuclear antigen), consequently not allowing it to complete
its purpose during Okazaki fragment maturation. Under
careful observation, cells homozygous for FFAA FEN1 mutations seem to display only partial defects in maturation,
meaning mice heterozygous for the mutation would be able
to survive into adulthood, despite sustaining multiple small
nicks in their genomes. Inevitably however, these nicks prevent future DNA replication because the break causes the
replication fork to collapse and causes double strand breaks
in the actual DNA sequence. In time, these nicks also cause

full chromosome breaks, which could lead to severe mutations and cancers. Other mutations have been implemented with altered versions of Polymerase α, leading to similar results.[19]

# 21.7   References

[1] Orsay, Jonathan (2014). *Biology 1: Molecules* (9th ed.). Examkrackers, Inc. p. 47. ISBN 9781893858701.

[2] Goldstein, Elliott, Lewin, Benjamin; Jocelyn E. Krebs; Stephen T. Kilpatrick (2011). *Lewin's genes X*. Boston: Jones and Bartlett. p. 329. ISBN 0-7637-7992-X.

[3] Sakabe K, Okazaki R (December 1966). "A unique property of the replicating region of chromosomal DNA". *Biochimica et Biophysica Acta*. **129** (3): 651–54. doi:10.1016/0005-2787(66)90088-8. PMID 5337977.

[4] Moitra, Karobi. *A Journey Through Genetics, Part I*. Biota Publishing. p. 49.

[5] Okazaki R, Okazaki T, Sakabe K, Sugimoto K (June 1967). "Mechanism of DNA replication possible discontinuity of DNA chain growth". *Japanese Journal of Medical Science & Biology*. **20** (3): 255–60. PMID 4861623.

[6] An American scientist, by the last name Shandel, discovered this mechanism prior to Okazaki, however he was never accredited with the discovery since the head of his research team decided the discovery was an erroneous interpretation of test results.

[7] Ogawa T, Okazaki T (1980). "Discontinuous DNA transcription". *Annual Review of Biochemistry*. **49**: 421–57. doi:10.1146/annurev.bi.49.070180.002225. PMID 6250445.

[8] Okazaki R, Okazaki T, Sakabe K, Sugimoto K, Sugino A (February 1968). "Mechanism of DNA chain growth. I. Possible discontinuity and unusual secondary structure of newly synthesized chains". *Proceedings of the National Academy of Sciences of the United States of America*. **59** (2): 598–605. doi:10.1073/pnas.59.2.598. PMC 224714. PMID 4967086.

[9] Sugimoto K, Okazaki T, Okazaki R (August 1968). "Mechanism of DNA chain growth. II. Accumulation of newly synthesized short chains in E. coli infected with ligase-defective T4 phages". *Proceedings of the National Academy of Sciences of the United States of America*. **60** (4): 1356–62. doi:10.1073/pnas.60.4.1356. PMC 224926. PMID 4299945.

[10] Henry RA, Balakrishnan L, Ying-Lin ST, Campbell JL, Bambara RA (September 2010). "Components of the secondary pathway stimulate the primary pathway of eukaryotic Okazaki fragment processing". *The Journal of Biological Chemistry*. **285** (37): 28496–505.

doi:10.1074/jbc.M110.131870. PMC 2937875. PMID 20628185.

[11] Pike JE, Henry RA, Burgers PM, Campbell JL, Bambara RA (December 2010). "An alternative pathway for Okazaki fragment processing: resolution of fold-back flaps by Pif1 helicase". *The Journal of Biological Chemistry*. **285** (53): 41712–23. doi:10.1074/jbc.M110.146894. PMC 3009898. PMID 20959454.

[12] Lee JB, Hite RK, Hamdan SM, Xie XS, Richardson CC, van Oijen AM (February 2006). "DNA primase acts as a molecular brake in DNA replication". *Nature*. **439** (7076): 621–4. doi:10.1038/nature04317. PMID 16452983.

[13] Jin YH, Obert R, Burgers PM, Kunkel TA, Resnick MA, Gordenin DA (April 2001). "The 3'-->5' exonuclease of DNA polymerase delta can substitute for the 5' flap endonuclease Rad27/Fen1 in processing Okazaki fragments and preventing genome instability". *Proceedings of the National Academy of Sciences of the United States of America*. **98** (9): 5122–7. doi:10.1073/pnas.091095198. PMC 33174. PMID 11309502.

[14] Jin YH, Ayyagari R, Resnick MA, Gordenin DA, Burgers PM (January 2003). "Okazaki fragment maturation in yeast. II. Cooperation between the polymerase and 3'–5'-exonuclease activities of Pol delta in the creation of a ligatable nick". *The Journal of Biological Chemistry*. **278** (3): 1626–33. doi:10.1074/jbc.M209803200. PMID 12424237.

[15] Levin DS, Bai W, Yao N, O'Donnell M, Tomkinson AE (November 1997). "An interaction between DNA ligase I and proliferating cell nuclear antigen: implications for Okazaki fragment synthesis and joining". *Proceedings of the National Academy of Sciences of the United States of America*. **94** (24): 12863–8. doi:10.1073/pnas.94.24.12863. PMC 24229. PMID 9371766.

[16] Levin DS, McKenna AE, Motycka TA, Matsumoto Y, Tomkinson AE (2000). "Interaction between PCNA and DNA ligase I is critical for joining of Okazaki fragments and long-patch base-excision repair". *Current Biology*. **10** (15): 919–22. doi:10.1016/S0960-9822(00)00619-9. PMID 10959839.

[17] Liu Y, Kao HI, Bambara RA (2004). "Flap endonuclease 1: a central component of DNA metabolism". *Annual Review of Biochemistry*. **73**: 589–615. doi:10.1146/annurev.biochem.73.012803.092453. PMID 15189154.

[18] Ayyagari R, Gomes XV, Gordenin DA, Burgers PM (January 2003). "Okazaki fragment maturation in yeast. I. Distribution of functions between FEN1 AND DNA2". *The Journal of Biological Chemistry*. **278** (3): 1618–25. doi:10.1074/jbc.M209801200. PMID 12424238.

[19] Zheng L, Shen B (February 2011). "Okazaki fragment maturation: nucleases take centre stage". *Journal of Molecular Cell Biology.* **3** (1): 23–30. doi:10.1093/jmcb/mjq048. PMC 3030970. PMID 21278448.

[20] "Eukaryotic DNA Replication." Molecular-Plant-Biotechnology. multilab.biz, n.d.Web. 29 Mar. 2011.

[21] Matsunaga F, Norais C, Forterre P, Myllykallio H (February 2003). "Identification of short 'eukaryotic' Okazaki fragments synthesized from a prokaryotic replication origin". *EMBO Reports.* **4** (2): 154–8. doi:10.1038/sj.embor.embor732. PMC 1315830. PMID 12612604.

**Notes**

- Inman RB, Schnos M, Structure of branch points in replicating DNA: Presence of single-stranded connections in lambda DNA branch points. *J. Mol Biol.* 56:319-625, 1971.

- Thommes P, Hubscher U. Eukaryotic DNA replication. Enzymes and proteins acting at the fork. *Eur. J. Biochem.* 194(3):699-712, 1990.

## 21.8 External links

- Okazaki fragments at the US National Library of Medicine Medical Subject Headings (MeSH)

- McGraw Hill Higher Education article discussing DNA synthesis

# Chapter 22

# Homologous recombination

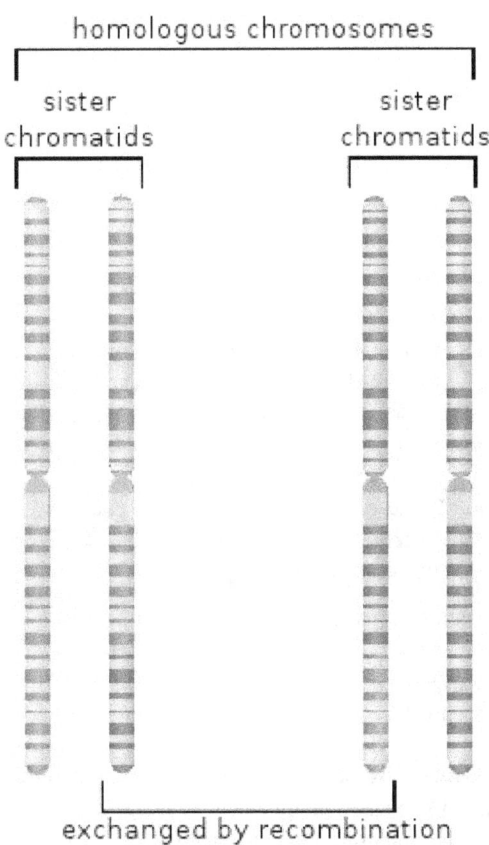

homologous chromosomes

sister chromatids          sister chromatids

exchanged by recombination

*Figure 1. During meiosis, homologous recombination can produce new combinations of genes as shown here between similar but not identical copies of human chromosome 1.*

**Homologous recombination** is a type of genetic recombination in which nucleotide sequences are exchanged between two similar or identical molecules of DNA. It is most widely used by cells to accurately repair harmful breaks that occur on both strands of DNA, known as double-strand breaks. Homologous recombination also produces new combinations of DNA sequences during meiosis, the process by which eukaryotes make gamete cells, like sperm and egg cells in animals. These new combinations of DNA represent genetic variation in offspring, which in turn enables populations to adapt during the course of evolution.[1] Homologous recombination is also used in horizontal gene transfer to exchange genetic material between different strains and species of bacteria and viruses.

Although homologous recombination varies widely among different organisms and cell types, most forms involve the same basic steps. After a double-strand break occurs, sections of DNA around the 5' ends of the break are cut away in a process called *resection*. In the *strand invasion* step that follows, an overhanging 3' end of the broken DNA molecule then "invades" a similar or identical DNA molecule that is not broken. After strand invasion, the further sequence of events may follow either of two main pathways discussed below (see Models); the DSBR (double-strand break repair) pathway or the SDSA (synthesis-dependent strand annealing) pathway. Homologous recombination that occurs during DNA repair tends to result in non-crossover products, in effect restoring the damaged DNA molecule as it existed before the double-strand break.

Homologous recombination is conserved across all three domains of life as well as viruses, suggesting that it is a nearly universal biological mechanism. The discovery of genes for homologous recombination in protists—a diverse group of eukaryotic microorganisms—has been interpreted as evidence that meiosis emerged early in the evolution of eukaryotes. Since their dysfunction has been strongly associated with increased susceptibility to several types of cancer, the proteins that facilitate homologous recombination are topics of active research. Homologous recombination is also used in gene targeting, a technique for introducing genetic changes into target organisms. For their development of this technique, Mario Capecchi, Martin Evans and Oliver Smithies were awarded the 2007 Nobel Prize for Physiology or Medicine; Capecchi[2] and Smithies[3] independently discovered applications to mouse embryonic stem cells, however the highly conserved mechanisms underlying the DSB repair model, including uniform homologous integration of transformed DNA (gene therapy), were first shown in plasmid experiments by Orr-Weaver,

Szostack and Rothstein.[4][5][6] Researching the plasmid-induced DSB, using γ-irradiation[7] in the 1970's-1980's, led to later experiments using endonucleases (e.g. I-SceI) to cut chromosomes for genetic engineering of mammalian cells, where nonhomologous recombination is more frequent than in yeast.[8]

## 22.1   History and discovery

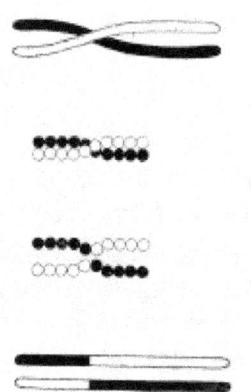

Fιο. 64.   Scheme to illustrate a method of crossing over of the chromosomes.

**Figure 2.** *An early illustration of crossing over from Thomas Hunt Morgan*

In the early 1900s, William Bateson and Reginald Punnett found an exception to one of the principles of inheritance originally described by Gregor Mendel in the 1860s. In contrast to Mendel's notion that traits are independently assorted when passed from parent to child—for example that a cat's hair color and its tail length are inherited independent of each other—Bateson and Punnett showed that certain genes associated with physical traits can be inherited together, or genetically linked.[9][10] In 1911, after observing that linked traits could on occasion be inherited separately, Thomas Hunt Morgan suggested that "crossovers" can occur between linked genes,[11] where one of the linked genes physically crosses over to a different chromosome. Two decades later, Barbara McClintock and Harriet Creighton demonstrated that chromosomal crossover occurs during meiosis,[12][13] the process of cell division by which sperm and egg cells are made. Within the same year as McClintock's discovery, Curt Stern showed that crossing over—later called "recombination"—could also occur in somatic cells like white blood cells and skin cells that divide through mitosis.[12][14]

In 1947, the microbiologist Joshua Lederberg showed that bacteria—which had been assumed to reproduce only asexually through binary fission—are capable of genetic recombination, which is more similar to sexual reproduction. This work established *E. coli* as a model organism in genetics,[15] and helped Lederberg win the 1958 Nobel Prize in Physiology or Medicine.[16] Building on studies in fungi, in 1964 Robin Holliday proposed a model for recombination in meiosis which introduced key details of how the process can work, including the exchange of material between chromosomes through Holliday junctions.[17] In 1983, Jack Szostak and colleagues presented a model now known as the DSBR pathway, which accounted for observations not explained by the Holliday model.[17][18] During the next decade, experiments in *Drosophila*, budding yeast and mammalian cells led to the emergence of other models of homologous recombination, called SDSA pathways, which do not always rely on Holliday junctions.[17]

## 22.2   In eukaryotes

Homologous recombination (HR) is essential to cell division in eukaryotes like plants, animals, fungi and protists. In cells that divide through mitosis, homologous recombination repairs double-strand breaks in DNA caused by ionizing radiation or DNA-damaging chemicals.[19] Left unrepaired, these double-strand breaks can cause large-scale rearrangement of chromosomes in somatic cells,[20] which can in turn lead to cancer.[21]

In addition to repairing DNA, homologous recombination also helps produce genetic diversity when cells divide in meiosis to become specialized gamete cells—sperm or egg cells in animals, pollen or ovules in plants, and spores in fungi. It does so by facilitating chromosomal crossover, in which regions of similar but not identical DNA are exchanged between homologous chromosomes.[22][23] This creates new, possibly beneficial combinations of genes, which can give offspring an evolutionary advantage.[24] Chromosomal crossover often begins when a protein called Spo11 makes a targeted double-strand break in DNA.[25] These sites are non-randomly located on the chromosomes; usually in intergenic promoter regions and preferentially in GC-rich domains[26] These double-strand break sites often occur at recombination hotspots, regions in chromosomes that are about 1,000–2,000 base pairs in length and have high rates of recombination. The absence of a recombination hotspot between two genes on the same chromosome often means that those genes will be inherited by future generations in equal proportion. This represents linkage between the two genes greater than would be expected from genes that independently assort during meiosis.[27]

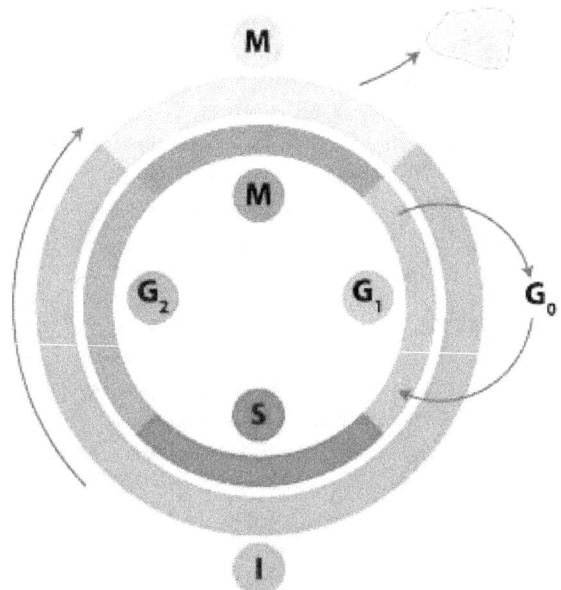

*Figure 3. Homologous recombination repairs DNA before the cell enters mitosis (M phase). It occurs only during and shortly after DNA replication, during the S and $G_2$ phases of the cell cycle.*

### 22.2.1 Timing within the mitotic cell cycle

Double-strand breaks can be repaired through homologous recombination or through non-homologous end joining (NHEJ). NHEJ is a DNA repair mechanism which, unlike homologous recombination, does not require a long homologous sequence to guide repair. Whether homologous recombination or NHEJ is used to repair double-strand breaks is largely determined by the phase of cell cycle. Homologous recombination repairs DNA before the cell enters mitosis (M phase). It occurs during and shortly after DNA replication, in the S and $G_2$ phases of the cell cycle, when sister chromatids are more easily available.[28] Compared to homologous chromosomes, which are similar to another chromosome but often have different alleles, sister chromatids are an ideal template for homologous recombination because they are an identical copy of a given chromosome. In contrast to homologous recombination, NHEJ is predominant in the $G_1$ phase of the cell cycle, when the cell is growing but not yet ready to divide. It occurs less frequently after the $G_1$ phase, but maintains at least some activity throughout the cell cycle. The mechanisms that regulate homologous recombination and NHEJ throughout the cell cycle vary widely between species.[29]

Cyclin-dependent kinases (CDKs), which modify the activity of other proteins by adding phosphate groups to (that is, phosphorylating) them, are important regulators of homologous recombination in eukaryotes.[29] When DNA replication begins in budding yeast, the cyclin-dependent kinase

Cdc28 begins homologous recombination by phosphorylating the Sae2 protein.[30] After being so activated by the addition of a phosphate, Sae2 uses its endonuclease activity to make a clean cut near a double-strand break in DNA. This allows a three-part protein known as the MRX complex to bind to DNA, and begins a series of protein-driven reactions that exchange material between two DNA molecules.[31]

### 22.2.2 Models

Two primary models for how homologous recombination repairs double-strand breaks in DNA are the double-strand break repair (DSBR) pathway (sometimes called the *double Holliday junction model*) and the synthesis-dependent strand annealing (SDSA) pathway.[32] The two pathways are similar in their first several steps. After a double-strand break occurs, the MRX complex (MRN complex in humans) binds to DNA on either side of the break. Next a resection, in which DNA around the 5' ends of the break is cut back, is carried out in two distinct steps. In the first step of resection, the MRX complex recruits the Sae2 protein. The two proteins then trim back the 5' ends on either side of the break to create short 3' overhangs of single-strand DNA. In the second step, 5'→3' resection is continued by the Sgs1 helicase and the Exo1 and Dna2 nucleases. As a helicase, Sgs1 "unzips" the double-strand DNA, while Exo1 and Dna2's nuclease activity allows them to cut the single-stranded DNA produced by Sgs1.[30]

The RPA protein, which has high affinity for single-stranded DNA, then binds the 3' overhangs.[33] With the help of several other proteins that mediate the process, the Rad51 protein (and Dmc1, in meiosis) then forms a filament of nucleic acid and protein on the single strand of DNA coated with RPA. This nucleoprotein filament then begins searching for DNA sequences similar to that of the 3' overhang. After finding such a sequence, the single-stranded nucleoprotein filament moves into (invades) the similar or identical recipient DNA duplex in a process called *strand invasion*. In cells that divide through mitosis, the recipient DNA duplex is generally a sister chromatid, which is identical to the damaged DNA molecule and provides a template for repair. In meiosis, however, the recipient DNA tends to be from a similar but not necessarily identical homologous chromosome.[32] A displacement loop (D-loop) is formed during strand invasion between the invading 3' overhang strand and the homologous chromosome. After strand invasion, a DNA polymerase extends the end of the invading 3' strand by synthesizing new DNA. This changes the D-loop to a cross-shaped structure known as a Holliday junction. Following this, more DNA synthesis occurs on the invading strand (i.e., one of the original 3' overhangs), effectively restoring the strand on the homologous chromosome that was displaced during strand invasion.[32]

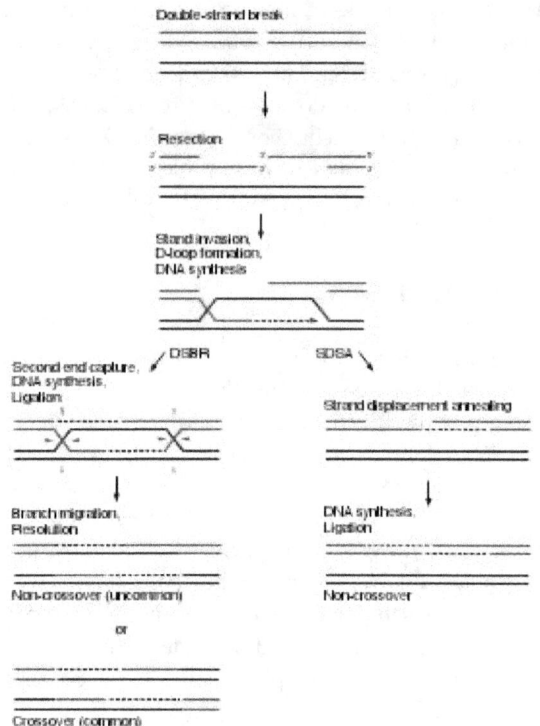

**Figure 4.** *The DSBR and SDSA pathways follow the same initial steps, but diverge thereafter. The DSBR pathway most often results in chromosomal crossover (bottom left), while SDSA always ends with non-crossover products (bottom right).*

## DSBR pathway

After the stages of resection, strand invasion and DNA synthesis, the DSBR and SDSA pathways become distinct.[32] The DSBR pathway is unique in that the second 3' overhang (which was not involved in strand invasion) also forms a Holliday junction with the homologous chromosome. The double Holliday junctions are then converted into recombination products by nicking endonucleases, a type of restriction endonuclease which cuts only one DNA strand. The DSBR pathway commonly results in crossover, though it can sometimes result in non-crossover products; the ability of a broken DNA molecule to collect sequences from separated donor loci was shown in mitotic budding yeast using plasmids or endonuclease induction of chromosomal events.[34][35] Because of this tendency for chromosomal crossover, the DSBR pathway is a likely model of how crossover homologous recombination occurs during meiosis.[22]

Whether recombination in the DSBR pathway results in chromosomal crossover is determined by how the double Holliday junction is cut, or "resolved". Chromosomal crossover will occur if one Holliday junction is cut on the

crossing strand and the other Holliday junction is cut on the non-crossing strand (in Figure 4, along the horizontal purple arrowheads at one Holliday junction and along the vertical orange arrowheads at the other). Alternatively, if the two Holliday junctions are cut on the crossing strands (along the horizontal purple arrowheads at both Holliday junctions in Figure 4), then chromosomes without crossover will be produced.[36]

## SDSA pathway

Homologous recombination via the SDSA pathway occurs in cells that divide through mitosis and meiosis and results in non-crossover products. In this model, the invading 3' strand is extended along the recipient DNA duplex by a DNA polymerase, and is released as the Holliday junction between the donor and recipient DNA molecules slides in a process called *branch migration*. The newly synthesized 3' end of the invading strand is then able to anneal to the other 3' overhang in the damaged chromosome through complementary base pairing. After the strands anneal, a small flap of DNA can sometimes remain. Any such flaps are removed, and the SDSA pathway finishes with the resealing, also known as *ligation*, of any remaining single-stranded gaps.[37]

During mitosis, the major homologous recombination pathway for repairing DNA double-strand breaks appears to be the SDSA pathway (rather than the DSBR pathway).[38] The SDSA pathway produces non-crossover recombinants (Figure 4). During meiosis non-crossover recombinants also occur frequently and these appear to arise mainly by the SDSA pathway as well.[38][39] Non-crossover recombination events occurring during meiosis likely reflect instances of repair of DNA double-strand damages or other types of DNA damages.

## SSA pathway

The single-strand annealing (SSA) pathway of homologous recombination repairs double-strand breaks between two repeat sequences. The SSA pathway is unique in that it does not require a separate similar or identical molecule of DNA, like the DSBR or SDSA pathways of homologous recombination. Instead, the SSA pathway only requires a single DNA duplex, and uses the repeat sequences as the identical sequences that homologous recombination needs for repair. The pathway is relatively simple in concept: after two strands of the same DNA duplex are cut back around the site of the double-strand break, the two resulting 3' overhangs then align and anneal to each other, restoring the DNA as a continuous duplex.[37][40]

As DNA around the double-strand break is cut back, the

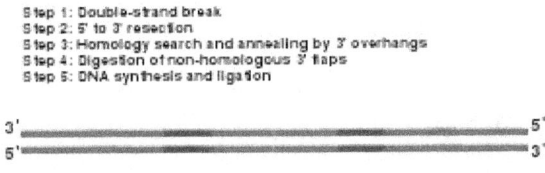

Step 1: Double-strand break
Step 2: 5' to 3' resection
Step 3: Homology search and annealing by 3' overhangs
Step 4: Digestion of non-homologous 3' flaps
Step 5: DNA synthesis and ligation

**Figure 5.** *Recombination via the SSA pathway occurs between two repeat elements (purple) on the same DNA duplex, and results in deletions of genetic material. (Click to view animated diagram in Firefox, Chrome, Safari, or Opera web browsers.)*

single-stranded 3' overhangs being produced are coated with the RPA protein, which prevents the 3' overhangs from sticking to themselves.[41] A protein called Rad52 then binds each of the repeat sequences on either side of the break, and aligns them to enable the two complementary repeat sequences to anneal.[41] After annealing is complete, leftover non-homologous flaps of the 3' overhangs are cut away by a set of nucleases, known as Rad1/Rad10, which are brought to the flaps by the Saw1 and Slx4 proteins.[41][42] New DNA synthesis fills in any gaps, and ligation restores the DNA duplex as two continuous strands.[43] The DNA sequence between the repeats is always lost, as is one of the two repeats. The SSA pathway is considered mutagenic since it results in such deletions of genetic material.[37]

### BIR pathway

During DNA replication, double-strand breaks can sometimes be encountered at replication forks as DNA helicase unzips the template strand. These defects are repaired in the *break-induced replication* (BIR) pathway of homologous recombination. The precise molecular mechanisms of the BIR pathway remain unclear. Three proposed mechanisms have strand invasion as an initial step, but they differ in how they model the migration of the D-loop and later phases of recombination.[44]

The BIR pathway can also help to maintain the length of telomeres (regions of DNA at the end of eukaryotic chromosomes) in the absence of (or in cooperation with) telomerase. Without working copies of the telomerase enzyme, telomeres typically shorten with each cycle of mitosis, which eventually blocks cell division and leads to senescence. In budding yeast cells where telomerase has been inactivated through mutations, two types of "sur-

vivor" cells have been observed to avoid senescence longer than expected by elongating their telomeres through BIR pathways.[44]

Maintaining telomere length is critical for cell immortalization, a key feature of cancer. Most cancers maintain telomeres by upregulating telomerase. However, in several types of human cancer, a BIR-like pathway helps to sustain some tumors by acting as an alternative mechanism of telomere maintenance.[45] This fact has led scientists to investigate whether such recombination-based mechanisms of telomere maintenance could thwart anti-cancer drugs like telomerase inhibitors.[46]

## 22.3  In bacteria

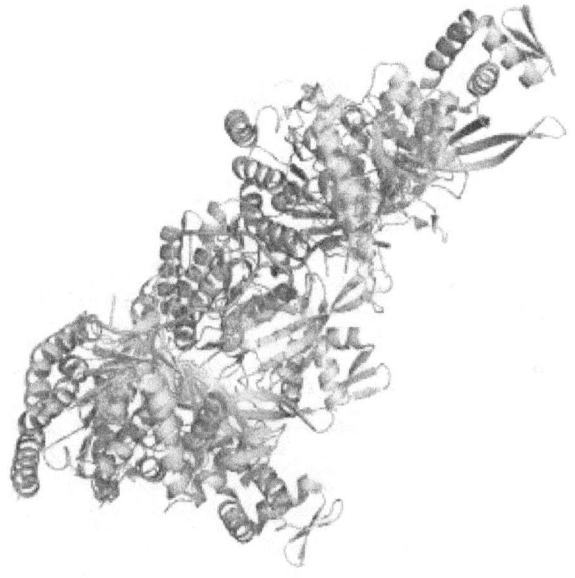

**Figure 6.** *Crystal structure of a RecA protein filament bound to DNA.[47] A 3' overhang is visible to the right of center.*

Homologous recombination is a major DNA repair process in bacteria. It is also important for producing genetic diversity in bacterial populations, although the process differs substantially from meiotic recombination, which repairs DNA damages and brings about diversity in eukaryotic genomes. Homologous recombination has been most studied and is best understood for *Escherichia coli*.[48] Double-strand DNA breaks in bacteria are repaired by the RecBCD pathway of homologous recombination. Breaks that occur on only one of the two DNA strands, known as single-strand gaps, are thought to be repaired by the RecF pathway.[49] Both the RecBCD and RecF pathways include a series of reactions known as *branch migration*, in which single DNA strands are exchanged between two intercrossed molecules of duplex DNA, and *resolution*, in which those two intercrossed molecules of DNA are cut apart and restored to their normal double-stranded state.

## 22.3.1   RecBCD pathway

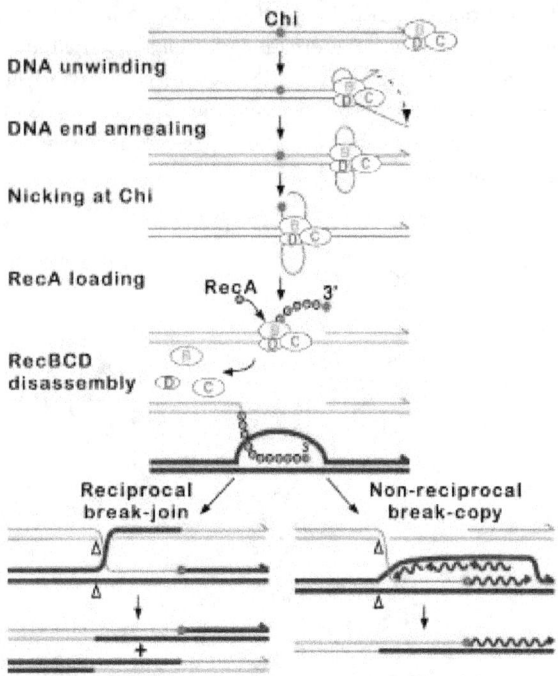

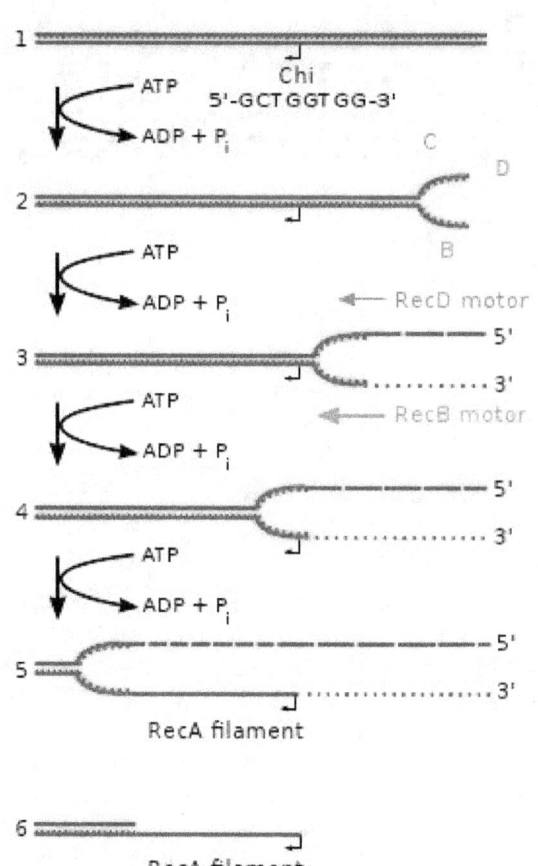

**Figure 7A.** *Molecular model for the RecBCD pathway of recombination. This model is based on reactions of DNA and RecBCD with ATP in excess over Mg2+ ions. Step 1: RecBCD binds to a double-stranded DNA end. Step 2: RecBCD unwinds DNA. RecD is a fast helicase on the 5'-ended strand, and RecB is a slower helicase on the 3'-ended strand (that with an arrowhead) [ref 46 in current Wiki version]. This produces two single-stranded (ss) DNA tails and one ss loop. The loop and tails enlarge as RecBCD moves along the DNA. Step 3: The two tails anneal to produce a second ss DNA loop, and both loops move and grow. Step 4: Upon reaching the Chi hotspot sequence (5' GCTGGTGG 3': red dot) RecBCD nicks the 3'-ended strand. Further unwinding produces a long 3'-ended ss tail with Chi near its end. Step 5: RecBCD loads RecA protein onto the Chi tail. At some undetermined point, the RecBCD subunits disassemble. Step 6: The RecA-ssDNA complex invades an intact homologous duplex DNA to produce a D-loop, which can be resolved into intact, recombinant DNA in two ways. Step 7: The D-loop is cut and anneals with the gap in the first DNA to produce a Holliday junction. Resolution of the Holliday junction (cutting, swapping of strands, and ligation) at the open arrowheads by some combination of RuvABC and RecG produces two recombinants of reciprocal type. Step 8: The 3' end of the Chi tail primes DNA synthesis, from which a replication fork can be generated. Resolution of the fork at the open arrowheads produces one recombinant (non-reciprocal) DNA, one parental-type DNA, and one DNA fragment.*[50]

**Figure 7B.** *Beginning of the RecBCD pathway. This model is based on reactions of DNA and RecBCD with $Mg^{2+}$ ions in excess over ATP. Step 1: RecBCD binds to a DNA double strand break. Step 2: RecBCD initiates unwinding of the DNA duplex through ATP-dependent helicase activity. Step 3: RecBCD continues its unwinding and moves down the DNA duplex, cleaving the 3' strand much more frequently than the 5' strand. Step 4: RecBCD encounters a Chi sequence and stops digesting the 3' strand; cleavage of the 5' strand is significantly increased. Step 5: RecBCD loads RecA onto the 3' strand. Step 6: RecBCD unbinds from the DNA duplex, leaving a RecA nucleoprotein filament on the 3' tail.*[51]

The RecBCD pathway is the main recombination pathway used in many bacteria to repair double-strand breaks in DNA, and the proteins are found in a broad array of bacteria.[52][53][54] These double-strand breaks can be caused by UV light and other radiation, as well as chemical mutagens. Double-strand breaks may also arise by DNA replication through a single-strand nick or gap. Such a situation causes what is known as a collapsed replication fork and is fixed by several pathways of homologous recombination including the RecBCD pathway.[55]

In this pathway, a three-subunit enzyme complex called RecBCD initiates recombination by binding to a blunt or nearly blunt end of a break in double-strand DNA. After RecBCD binds the DNA end, the RecB and RecD subunits begin unzipping the DNA duplex through helicase activity. The RecB subunit also has a nuclease domain, which cuts the single strand of DNA that emerges from the unzipping process. This unzipping continues until RecBCD encounters a specific nucleotide sequence (5'-GCTGGTGG-3') known as a Chi site.[54]

Upon encountering a Chi site, the activity of the RecBCD enzyme changes drastically.[53][56][57] DNA unwinding pauses for a few seconds and then resumes at roughly half the initial speed. This is likely because the slower RecB helicase unwinds the DNA after Chi, rather than the faster RecD helicase, which unwinds the DNA before Chi.[58][59] Recognition of the Chi site also changes the RecBCD enzyme so that it cuts the DNA strand with Chi and begins loading multiple RecA proteins onto the single-stranded DNA with the newly generated 3' end. The resulting RecA-coated nucleoprotein filament then searches out similar sequences of DNA on a homologous chromosome. The search process induces stretching of the DNA duplex, which enhances homology recognition (a mechanism termed conformational proofreading [60][61][62]). Upon finding such a sequence, the single-stranded nucleoprotein filament moves into the homologous recipient DNA duplex in a process called strand invasion.[63] The invading 3' overhang causes one of the strands of the recipient DNA duplex to be displaced, to form a D-loop. If the D-loop is cut, another swapping of strands forms a cross-shaped structure called a Holliday junction.[54] Resolution of the Holliday junction by some combination of RuvABC or RecG can produce two recombinant DNA molecules with reciprocal genetic types, if the two interacting DNA molecules differ genetically. Alternatively, the invading 3' end near Chi can prime DNA synthesis and form a replication fork. This type of resolution produces only one type of recombinant (non-reciprocal).

## 22.3.2   RecF pathway

Further information: RecF pathway

Bacteria appear to use the RecF pathway of homologous recombination to repair single-strand gaps in DNA. When the RecBCD pathway is inactivated by mutations and additional mutations inactivate the SbcCD and ExoI nucleases, the RecF pathway can also repair DNA double-strand breaks.[64] In the RecF pathway the RecQ helicase unwinds the DNA and the RecJ nuclease degrades the strand with a 5' end, leaving the strand with the 3' end intact. RecA protein binds to this strand and is either aided by the RecF, RecO, and RecR proteins or stabilized by them. The

RecA nucleoprotein filament then searches for a homologous DNA and exchanges places with the identical or nearly identical strand in the homologous DNA.

Although the proteins and specific mechanisms involved in their initial phases differ, the two pathways are similar in that they both require single-stranded DNA with a 3' end and the RecA protein for strand invasion. The pathways are also similar in their phases of branch migration, in which the Holliday junction slides in one direction, and resolution, in which the Holliday junctions are cleaved apart by enzymes.[65][66] The alternative, non-reciprocal type of resolution may also occur by either pathway.

## 22.3.3   Branch migration

Immediately after strand invasion, the Holliday junction moves along the linked DNA during the branch migration process. It is in this movement of the Holliday junction that base pairs between the two homologous DNA duplexes are exchanged. To catalyze branch migration, the RuvA protein first recognizes and binds to the Holliday junction and recruits the RuvB protein to form the RuvAB complex. Two sets of the RuvB protein, which each form a ring-shaped ATPase, are loaded onto opposite sides of the Holliday junction, where they act as twin pumps that provide the force for branch migration. Between those two rings of RuvB, two sets of the RuvA protein assemble in the center of the Holliday junction such that the DNA at the junction is sandwiched between each set of RuvA. The strands of both DNA duplexes—the "donor" and the "recipient" duplexes—are unwound on the surface of RuvA as they are guided by the protein from one duplex to the other.[67][68]

## 22.3.4   Resolution

In the resolution phase of recombination, any Holliday junctions formed by the strand invasion process are cut, thereby restoring two separate DNA molecules. This cleavage is done by RuvAB complex interacting with RuvC, which together form the RuvABC complex. RuvC is an endonuclease that cuts the degenerate sequence 5'-(A/T)TT(G/C)–3'. The sequence is found frequently in DNA, about once every 64 nucleotides.[68] Before cutting, RuvC likely gains access to the Holliday junction by displacing one of the two RuvA tetramers covering the DNA there.[67] Recombination results in either "splice" or "patch" products, depending on how RuvC cleaves the Holliday junction.[68] Splice products are crossover products, in which there is a rearrangement of genetic material around the site of recombination. Patch products, on the other hand, are non-crossover products in which there is no such rearrangement and there is only a "patch" of hybrid DNA in the recombination product.[69]

### 22.3.5 Facilitating genetic transfer

Homologous recombination is an important method of integrating donor DNA into a recipient organism's genome in horizontal gene transfer, the process by which an organism incorporates foreign DNA from another organism without being the offspring of that organism. Homologous recombination requires incoming DNA to be highly similar to the recipient genome, and so horizontal gene transfer is usually limited to similar bacteria.[70] Studies in several species of bacteria have established that there is a log-linear decrease in recombination frequency with increasing difference in sequence between host and recipient DNA.[71][72][73]

In bacterial conjugation, where DNA is transferred between bacteria through direct cell-to-cell contact, homologous recombination helps integrate foreign DNA into the host genome via the RecBCD pathway. The RecBCD enzyme promotes recombination after DNA is converted from single-strand DNA–in which form it originally enters the bacterium–to double-strand DNA during replication. The RecBCD pathway is also essential for the final phase of transduction, a type of horizontal gene transfer in which DNA is transferred from one bacterium to another by a virus. Foreign, bacterial DNA is sometimes misincorporated in the capsid head of bacteriophage virus particles as DNA is packaged into new bacteriophages during viral replication. When these new bacteriophages infect other bacteria, DNA from the previous host bacterium is injected into the new bacterial host as double-strand DNA. The RecBCD enzyme then incorporates this double-strand DNA into the genome of the new bacterial host.[54]

### 22.3.6 Bacterial transformation

Natural bacterial transformation involves the transfer of DNA from a donor bacterium to a recipient bacterium, where both donor and recipient are ordinarily of the same species. Transformation, unlike bacterial conjugation and transduction, depends on numerous bacterial gene products that specifically interact to perform this process.[74] Thus transformation is clearly a bacterial adaptation for DNA transfer. In order for a bacterium to bind, take up and integrate donor DNA into its resident chromosome by homologous recombination, it must first enter a special physiological state termed competence. The *RecA/Rad51/DMC1* gene family plays a central role in homologous recombination during bacterial transformation as it does during eukaryotic meiosis and mitosis. For instance, the RecA protein is essential for transformation in *Bacillus subtilis* and

*Streptococcus pneumoniae*,[75] and expression of the RecA gene is induced during the development of competence for transformation in these organisms.

As part of the transformation process, the RecA protein interacts with entering single-stranded DNA (ssDNA) to form RecA/ssDNA nucleofilaments that scan the resident chromosome for regions of homology and bring the entering ssDNA to the corresponding region, where strand exchange and homologous recombination occur.[76] Thus the process of homologous recombination during bacterial transformation has fundamental similarities to homologous recombination during meiosis.

## 22.4  In viruses

Homologous recombination occurs in several groups of viruses. In DNA viruses such as herpesvirus, recombination occurs through a break-and-rejoin mechanism like in bacteria and eukaryotes.[77] There is also evidence for recombination in some RNA viruses, specifically positive-sense ssRNA viruses like retroviruses, picornaviruses, and coronaviruses. There is controversy over whether homologous recombination occurs in negative-sense ssRNA viruses like influenza.[78]

In RNA viruses, homologous recombination can be either precise or imprecise. In the precise type of RNA-RNA recombination, there is no difference between the two parental RNA sequences and the resulting crossover RNA region. Because of this, it is often difficult to determine the location of crossover events between two recombining RNA sequences. In imprecise RNA homologous recombination, the crossover region has some difference with the parental RNA sequences – caused by either addition, deletion, or other modification of nucleotides. The level of precision in crossover is controlled by the sequence context of the two recombining strands of RNA: sequences rich in adenine and uracil decrease crossover precision.[79][80]

Homologous recombination is important in facilitating viral evolution.[79][81] For example, if the genomes of two viruses with different disadvantageous mutations undergo recombination, then they may be able to regenerate a fully functional genome. Alternatively, if two similar viruses have infected the same host cell, homologous recombination can allow those two viruses to swap genes and thereby evolve more potent variations of themselves.[81]

Homologous recombination is the proposed mechanism whereby the DNA virus *human herpesvirus-6* integrates into human telomeres.[82]

When two or more viruses, each containing lethal genomic damage, infect the same host cell, the virus genomes can often pair with each other and undergo homologous recombinational repair to produce viable progeny. This process, known as multiplicity reactivation, has been studied in several bacteriophages, including phage T4.[83] Enzymes employed in recombinational repair in phage T4 are functionally homologous to enzymes employed in bacterial and eukaryotic recombinational repair.[84] In particular, with regard to a gene necessary for the strand exchange reaction, a key step in homologous recombinational repair, there is functional homology from viruses to humans (i. e. *uvsX* in phage T4; *recA* in E. coli and other bacteria, and *rad51* and *dmc1* in yeast and other eukaryotes, including humans).[85] Multiplicity reactivation has also been demonstrated in numerous pathogenic viruses.[86]

## 22.5    Effects of dysfunction

Without proper homologous recombination, chromosomes often incorrectly align for the first phase of cell division in meiosis. This causes chromosomes to fail to properly segregate in a process called nondisjunction. In turn, nondisjunction can cause sperm and ova to have too few or too many chromosomes. Down's syndrome, which is caused by an extra copy of chromosome 21, is one of many abnormalities that result from such a failure of homologous recombination in meiosis.[68][87]

Deficiencies in homologous recombination have been strongly linked to cancer formation in humans. For example, each of the cancer-related diseases Bloom's syndrome, Werner's syndrome and Rothmund-Thomson syndrome are caused by malfunctioning copies of RecQ helicase genes involved in the regulation of homologous recombination: *BLM*, *WRN* and *RECQL4*, respectively.[88] In the cells of Bloom's syndrome patients, who lack a working copy of the BLM protein, there is an elevated rate of homologous recombination.[89] Experiments in mice deficient in BLM have suggested that the mutation gives rise to cancer through a loss of heterozygosity caused by increased homologous recombination.[90] A loss in heterozygosity refers to the loss of one of two versions—or alleles—of a gene. If one of the lost alleles helps to suppress tumors, like the gene for the retinoblastoma protein for example, then the loss of heterozygosity can lead to cancer.[91]:1236

Decreased rates of homologous recombination cause inefficient DNA repair,[91]:310 which can also lead to cancer.[92] This is the case with BRCA1 and BRCA2, two similar tumor suppressor genes whose malfunctioning has been linked with considerably increased risk for breast and ovarian cancer. Cells missing BRCA1 and BRCA2 have a decreased rate of homologous recombination and increased sensitivity to ionizing radiation, suggesting that decreased homologous recombination leads to increased susceptibility

to cancer.[92] Because the only known function of BRCA2 is to help initiate homologous recombination, researchers have speculated that more detailed knowledge of BRCA2's role in homologous recombination may be the key to understanding the causes of breast and ovarian cancer.[92]

Tumours with a homologous recombination deficiency (including BRCA defects) are described as HRD-positive.[93]

## 22.6    Evolutionary conservation

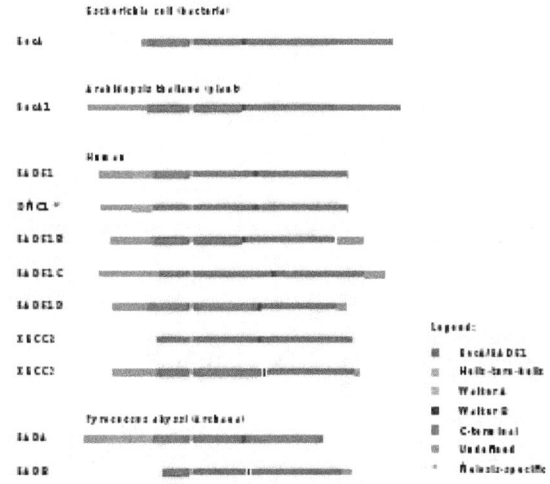

**Figure 8.** *Protein domains in homologous recombination-related proteins are conserved across the three main groups of life: archaea, bacteria and eukaryotes.*

While the pathways can mechanistically vary, the ability of organisms to perform homologous recombination is universally conserved across all domains of life.[94] Based on the similarity of their amino acid sequences, homologs of a number of proteins can be found in multiple domains of life indicating that they evolved a long time ago, and have since diverged from common ancestral proteins.[94]

RecA recombinase family members are found in almost all organisms with RecA in bacteria, Rad51 and DMC1 in eukaryotes, RadA in archaea, and UvsX in T4 phage.[95]

Related single stranded binding proteins that are important for homologous recombination, and many other processes, are also found in all domains of life.[96]

Rad54, Mre11, Rad50, and a number of other proteins are also found in both archaea and eukaryotes.[94][95][97]

### 22.6.1   The RecA recombinase family

The proteins of the RecA recombinase family of proteins are thought to be descended from a common ancestral recombinase.[94] The RecA recombinase family contains RecA protein from bacteria, the Rad51 and Dmc1 proteins from eukaryotes, and RadA from archaea, and the recombinase paralog proteins. Studies modeling the evolutionary relationships between the Rad51, Dmc1 and RadA proteins indicate that they are monophyletic, or that they share a common molecular ancestor.[94] Within this protein family, Rad51 and Dmc1 are grouped together in a separate clade from RadA. One of the reasons for grouping these three proteins together is that they all possess a modified helix-turn-helix motif, which helps the proteins bind to DNA, toward their N-terminal ends.[94] An ancient gene duplication event of a eukaryotic RecA gene and subsequent mutation has been proposed as a likely origin of the modern RAD51 and DMC1 genes.[94]

The proteins generally share a long conserved region known as the RecA/Rad51 domain. Within this protein domain are two sequence motifs, Walker A motif and Walker B motif. The Walker A and B motifs allow members of the RecA/Rad51 protein family to engage in ATP binding and ATP hydrolysis.[94][98]

*Figure 9. As a developing embryo, this chimeric mouse had the agouti coat color gene introduced into its DNA via gene targeting. Its offspring are homozygous for the agouti gene.*

### 22.6.2   Meiosis-specific proteins

The discovery of Dmc1 in several species of *Giardia*, one of the earliest protists to diverge as a eukaryote, suggests that meiotic homologous recombination—and thus meiosis itself—emerged very early in eukaryotic evolution.[99] In addition to research on Dmc1, studies on the Spo11 protein have provided information on the origins of meiotic recombination.[100] Spo11, a type II topoisomerase, can initiate homologous recombination in meiosis by making targeted double-strand breaks in DNA.[25] Phylogenetic trees based on the sequence of genes similar to SPO11 in animals, fungi, plants, protists and archaea have led scientists to believe that the version Spo11 currently in eukaryotes emerged in the last common ancestor of eukaryotes and archaea.[100]

## 22.7   Technological applications

### 22.7.1   Gene targeting

Main article: Gene targeting

Many methods for introducing DNA sequences into organisms to create recombinant DNA and genetically modified organisms use the process of homologous recombination.[101] Also called gene targeting, the method is especially common in yeast and mouse genetics. The gene targeting method in knockout mice uses mouse embryonic stem cells to deliver artificial genetic material (mostly of therapeutic interest), which represses the target gene of the mouse by the principle of homologous recombination. The mouse thereby acts as a working model to understand the effects of a specific mammalian gene. In recognition of their discovery of how homologous recombination can be used to introduce genetic modifications in mice through embryonic stem cells, Mario Capecchi, Martin Evans and Oliver Smithies were awarded the 2007 Nobel Prize for Physiology or Medicine.[102]

Advances in gene targeting technologies which hijack the homologous recombination mechanics of cells are now leading to the development of a new wave of more accurate, isogenic human disease models. These engineered human cell models are thought to more accurately reflect the genetics of human diseases than their mouse model predecessors. This is largely because mutations of interest are introduced into endogenous genes, just as they occur in the real patients, and because they are based on human genomes rather than rat genomes. Furthermore, certain technologies enable the knock-in of a particular mutation rather than just knock-outs associated with older gene targeting technologies.

## 22.7.2   Protein engineering

Protein engineering with homologous recombination develops chimeric proteins by swapping fragments between two parental proteins. These techniques exploit the fact that recombination can introduce a high degree of sequence diversity while preserving a protein's ability to fold into its tertiary structure, or three-dimensional shape.[103] This stands in contrast to other protein engineering techniques, like random point mutagenesis, in which the probability of maintaining protein function declines exponentially with increasing amino acid substitutions.[104] The chimeras produced by recombination techniques are able to maintain their ability to fold because their swapped parental fragments are structurally and evolutionarily conserved. These recombinable "building blocks" preserve structurally important interactions like points of physical contact between different amino acids in the protein's structure. Computational methods like SCHEMA and statistical coupling analysis can be used to identify structural subunits suitable for recombination.[105][106][107]

Techniques that rely on homologous recombination have been used to engineer new proteins.[105] In a study published in 2007, researchers were able to create chimeras of two enzymes involved in the biosynthesis of isoprenoids, a diverse class of compounds including hormones, visual pigments and certain pheromones. The chimeric proteins acquired an ability to catalyze an essential reaction in isoprenoid biosynthesis—one of the most diverse pathways of biosynthesis found in nature—that was absent in the parent proteins.[108] Protein engineering through recombination has also produced chimeric enzymes with new function in members of a group of proteins known as the cytochrome P450 family,[109] which in humans is involved in detoxifying foreign compounds like drugs, food additives and preservatives.[22]

## 22.7.3   Cancer therapy

Cancer cells with BRCA mutations have deficiencies in homologous recombination, and drugs to exploit those deficiencies have been developed and used successfully in clinical trials.[110][111] Olaparib, a PARP1 inhibitor, shrunk or stopped the growth of tumors from breast, ovarian and prostate cancers caused by mutations in the BRCA1 or BRCA2 genes, which are necessary for HR. When BRCA1 or BRCA2 is absent, other types of DNA repair mechanisms must compensate for the deficiency of HR, such as base-excision repair (BER) for stalled replication forks or non-homologous end joining (NHEJ) for double strand breaks.[110] By inhibiting BER in an HR-deficient cell, olaparib applies the concept of synthetic lethality to specifically target cancer cells. While PARP1 inhibitors represent

a novel approach to cancer therapy, researchers have cautioned that they may prove insufficient for treating late-stage metastatic cancers.[110] Cancer cells can become resistant to a PARP1 inhibitor if they undergo deletions of mutations in BRCA2, undermining the drug's synthetic lethality by restoring cancer cells' ability to repair DNA by HR.[112]

## 22.8   References

[1] Alberts B, Johnson A, Lewis J, Raff M, Roberts K, Walter P, et al. (2002). "Chapter 5: DNA Replication, Repair, and Recombination". *Molecular Biology of the Cell* (4th ed.). New York: Garland Science. p. 845. ISBN 0-8153-3218-1. OCLC 145080076.

[2] Capecchi MR (June 1989). "Altering the genome by homologous recombination". *Science*. **244** (4910): 1288–92. doi:10.1126/science.2660260. PMID 2660260.

[3] Smithies O, Gregg RG, Boggs SS, Koralewski MA, Kucherlapati RS (1985-09-19). "Insertion of DNA sequences into the human chromosomal beta-globin locus by homologous recombination". *Nature*. **317** (6034): 230–4. doi:10.1038/317230a0. PMID 2995814.

[4] Orr-Weaver TL, Szostak JW, Rothstein RJ (October 1981). "Yeast transformation: a model system for the study of recombination". *Proceedings of the National Academy of Sciences of the United States of America*. **78** (10): 6354–8. doi:10.1073/pnas.78.10.6354. PMC 349037. PMID 6273866.

[5] Orr-Weaver TL, Szostak JW (July 1983). "Yeast recombination: the association between double-strand gap repair and crossing-over". *Proceedings of the National Academy of Sciences of the United States of America*. **80** (14): 4417–21. doi:10.1073/pnas.80.14.4417. PMC 384049. PMID 6308623.

[6] Szostak JW, Orr-Weaver TL, Rothstein RJ, Stahl FW (May 1983). "The double-strand-break repair model for recombination". *Cell*. **33** (1): 25–35. doi:10.1016/0092-8674(83)90331-8. PMID 6380756.

[7] Resnick MA (June 1976). "The repair of double-strand breaks in DNA; a model involving recombination". *Journal of Theoretical Biology*. **59** (1): 97–106. doi:10.1016/s0022-5193(76)80025-2. PMID 940351.

[8] Jasin M, Rothstein R (November 2013). "Repair of strand breaks by homologous recombination". *Cold Spring Harbor Perspectives in Biology*. **5** (11): a012740. doi:10.1101/cshperspect.a012740. PMC 3809576. PMID 24097900.

[9] Bateson P (August 2002). "William Bateson: a biologist ahead of his time" (PDF). *Journal of Genetics*. **81** (2): 49–58. doi:10.1007/BF02715900. PMID 12532036.

[10] "Reginald Crundall Punnett". NAHSTE. University of Edinburgh. Retrieved 3 July 2010.

[11] Lobo I, Shaw K (2008). "Thomas Hunt Morgan, genetic recombination, and gene mapping". *Nature Education.* **1** (1).

[12] Coe E, Kass LB (May 2005). "Proof of physical exchange of genes on the chromosomes". *Proceedings of the National Academy of Sciences of the United States of America.* **102** (19): 6641–6. doi:10.1073/pnas.0407340102. PMC 1100733. PMID 15867161.

[13] Creighton HB, McClintock B (August 1931). "A Correlation of Cytological and Genetical Crossing-Over in Zea Mays". *Proceedings of the National Academy of Sciences of the United States of America.* **17** (8): 492–7. doi:10.1073/pnas.17.8.492. PMC 1076098. PMID 16587654.

[14] Stern, C (1931). "Zytologisch-genetische untersuchungen alsbeweise fur die Morgansche theorie des faktoraustauschs". *Biol. Zentbl.* **51**: 547–587.

[15] "The development of bacterial genetics". US National Library of Medicine. Retrieved 3 July 2010.

[16] "The Nobel Prize in Physiology or Medicine 1958". Nobelprize.org. Retrieved 3 July 2010.

[17] Haber JE, Ira G, Malkova A, Sugawara N (January 2004). "Repairing a double-strand chromosome break by homologous recombination: revisiting Robin Holliday's model". *Philosophical Transactions of the Royal Society of London. Series B, Biological Sciences.* **359** (1441): 79–86. doi:10.1098/rstb.2003.1367. PMC 1693306. PMID 15065659.

[18] Szostak JW, Orr-Weaver TL, Rothstein RJ, Stahl FW (May 1983). "The double-strand-break repair model for recombination". *Cell.* **33** (1): 25–35. doi:10.1016/0092-8674(83)90331-8. PMID 6380756.

[19] Lodish H, Berk A, Zipursky SL, Matsudaira P, Baltimore D, Darnell J (2000). "12.5: Recombination between Homologous DNA Sites: Double-Strand Breaks in DNA Initiate Recombination". *Molecular Cell Biology* (4th ed.). W. H. Freeman and Company. ISBN 0-7167-3136-3.

[20] Griffiths A, et al. (1999). "8: Chromosome Mutations: Chromosomal Rearrangements". *Modern Genetic Analysis.* W. H. Freeman and Company. ISBN 0-7167-3118-5.

[21] Khanna KK, Jackson SP (March 2001). "DNA double-strand breaks: signaling, repair and the cancer connection". *Nature Genetics.* **27** (3): 247–54. doi:10.1038/85798. PMID 11242102.

[22] Nelson DL, Cox MM (2005). *Principles of Biochemistry* (4th ed.). Freeman. pp. 980–981. ISBN 978-0-7167-4339-2.

[23] Marcon E, Moens PB (August 2005). "The evolution of meiosis: recruitment and modification of somatic DNA-repair proteins". *BioEssays.* **27** (8): 795–808. doi:10.1002/bies.20264. PMID 16015600.

[24] Alberts B, Johnson A, Lewis J, Raff M, Roberts K, Walter P (2008). *Molecular Biology of the Cell* (5th ed.). Garland Science. p. 305. ISBN 978-0-8153-4105-5.

[25] Keeney S, Giroux CN, Kleckner N (February 1997). "Meiosis-specific DNA double-strand breaks are catalyzed by Spo11, a member of a widely conserved protein family". *Cell.* **88** (3): 375–84. doi:10.1016/S0092-8674(00)81876-0. PMID 9039264.

[26] Longhese MP, Bonetti D, Guerini I, Manfrini N, Clerici M (September 2009). "DNA double-strand breaks in meiosis: checking their formation, processing and repair". *DNA Repair.* **8** (9): 1127–38. doi:10.1016/j.dnarep.2009.04.005. PMID 19464965.

[27] Cahill LP, Mariana JC, Mauléon P (January 1979). "Total follicular populations in ewes of high and low ovulation rates". *Journal of Reproduction and Fertility.* **55** (1): 27–36. doi:10.1371/journal.pbio.0020192. PMC 423159.

[28] Alberts B, Johnson A, Lewis J, Raff M, Roberts K, Walter P (2008). *Molecular Biology of the Cell* (5th ed.). Garland Science. p. 303. ISBN 978-0-8153-4105-5.

[29] Shrivastav M, De Haro LP, Nickoloff JA (January 2008). "Regulation of DNA double-strand break repair pathway choice". *Cell Research.* **18** (1): 134–47. doi:10.1038/cr.2007.111. PMID 18157161.

[30] Mimitou EP, Symington LS (May 2009). "Nucleases and helicases take center stage in homologous recombination". *Trends in Biochemical Sciences.* **34** (5): 264–72. doi:10.1016/j.tibs.2009.01.010. PMID 19375328.

[31] Huertas P, Cortés-Ledesma F, Sartori AA, Aguilera A, Jackson SP (October 2008). "CDK targets Sae2 to control DNA-end resection and homologous recombination". *Nature.* **455** (7213): 689–92. doi:10.1038/nature07215. PMC 2635538. PMID 18716619.

[32] Sung P, Klein H (October 2006). "Mechanism of homologous recombination: mediators and helicases take on regulatory functions". *Nature Reviews. Molecular Cell Biology.* **7** (10): 739–50. doi:10.1038/nrm2008. PMID 16926856.

[33] Wold MS (1997). "Replication protein A: a heterotrimeric, single-stranded DNA-binding protein required for eukaryotic DNA metabolism". *Annual Review of Biochemistry.* **66**: 61–92. doi:10.1146/annurev.biochem.66.1.61. PMID 9242902.

[34] McMahill MS, Sham CW, Bishop DK (November 2007). "Synthesis-dependent strand annealing in meiosis". *PLoS Biology.* **5** (11): e299. doi:10.1371/journal.pbio.0050299. PMC 2062477. PMID 17988174.

[35] Bärtsch S, Kang LE, Symington LS (February 2000). "RAD51 is required for the repair of plasmid double-stranded DNA gaps from either plasmid or chromosomal templates". *Molecular and Cellular Biology*. **20** (4): 1194–205. doi:10.1128/MCB.20.4.1194-1205.2000. PMC 85244. PMID 10648605.

[36] Alberts B, Johnson A, Lewis J, Raff M, Roberts K, Walter P (2008). *Molecular Biology of the Cell* (5th ed.). Garland Science. pp. 312–313. ISBN 978-0-8153-4105-5.

[37] Helleday T, Lo J, van Gent DC, Engelward BP (July 2007). "DNA double-strand break repair: from mechanistic understanding to cancer treatment". *DNA Repair*. **6** (7): 923–35. doi:10.1016/j.dnarep.2007.02.006. PMID 17363343.

[38] Andersen SL, Sekelsky J (December 2010). "Meiotic versus mitotic recombination: two different routes for double-strand break repair: the different functions of meiotic versus mitotic DSB repair are reflected in different pathway usage and different outcomes". *BioEssays*. **32** (12): 1058–66. doi:10.1002/bies.201000087. PMC 3090628. PMID 20967781.

[39] Allers T, Lichten M (July 2001). "Differential timing and control of noncrossover and crossover recombination during meiosis". *Cell*. **106** (1): 47–57. doi:10.1016/s0092-8674(01)00416-0. PMID 11461701.

[40] Haber lab. "Single-strand annealing". "Brandeis University". Retrieved 3 July 2010.

[41] Lyndaker AM, Alani E (March 2009). "A tale of tails: insights into the coordination of 3' end processing during homologous recombination". *BioEssays*. **31** (3): 315–21. doi:10.1002/bies.200800195. PMC 2958051. PMID 19260026.

[42] Mimitou EP, Symington LS (September 2009). "DNA end resection: many nucleases make light work". *DNA Repair*. **8** (9): 983–95. doi:10.1016/j.dnarep.2009.04.017. PMC 2760233. PMID 19473888.

[43] Pâques F, Haber JE (June 1999). "Multiple pathways of recombination induced by double-strand breaks in Saccharomyces cerevisiae". *Microbiology and Molecular Biology Reviews*. **63** (2): 349–404. PMC 98970. PMID 10357855.

[44] McEachern MJ, Haber JE (2006). "Break-induced replication and recombinational telomere elongation in yeast". *Annual Review of Biochemistry*. **75**: 111–35. doi:10.1146/annurev.biochem.74.082803.133234. PMID 16756487.

[45] Morrish TA, Greider CW (January 2009). Haber JE. ed. "Short telomeres initiate telomere recombination in primary and tumor cells". *PLoS Genetics*. **5** (1): e1000357. doi:10.1371/journal.pgen.1000357. PMC 2627939. PMID 19180191.

[46] Muntoni A, Reddel RR (October 2005). "The first molecular details of ALT in human tumor cells". *Human Molecular Genetics*. 14 Spec No. 2 (Review Issue 2): R191–6. doi:10.1093/hmg/ddi266. PMID 16244317.

[47] PDB: 3cmt; Chen Z, Yang H, Pavletich NP (May 2008). "Mechanism of homologous recombination from the RecA-ssDNA/dsDNA structures". *Nature*. **453** (7194): 489–4. doi:10.1038/nature06971. PMID 18497818.

[48] Kowalczykowski SC, Dixon DA, Eggleston AK, Lauder SD, Rehrauer WM (September 1994). "Biochemistry of homologous recombination in Escherichia coli". *Microbiological Reviews*. **58** (3): 401–65. PMC 372975. PMID 7968921.

[49] Rocha EP, Cornet E, Michel B (August 2005). "Comparative and evolutionary analysis of the bacterial homologous recombination systems". *PLoS Genetics*. **1** (2): e15. doi:10.1371/journal.pgen.0010015. PMC 1193525. PMID 16132081.

[50] Amundsen SK, Taylor AF, Reddy M, Smith GR (December 2007). "Intersubunit signaling in RecBCD enzyme, a complex protein machine regulated by Chi hot spots". *Genes & Development*. **21** (24): 3296–307. doi:10.1101/gad.1605807. PMC 2113030. PMID 18079176.

[51] Singleton MR, Dillingham MS, Gaudier M, Kowalczykowski SC, Wigley DB (November 2004). "Crystal structure of RecBCD enzyme reveals a machine for processing DNA breaks" (PDF). *Nature*. **432** (7014): 187–93. doi:10.1038/nature02988. PMID 15538360.

[52] Cromie GA (August 2009). "Phylogenetic ubiquity and shuffling of the bacterial RecBCD and AddAB recombination complexes". *Journal of Bacteriology*. **191** (16): 5076–84. doi:10.1128/JB.00254-09. PMC 2725590. PMID 19542287.

[53] Smith GR (June 2012). "How RecBCD enzyme and Chi promote DNA break repair and recombination: a molecular biologist's view". *Microbiology and Molecular Biology Reviews*. **76** (2): 217–28. doi:10.1128/MMBR.05026-11. PMID 22688812.

[54] Dillingham MS, Kowalczykowski SC (December 2008). "RecBCD enzyme and the repair of double-stranded DNA breaks". *Microbiology and Molecular Biology Reviews*. **72** (4): 642–71, Table of Contents. doi:10.1128/MMBR.00020-08. PMC 2593567. PMID 19052323.

[55] Michel B, Boubakri H, Baharoglu Z, LeMasson M, Lestini R (July 2007). "Recombination proteins and rescue of arrested replication forks". *DNA Repair*. **6** (7): 967–80. doi:10.1016/j.dnarep.2007.02.016. PMID 17395553.

[56] Amundsen SK, Taylor AF, Reddy M, Smith GR (December 2007). "Intersubunit signaling in RecBCD enzyme, a complex protein machine regulated by Chi hot

spots". *Genes & Development*. **21** (24): 3296–307. doi:10.1101/gad.1605807. PMC 2113030⊖. PMID 18079176.

[57] Spies M, Bianco PR, Dillingham MS, Handa N, Baskin RJ, Kowalczykowski SC (September 2003). "A molecular throttle: the recombination hotspot chi controls DNA translocation by the RecBCD helicase". *Cell*. **114** (5): 647–54. doi:10.1016/S0092-8674(03)00681-0. PMID 13678587.

[58] Taylor AF, Smith GR (June 2003). "RecBCD enzyme is a DNA helicase with fast and slow motors of opposite polarity". *Nature*. **423** (6942): 889–93. doi:10.1038/nature01674. PMID 12815437.

[59] Spies M, Amitani I, Baskin RJ, Kowalczykowski SC (November 2007). "RecBCD enzyme switches lead motor subunits in response to chi recognition". *Cell*. **131** (4): 694–705. doi:10.1016/j.cell.2007.09.023. PMC 2151923⊖. PMID 18022364.

[60] Savir Y, Tlusty T (November 2010). "RecA-mediated homology search as a nearly optimal signal detection system" (PDF). *Molecular Cell*. **40** (3): 388–96. doi:10.1016/j.molcel.2010.10.020. PMID 21070965.

[61] Rambo RP, Williams GJ, Tainer JA (November 2010). "Achieving fidelity in homologous recombination despite extreme complexity: informed decisions by molecular profiling" (PDF). *Molecular Cell*. **40** (3): 347–8. doi:10.1016/j.molcel.2010.10.032. PMC 3003302⊖. PMID 21070960.

[62] De Vlaminck I, van Loenhout MT, Zweifel L, den Blanken J, Hooning K, Hage S, Kerssemakers J, Dekker C (June 2012). "Mechanism of homology recognition in DNA recombination from dual-molecule experiments". *Molecular Cell*. **46** (5): 616–24. doi:10.1016/j.molcel.2012.03.029. PMID 22560720.

[63] Alberts B, Johnson A, Lewis J, Raff M, Roberts K, Walter P (2008). *Molecular Biology of the Cell* (5th ed.). Garland Science. p. 307. ISBN 978-0-8153-4105-5.

[64] Morimatsu K, Kowalczykowski SC (May 2003). "RecFOR proteins load RecA protein onto gapped DNA to accelerate DNA strand exchange: a universal step of recombinational repair". *Molecular Cell*. **11** (5): 1337–47. doi:10.1016/S1097-2765(03)00188-6. PMID 12769856.

[65] Hiom K (July 2009). "DNA repair: common approaches to fixing double-strand breaks". *Current Biology*. **19** (13): R523–5. doi:10.1016/j.cub.2009.06.009. PMID 19602417.

[66] Handa N, Morimatsu K, Lovett ST, Kowalczykowski SC (May 2009). "Reconstitution of initial steps of dsDNA break repair by the RecF pathway of E. coli". *Genes & Development*. **23** (10): 1234–45. doi:10.1101/gad.1780709. PMC 2685532⊖. PMID 19451222.

[67] West SC (June 2003). "Molecular views of recombination proteins and their control". *Nature Reviews. Molecular Cell Biology*. **4** (6): 435–45. doi:10.1038/nrm1127. PMID 12778123.

[68] Watson JD, Baker TA, Bell SP, Gann A, Levine M, Losick R (2003). *Molecular Biology of the Gene* (5th ed.). Pearson/Benjamin Cummings. pp. 259–291. ISBN 978-0-8053-4635-0.

[69] Gumbiner-Russo LM, Rosenberg SM (28 November 2007). Sandler S, ed. "Physical analyses of E. coli heteroduplex recombination products in vivo: on the prevalence of 5' and 3' patches". *PloS One*. **2** (11): e1242. doi:10.1371/journal.pone.0001242. PMC 2082072⊖. PMID 18043749. ⊖

[70] Thomas CM, Nielsen KM (September 2005). "Mechanisms of, and barriers to, horizontal gene transfer between bacteria" (PDF). *Nature Reviews. Microbiology*. **3** (9): 711–21. doi:10.1038/nrmicro1234. PMID 16138099.

[71] Vulić M, Dionisio F, Taddei F, Radman M (September 1997). "Molecular keys to speciation: DNA polymorphism and the control of genetic exchange in enterobacteria". *Proceedings of the National Academy of Sciences of the United States of America*. **94** (18): 9763–7. doi:10.1073/pnas.94.18.9763. PMC 23264⊖. PMID 9275198.

[72] Majewski J, Cohan FM (January 1998). "The effect of mismatch repair and heteroduplex formation on sexual isolation in Bacillus". *Genetics*. **148** (1): 13–8. PMC 1459767⊖. PMID 9475717.

[73] Majewski J, Zawadzki P, Pickerill P, Cohan FM, Dowson CG (February 2000). "Barriers to genetic exchange between bacterial species: Streptococcus pneumoniae transformation". *Journal of Bacteriology*. **182** (4): 1016–23. doi:10.1128/JB.182.4.1016-1023.2000. PMC 94378⊖. PMID 10648528.

[74] Chen I, Dubnau D (March 2004). "DNA uptake during bacterial transformation". *Nature Reviews. Microbiology*. **2** (3): 241–9. doi:10.1038/nrmicro844. PMID 15083159.

[75] Claverys JP, Martin B, Polard P (May 2009). "The genetic transformation machinery: composition, localization, and mechanism". *FEMS Microbiology Reviews*. **33** (3): 643–56. doi:10.1111/j.1574-6976.2009.00164.x. PMID 19228200.

[76] Kidane D, Graumann PL (July 2005). "Intracellular protein and DNA dynamics in competent Bacillus subtilis cells". *Cell*. **122** (1): 73–84. doi:10.1016/j.cell.2005.04.036. PMID 16009134.

[77] Fleischmann Jr WR (1996). "43". *Medical Microbiology* (4th ed.). University of Texas Medical Branch at Galveston. ISBN 0-9631172-1-1.

[78] Boni MF, de Jong MD, van Doorn HR, Holmes EC (3 May 2010). Martin DP, ed. "Guidelines for identifying homologous recombination events in influenza A virus". *PloS One.* **5** (5): e10434. doi:10.1371/journal.pone.0010434. PMC 2862710. PMID 20454662.

[79] Nagy PD, Bujarski JJ (January 1996). "Homologous RNA recombination in brome mosaic virus: AU-rich sequences decrease the accuracy of crossovers". *Journal of Virology.* **70** (1): 415–26. PMC 189831. PMID 8523555.

[80] Chetverin AB (October 1999). "The puzzle of RNA recombination". *FEBS Letters.* **460** (1): 1–5. doi:10.1016/S0014-5793(99)01282-X. PMID 10571050.

[81] Roossinck MJ (September 1997). "Mechanisms of plant virus evolution". *Annual Review of Phytopathology.* **35**: 191–209. doi:10.1146/annurev.phyto.35.1.191. PMID 15012521.

[82] Arbuckle JH, Medveczky PG (August 2011). "The molecular biology of human herpesvirus-6 latency and telomere integration". *Microbes and Infection / Institut Pasteur.* **13** (8-9): 731–41. doi:10.1016/j.micinf.2011.03.006. PMC 3130849. PMID 21458587.

[83] Bernstein C (March 1981). "Deoxyribonucleic acid repair in bacteriophage". *Microbiological Reviews.* **45** (1): 72–98. PMC 281499. PMID 6261109.

[84] Bernstein C, Bernstein H (2001). DNA repair in bacteriophage. In: Nickoloff JA, Hoekstra MF (Eds.) DNA Damage and Repair, Vol.3. Advances from Phage to Humans. Humana Press, Totowa, NJ, pp. 1–19. ISBN 978-0896038035

[85] Story RM, Bishop DK, Kleckner N, Steitz TA (March 1993). "Structural relationship of bacterial RecA proteins to recombination proteins from bacteriophage T4 and yeast". *Science.* **259** (5103): 1892–6. doi:10.1126/science.8456313. PMID 8456313.

[86] Michod RE, Bernstein H, Nedelcu AM (May 2008). "Adaptive value of sex in microbial pathogens". *Infection, Genetics and Evolution.* **8** (3): 267–85. doi:10.1016/j.meegid.2008.01.002. PMID 18295550.http://www.hummingbirds.arizona.edu/Faculty/Michod/Downloads/IGE%20review%20sex.pdf

[87] Lamb NE, Yu K, Shaffer J, Feingold E, Sherman SL (January 2005). "Association between maternal age and meiotic recombination for trisomy 21". *American Journal of Human Genetics.* **76** (1): 91–9. doi:10.1086/427266. PMC 1196437. PMID 15551222.

[88] Cold Spring Harbor Laboratory (2007). "Human RecQ Helicases, Homologous Recombination And Genomic Instability". ScienceDaily. Retrieved 3 July 2010.

[89] Modesti M, Kanaar R (2001). "Homologous recombination: from model organisms to human disease". *Genome Biology.* **2** (5): REVIEWS1014. doi:10.1186/gb-2001-2-5-reviews1014. PMC 138934. PMID 11387040.

[90] Luo G, Santoro IM, McDaniel LD, Nishijima I, Mills M, Youssoufian H, Vogel H, Schultz RA, Bradley A (December 2000). "Cancer predisposition caused by elevated mitotic recombination in Bloom mice". *Nature Genetics.* **26** (4): 424–9. doi:10.1038/82548. PMID 11101838.

[91] Alberts B, Johnson A, Lewis J, Raff M, Roberts K, Walter P (2007). *Molecular Biology of the Cell* (5th ed.). Garland Science. ISBN 978-0-8153-4110-9.

[92] Powell SN, Kachnic LA (September 2003). "Roles of BRCA1 and BRCA2 in homologous recombination, DNA replication fidelity and the cellular response to ionizing radiation". *Oncogene.* **22** (37): 5784–91. doi:10.1038/sj.onc.1206678. PMID 12947386.

[93] Use of homologous recombination deficiency (HRD) score to enrich for niraparib sensitive high grade ovarian tumors.

[94] Lin Z, Kong H, Nei M, Ma H (July 2006). "Origins and evolution of the recA/RAD51 gene family: evidence for ancient gene duplication and endosymbiotic gene transfer". *Proceedings of the National Academy of Sciences of the United States of America.* **103** (27): 10328–33. doi:10.1073/pnas.0604232103. PMC 1502457. PMID 16798872.

[95] Haseltine CA, Kowalczykowski SC (May 2009). "An archaeal Rad54 protein remodels DNA and stimulates DNA strand exchange by RadA". *Nucleic Acids Research.* **37** (8): 2757–70. doi:10.1093/nar/gkp068. PMC 2677860. PMID 19282450.

[96] Rolfsmeier ML, Haseltine CA (March 2010). "The single-stranded DNA binding protein of Sulfolobus solfataricus acts in the presynaptic step of homologous recombination". *Journal of Molecular Biology.* **397** (1): 31–45. doi:10.1016/j.jmb.2010.01.004. PMID 20080104.

[97] Huang Q, Liu L, Liu J, Ni J, She Q, Shen Y (2015). "Efficient 5′-3′ DNA end resection by HerA and NurA is essential for cell viability in the crenarchaeon Sulfolobus islandicus". *BMC Molecular Biology.* **16**: 2. doi:10.1186/s12867-015-0030-z. PMC 4351679. PMID 25880130.

[98] Jain SK, Cox MM, Inman RB (August 1994). "On the role of ATP hydrolysis in RecA protein-mediated DNA strand exchange. III. Unidirectional branch migration and extensive hybrid DNA formation". *The Journal of Biological Chemistry.* **269** (32): 20653–61. PMID 8051165.

[99] Ramesh MA, Malik SB, Logsdon JM (January 2005). "A phylogenomic inventory of meiotic genes; evidence for sex in Giardia and an early eukaryotic origin of meiosis". *Current Biology.* **15** (2): 185–91. doi:10.1016/j.cub.2005.01.003. PMID 15668177.

[100] Malik SB, Ramesh MA, Hulstrand AM, Logsdon JM (December 2007). "Protist homologs of the meiotic

Spo11 gene and topoisomerase VI reveal an evolutionary history of gene duplication and lineage-specific loss". *Molecular Biology and Evolution*. **24** (12): 2827–41. doi:10.1093/molbev/msm217. PMID 17921483.

[101] Lodish H, Berk A, Zipursky SL, Matsudaira P, Baltimore D, Darnell J (2000). "Chapter 8.5: Gene Replacement and Transgenic Animals: DNA Is Transferred into Eukaryotic Cells in Various Ways". *Molecular Cell Biology* (4th ed.). W. H. Freeman and Company. ISBN 0-7167-3136-3.

[102] "The Nobel Prize in Physiology or Medicine 2007". The Nobel Foundation. Retrieved December 15, 2008.

[103] Drummond DA, Silberg JJ, Meyer MM, Wilke CO, Arnold FH (April 2005). "On the conservative nature of intragenic recombination". *Proceedings of the National Academy of Sciences of the United States of America*. **102** (15): 5380–5. doi:10.1073/pnas.0500729102. PMC 556249⊙. PMID 15809422.

[104] Bloom JD, Silberg JJ, Wilke CO, Drummond DA, Adami C, Arnold FH (January 2005). "Thermodynamic prediction of protein neutrality". *Proceedings of the National Academy of Sciences of the United States of America*. **102** (3): 606–11. doi:10.1073/pnas.0406744102. PMC 545518⊙. PMID 15644440.

[105] Carbone MN, Arnold FH (August 2007). "Engineering by homologous recombination: exploring sequence and function within a conserved fold". *Current Opinion in Structural Biology*. **17** (4): 454–9. doi:10.1016/j.sbi.2007.08.005. PMID 17884462.

[106] Otey CR, Landwehr M, Endelman JB, Hiraga K, Bloom JD, Arnold FH (May 2006). "Structure-guided recombination creates an artificial family of cytochromes P450". *PLoS Biology*. **4** (5): e112. doi:10.1371/journal.pbio.0040112. PMC 1431580⊙. PMID 16594730. ⊙

[107] Socolich M, Lockless SW, Russ WP, Lee H, Gardner KH, Ranganathan R (September 2005). "Evolutionary information for specifying a protein fold". *Nature*. **437** (7058): 512–8. doi:10.1038/nature03991. PMID 16177782.

[108] Thulasiram HV, Erickson HK, Poulter CD (April 2007). "Chimeras of two isoprenoid synthases catalyze all four coupling reactions in isoprenoid biosynthesis". *Science*. **316** (5821): 73–6. doi:10.1126/science.1137786. PMID 17412950.

[109] Landwehr M, Carbone M, Otey CR, Li Y, Arnold FH (March 2007). "Diversification of catalytic function in a synthetic family of chimeric cytochrome p450s". *Chemistry & Biology*. **14** (3): 269–78. doi:10.1016/j.chembiol.2007.01.009. PMC 1991292⊙. PMID 17379142.

[110] Iglehart JD, Silver DP (July 2009). "Synthetic lethality--a new direction in cancer-drug development". *The New England Journal of Medicine*. **361** (2): 189–91. doi:10.1056/NEJMe0903044. PMID 19553640.

[111] Fong PC, Boss DS, Yap TA, Tutt A, Wu P, Mergui-Roelvink M, Mortimer P, Swaisland H, Lau A, O'Connor MJ, Ashworth A, Carmichael J, Kaye SB, Schellens JH, de Bono JS (July 2009). "Inhibition of poly(ADP-ribose) polymerase in tumors from BRCA mutation carriers". *The New England Journal of Medicine*. **361** (2): 123–34. doi:10.1056/NEJMoa0900212. PMID 19553641.

[112] Edwards SL, Brough R, Lord CJ, Natrajan R, Vatcheva R, Levine DA, Boyd J, Reis-Filho JS, Ashworth A (February 2008). "Resistance to therapy caused by intragenic deletion in BRCA2". *Nature*. **451** (7182): 1111–5. doi:10.1038/nature06548. PMID 18264088.

## 22.9 External links

- Animations – homologous recombination: Animations showing several models of homologous recombination

- Homologous recombination: Tempy & Trun: Animation of the bacterial RecBCD pathway of homologous recombination

# Chapter 23

# Non-homologous end joining

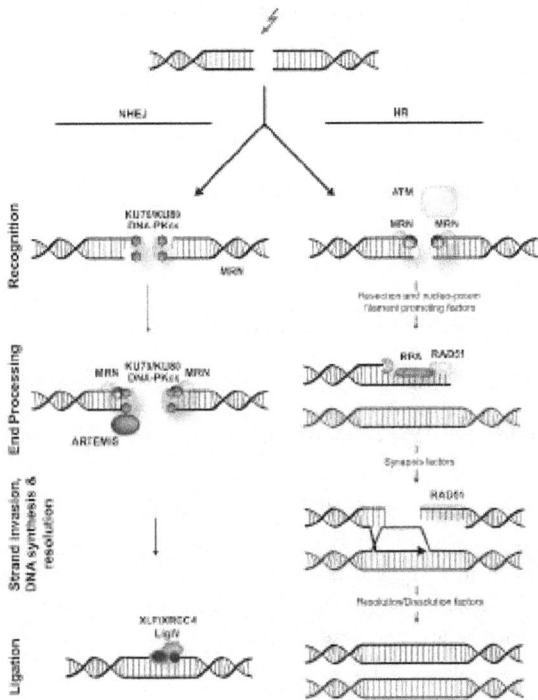

*Non-homologous end joining (NHEJ) and homologous recombination (HR) in mammals during DNA double-strand break*

**Non-homologous end joining** (**NHEJ**) is a pathway that repairs double-strand breaks in DNA. NHEJ is referred to as "non-homologous" because the break ends are directly ligated without the need for a homologous template, in contrast to homology directed repair, which requires a homologous sequence to guide repair. The term "non-homologous end joining" was coined in 1996 by Moore and Haber.[1]

NHEJ typically utilizes short homologous DNA sequences called microhomologies to guide repair. These microhomologies are often present in single-stranded overhangs on the ends of double-strand breaks. When the overhangs are perfectly compatible, NHEJ usually repairs the break accurately.[1][2][3][4] Imprecise repair leading to loss of nucleotides can also occur, but is much more common when

the overhangs are not compatible. Inappropriate NHEJ can lead to translocations and telomere fusion, hallmarks of tumor cells.[5]

NHEJ is evolutionarily conserved throughout all kingdoms of life and is the predominant double-strand break repair pathway in mammalian cells.[6] In budding yeast (*Saccharomyces cerevisiae*), however, homologous recombination dominates when the organism is grown under common laboratory conditions.

When the NHEJ pathway is inactivated, double-strand breaks can be repaired by a more error-prone pathway called microhomology-mediated end joining (MMEJ). In this pathway, end resection reveals short microhomologies on either side of the break, which are then aligned to guide repair.[7] This contrasts with classical NHEJ, which typically uses microhomologies already exposed in single-stranded overhangs on the DSB ends. Repair by MMEJ therefore leads to deletion of the DNA sequence between the microhomologies.

## 23.1   In bacteria

Many species of bacteria, including *Escherichia coli*, lack an end joining pathway and thus rely completely on homologous recombination to repair double-strand breaks. NHEJ proteins have been identified in a number of bacteria, however, including *Bacillus subtilis*, *Mycobacterium tuberculosis*, and *Mycobacterium smegmatis*.[8][9] Bacteria utilize a remarkably compact version of NHEJ in which all of the required activities are contained in only two proteins: a Ku homodimer and the multifunctional ligase/polymerase/nuclease LigD.[10] In mycobacteria, NHEJ is much more error prone than in yeast, with bases often added to and deleted from the ends of double-strand breaks during repair.[9] Many of the bacteria that possess NHEJ proteins spend a significant portion of their life cycle in a stationary haploid phase, in which a template for recombination is not available.[8] NHEJ may have evolved to help these organisms survive DSBs induced during desiccation.[11] Corndog and Omega, two related mycobacteriophages of *Mycobacterium smegmatis*, also encode Ku homologs and exploit the NHEJ pathway to recircularize their genomes during infection.[12] Unlike homologous recombination, which has been studied extensively in bacteria, NHEJ was originally discovered in eukaryotes and was only identified in prokaryotes in the past decade.

## 23.2 In eukaryotes

In contrast to bacteria, NHEJ in eukaryotes utilizes a number of proteins, which participate in the following steps:

### 23.2.1 End binding and tethering

In yeast, the Mre11-Rad50-Xrs2 (MRX) complex is recruited to DSBs early and is thought to promote bridging of the DNA ends.[13] The corresponding mammalian complex of Mre11-Rad50-Nbs1 (MRN) is also involved in NHEJ, but it may function at multiple steps in the pathway beyond simply holding the ends in proximity.[14] DNA-PKcs is also thought to participate in end bridging during mammalian NHEJ.[15]

Eukaryotic Ku is a heterodimer consisting of Ku70 and Ku80, and forms a complex with DNA-PKcs, which is present in mammals but absent in yeast. Ku is a basket-shaped molecule that slides onto the DNA end and translocates inward. Ku may function as a docking site for other NHEJ proteins, and is known to interact with the DNA ligase IV complex and XLF.[16][17]

### 23.2.2 End processing

End processing involves removal of damaged or mismatched nucleotides by nucleases and resynthesis by DNA polymerases. This step is not necessary if the ends are already compatible and have 3' hydroxyl and 5' phosphate termini.

Little is known about the function of nucleases in NHEJ. Artemis is required for opening the hairpins that are formed on DNA ends during V(D)J recombination, a specific type of NHEJ, and may also participate in end trimming during general NHEJ.[18] Mre11 has nuclease activity, but it seems to be involved in homologous recombination, not NHEJ.

The X family DNA polymerases Pol λ and Pol μ (Pol4 in yeast) fill gaps during NHEJ.[3][19][20] Yeast lacking Pol4 are unable to join 3' overhangs that require gap filling, but remain proficient for gap filling at 5' overhangs.[21] This is

because the primer terminus used to initiate DNA synthesis is less stable at 3' overhangs, necessitating a specialized NHEJ polymerase.

### 23.2.3 Ligation

The DNA ligase IV complex, consisting of the catalytic subunit DNA ligase IV and its cofactor XRCC4 (Dnl4 and Lif1 in yeast), performs the ligation step of repair.[22] XLF, also known as Cernunnos, is homologous to yeast Nej1 and is also required for NHEJ.[23][24] While the precise role of XLF is unknown, it interacts with the XRCC4/DNA ligase IV complex and likely participates in the ligation step.[25] Recent evidence suggests that XLF promotes re-adenylation of DNA ligase IV after ligation, recharging the ligase and allowing it to catalyze a second ligation.[26]

### 23.2.4 Other

In yeast, Sir2 was originally identified as an NHEJ protein, but is now known to be required for NHEJ only because it is required for the transcription of Nej1.[27]

## 23.3 Regulation

The choice between NHEJ and homologous recombination for repair of a double-strand break is regulated at the initial step in recombination, 5' end resection. In this step, the 5' strand of the break is degraded by nucleases to create long 3' single-stranded tails. DSBs that have not been resected can be rejoined by NHEJ, but resection of even a few nucleotides strongly inhibits NHEJ and effectively commits the break to repair by recombination.[20] NHEJ is active throughout the cell cycle, but is most important during G1 when no homologous template for recombination is available. This regulation is accomplished by the cyclin-dependent kinase Cdk1 (Cdc28 in yeast), which is turned off in G1 and expressed in S and G2. Cdk1 phosphorylates the nuclease Sae2, allowing resection to initiate.[28]

## 23.4 V(D)J recombination

NHEJ plays a critical role in V(D)J recombination, the process by which B-cell and T-cell receptor diversity is generated in the vertebrate immune system.[29] In V(D)J recombination, hairpin-capped double-strand breaks are created by the RAG1/RAG2 nuclease, which cleaves the DNA at recombination signal sequences.[30] These hairpins are then opened by the Artemis nuclease and joined by NHEJ.[18]

A specialized DNA polymerase called terminal deoxynucleotidyl transferase (TdT), which is only expressed in lymph tissue, adds nontemplated nucleotides to the ends before the break is joined.[31][32] This process couples "variable" (V), "diversity" (D), and "joining" (J) regions, which when assembled together create the variable region of a B-cell or T-cell receptor gene. Unlike typical cellular NHEJ, in which accurate repair is the most favorable outcome, error-prone repair in V(D)J recombination is beneficial in that it maximizes diversity in the coding sequence of these genes. Patients with mutations in NHEJ genes are unable to produce functional B cells and T cells and suffer from severe combined immunodeficiency (SCID).

## 23.5   At telomeres

Telomeres are normally protected by a "cap" that prevents them from being recognized as double-strand breaks. Loss of capping proteins causes telomere shortening and inappropriate joining by NHEJ, producing dicentric chromosomes which are then pulled apart during mitosis. Paradoxically, some NHEJ proteins are involved in telomere capping. For example, Ku localizes to telomeres and its deletion leads to shortened telomeres.[33] Ku is also required for subtelomeric silencing, the process by which genes located near telomeres are turned off.

## 23.6   Consequences of dysfunction

Several human syndromes are associated with dysfunctional NHEJ.[34] Hypomorphic mutations in LIG4 and XLF cause LIG4 syndrome and XLF-SCID, respectively. These syndromes share many features including cellular radiosensitivity, microcephaly and severe combined immunodeficiency (SCID) due to defective V(D)J recombination. Loss-of-function mutations in Artemis also cause SCID, but these patients do not show the neurological defects associated with LIG4 or XLF mutations. The difference in severity may be explained by the roles of the mutated proteins. Artemis is a nuclease and is thought to be required only for repair of DSBs with damaged ends, whereas DNA Ligase IV and XLF are required for all NHEJ events.

Many NHEJ genes have been knocked out in mice. Deletion of XRCC4 or LIG4 causes embryonic lethality in mice, indicating that NHEJ is essential for viability in mammals. In contrast, mice lacking Ku or DNA-PKcs are viable, probably because low levels of end joining can still occur in the absence of these components.[35] All NHEJ mutant mice show a SCID phenotype, sensitivity to ionizing radiation, and neuronal apoptosis.

## 23.7   Aging

A system was developed for measuring NHEJ efficiency in the mouse.[36] NHEJ efficiency could be compared across tissues of the same mouse and in mice of different age. Efficiency was higher in the skin, lung and kidney fibroblasts, and lower in heart fibroblasts and brain astrocytes. Furthermore, NHEJ efficiency declined with age. The decline was 1.8 to 3.8-fold, depending on the tissue, in the 5 month old compared to the 24-month old mice. Reduced capability for NHEJ can lead to an increase in the number of unrepaired or faultily repaired DNA double-strand breaks that may then contribute to aging.[37] (Also see DNA damage theory of aging.) An analysis of the level of NHEJ protein Ku80 in human, cow, and mouse indicated that Ku80 levels vary dramatically between species, and that these levels are strongly correlated with species longevity.[38]

## 23.8   List of proteins involved in NHEJ in human cells

- Ku70/80
- DNA-PKcs
- DNA Ligase IV
- XRCC4
- XLF
- Artemis
- DNA polymerase mu
- DNA polymerase lambda
- PNKP
- Aprataxin
- APLF
- BRCA1
- BRCA2

## 23.9   References

[1] Moore JK, Haber JE (May 1996). "Cell cycle and genetic requirements of two pathways of nonhomologous end-joining repair of double-strand breaks in Saccharomyces cerevisiae". Molecular and Cellular Biology. 16 (5): 2164–73. doi:10.1128/mcb.16.5.2164. PMC 231204. PMID 8628283.

[2] Boulton SJ, Jackson SP (September 1996). "Saccharomyces cerevisiae Ku70 potentiates illegitimate DNA double-strand break repair and serves as a barrier to error-prone DNA repair pathways". *EMBO J.* **15** (18): 5093–103. PMC 452249⊙. PMID 8890183.

[3] Wilson TE, Lieber MR (1999). "Efficient processing of DNA ends during yeast nonhomologous end joining. Evidence for a DNA polymerase beta (Pol4)-dependent pathway". *J. Biol. Chem.* **274**: 23599–23609. doi:10.1074/jbc.274.33.23599. PMID 10438542.

[4] Budman J, Chu G (Feb 2005). "Processing of DNA for nonhomologous end-joining by cell-free extract". *EMBO J.* **24** (4): 849–60. doi:10.1038/sj.emboj.7600563. PMC 549622⊙. PMID 15692565.

[5] Espejel S, Franco S, Rodríguez-Perales S, Bouffler SD, Cigudosa JC, Blasco MA (May 2002). "Mammalian Ku86 mediates chromosomal fusions and apoptosis caused by critically short telomeres". *The EMBO Journal.* **21** (9): 2207–19. doi:10.1093/emboj/21.9.2207. PMC 125978⊙. PMID 11980718.

[6] Guirouilh-Barbat J, Huck S, Bertrand P, et al. (June 2004). "Impact of the KU80 pathway on NHEJ-induced genome rearrangements in mammalian cells". *Mol. Cell.* **14** (5): 611–23. doi:10.1016/j.molcel.2004.05.008. PMID 15175156.

[7] McVey M, Lee SE (November 2008). "MMEJ repair of double-strand breaks (director's cut): deleted sequences and alternative endings". *Trends Genet.* **24** (11): 529–38. doi:10.1016/j.tig.2008.08.007. PMID 18809224.

[8] Weller GR, Kysela B, Roy R, et al. (September 2002). "Identification of a DNA nonhomologous end-joining complex in bacteria". *Science.* **297** (5587): 1686–9. doi:10.1126/science.1074584. PMID 12215643.

[9] Gong C, Bongiorno P, Martins A, et al. (April 2005). "Mechanism of nonhomologous end-joining in mycobacteria: a low-fidelity repair system driven by Ku, ligase D and ligase C". *Nat. Struct. Mol. Biol.* **12** (4): 304–12. doi:10.1038/nsmb915. PMID 15778718.

[10] Della M, Palmbos PL, Tseng HM, et al. (October 2004). "Mycobacterial Ku and ligase proteins constitute a two-component NHEJ repair machine". *Science.* **306** (5696): 683–5. doi:10.1126/science.1099824. PMID 15499016.

[11] Pitcher RS, Green AJ, Brzostek A, Korycka-Machala M, Dziadek J, Doherty AJ (September 2007). "NHEJ protects mycobacteria in stationary phase against the harmful effects of desiccation". *DNA Repair (Amst.).* **6** (9): 1271–6. doi:10.1016/j.dnarep.2007.02.009. PMID 17360246.

[12] Pitcher RS, Tonkin LM, Daley JM, et al. (September 2006). "Mycobacteriophage exploit NHEJ to facilitate genome circularization". *Mol. Cell.* **23** (5): 743–8. doi:10.1016/j.molcel.2006.07.009. PMID 16949369.

[13] Chen, L., Trujillo, K., Ramos, W., Sung, P., and Tomkinson, A. E. Promotion of Dnl4-catalyzed DNA end-joining by the Rad50/Mre11/Xrs2 and Hdf1/Hdf2 complexes. (2001) Mol" *Cell* 8, 1105–1115. PMID 11741545

[14] Zha S, Boboila C, Alt FW (August 2009). "Mre11: roles in DNA repair beyond homologous recombination". *Nat. Struct. Mol. Biol.* **16** (8): 798–800. doi:10.1038/nsmb0809-798. PMID 19654615.

[15] DeFazio LG, Stansel RM, Griffith JD, Chu G (June 2002). "Synapsis of DNA ends by DNA-dependent protein kinase". *The EMBO Journal.* **21** (12): 3192–200. doi:10.1093/emboj/cdf299. PMC 126055⊙. PMID 12065431.

[16] Palmbos PL, Wu D, Daley JM, Wilson TE (December 2008). "Recruitment of Saccharomyces cerevisiae Dnl4-Lif1 complex to a double-strand break requires interactions with Yku80 and the Xrs2 FHA domain". *Genetics.* **180** (4): 1809–19. doi:10.1534/genetics.108.095539. PMC 2600923⊙. PMID 18832348.

[17] Yano K, Morotomi-Yano K, Wang SY, et al. (January 2008). "Ku recruits XLF to DNA double-strand breaks". *EMBO Rep.* **9** (1): 91–6. doi:10.1038/sj.embor.7401137. PMC 2246615⊙. PMID 18064046.

[18] Ma, Y., Pannicke, U., Schwarz, K., and Lieber, M. R. Hairpin opening and overhang processing by an Artemis/DNA-dependent protein kinase complex in nonhomologous end joining and V(D)J recombination. (2002) *Cell* 108, 781–794. PMID 11955432

[19] Nick McElhinny SA, Ramsden DA (August 2004). "Sibling rivalry: competition between Pol X family members in V(D)J recombination and general double strand break repair". *Immunol. Rev.* **200**: 156–64. doi:10.1111/j.0105-2896.2004.00160.x. PMID 15242403.

[20] Daley JM, Laan RL, Suresh A, Wilson TE (August 2005). "DNA joint dependence of pol X family polymerase action in nonhomologous end joining". *J. Biol. Chem.* **280** (32): 29030–7. doi:10.1074/jbc.M505277200. PMID 15964833.

[21] Daley JM, Laan RL, Suresh A, Wilson TE (August 2005). "DNA joint dependence of pol X family polymerase action in nonhomologous end joining". *J. Biol. Chem.* **280** (32): 29030–7. doi:10.1074/jbc.M505277200. PMID 15964833.

[22] Wilson T. E.; Grawunder U.; Lieber M. R. (1997). "Yeast DNA ligase IV mediates non-homologous DNA end joining". *Nature.* **388**: 495–498. doi:10.1038/41365. PMID 9242411.

[23] Ahnesorg P, Smith P, Jackson SP (Jan 2006). "XLF interacts with the XRCC4-DNA ligase IV complex to promote DNA nonhomologous end-joining". *Cell.* **124** (2): 301–13. doi:10.1016/j.cell.2005.12.031. PMID 16439205.

[24] Buck D, Malivert L, de Chasseval R, Barraud A, Fondaneche MC, Sanal O, Plebani A, Stephan JL, Hufnagel M, et al. (Jan 2006). "Cernunnos, a novel nonhomologous end-joining factor, is mutated in human immunodeficiency with microcephaly". *Cell*. **124** (2): 287–99. doi:10.1016/j.cell.2005.12.030. PMID 16439204.

[25] Callebaut I, Malivert L, Fischer A, Mornon JP, Revy P, de Villartay JP (2006). "Cernunnos Interacts with the XRCC4{middle dot}DNA-ligase IV Complex and Is Homologous to the Yeast Nonhomologous End-joining Factor Nej1". *J Biol Chem*. **281** (20): 13857–60. doi:10.1074/jbc.C500473200. PMID 16571728.

[26] Riballo E, Woodbine L, Stiff T, Walker SA, Goodarzi AA, Jeggo PA (February 2009). "XLF-Cernunnos promotes DNA ligase IV-XRCC4 re-adenylation following ligation". *Nucleic Acids Res*. **37** (2): 482–92. doi:10.1093/nar/gkn957. PMC 2632933. PMID 19056826.

[27] Lee SE, Pâques F, Sylvan J, Haber JE (July 1999). "Role of yeast SIR genes and mating type in directing DNA double-strand breaks to homologous and non-homologous repair paths". *Curr. Biol*. **9** (14): 767–70. doi:10.1016/s0960-9822(99)80339-x. PMID 10421582.

[28] Mimitou EP, Symington LS (September 2009). "DNA end resection: Many nucleases make light work". *DNA Repair (Amst.)*. **8** (9): 983–95. doi:10.1016/j.dnarep.2009.04.017. PMC 2760233. PMID 19473888.

[29] Jung D, Alt FW (Jan 2004). "Unraveling V(D)J recombination: insights into gene regulation". *Cell*. **116** (2): 299–311. doi:10.1016/S0092-8674(04)00039-X. PMID 14744439.

[30] Schatz DG, Baltimore D (April 1988). "Stable expression of immunoglobulin gene V(D)J recombinase activity by gene transfer into 3T3 fibroblasts". *Cell*. **53** (1): 107–15. doi:10.1016/0092-8674(88)90492-8. PMID 3349523.

[31] Gilfillan S, Dierich A, Lemeur M, Benoist C, Mathis D (August 1993). "Mice lacking TdT: mature animals with an immature lymphocyte repertoire". *Science*. **261** (5125): 1175–8. doi:10.1126/science.8356452. PMID 8356452.

[32] Komori T, Okada A, Stewart V, Alt FW (August 1993). "Lack of N regions in antigen receptor variable region genes of TdT-deficient lymphocytes". *Science*. **261** (5125): 1171–5. doi:10.1126/science.8356451. PMID 8356451.

[33] Boulton SJ, Jackson SP (1998). "Components of the Ku-dependent non-homologous endjoining pathway are involved in telomeric length maintenance and telomeric silencing". *EMBO J*. **17**: 1819–28. doi:10.1093/emboj/17.6.1819. PMC 1170529. PMID 9501103.

[34] Kerzendorfer C, O'Driscoll M (September 2009). "Human DNA damage response and repair deficiency syndromes: Linking genomic instability and cell cycle checkpoint proficiency". *DNA Repair (Amst.)*. **8** (9): 1139–52. doi:10.1016/j.dnarep.2009.04.018. PMID 19473885.

[35] Li H, Vogel H, Holcomb VB, Gu Y, Hasty P (December 2007). "Deletion of Ku70, Ku80, or both causes early aging without substantially increased cancer". *Mol. Cell. Biol*. **27** (23): 8205–14. doi:10.1128/MCB.00785-07. PMC 2169178. PMID 17875923.

[36] Vaidya A, Mao Z, Tian X, Spencer B, Seluanov A, Gorbunova V (2014). "Knock-in reporter mice demonstrate that DNA repair by non-homologous end joining declines with age". *PLoS Genet*. **10** (7): e1004511. doi:10.1371/journal.pgen.1004511. PMC 4102425. PMID 25033455.

[37] Bernstein H, Payne CM, Bernstein C, Garewal H, Dvorak K (2008). Cancer and aging as consequences of unrepaired DNA damage. In: New Research on DNA Damages (Editors: Honoka Kimura and Aoi Suzuki) Nova Science Publishers, Inc., New York, Chapter 1, pp. 1-47. open access, but read only https://www.novapublishers.com/catalog/product_info.php?products_id=43247 ISBN 978-1604565812

[38] Lorenzini A, Johnson FB, Oliver A, Tresini M, Smith JS, Hdeib M, Sell C, Cristofalo VJ, Stamato TD (2009). "Significant correlation of species longevity with DNA double strand break recognition but not with telomere length". *Mech. Ageing Dev*. **130** (11-12): 784–92. doi:10.1016/j.mad.2009.10.004. PMC 2799038. PMID 19896964.

# 23.10 Text and image sources, contributors, and licenses

## 23.10.1 Text

- **Genetics of aging** *Source:* https://en.wikipedia.org/wiki/Genetics_of_aging?oldid=754991669 *Contributors:* Benbest, Rjwilmsi, McGeddon, Cydebot, Hebrides, Headbomb, Jytdog, RjwilmsiBot, Dcirovic, Mazdarin and Anonymous: 3

- **Maximum life span** *Source:* https://en.wikipedia.org/wiki/Maximum_life_span?oldid=767323464 *Contributors:* Taw, Miguel~enwiki, William Avery, Patrick, Michael Hardy, Lexor, TakuyaMurata, Ike9898, David Latapie, Furrykef, SEWilco, Rhys~enwiki, Donreed, TimR, Tim-rollpickering, Intangir, Xanzzibar, MitkMiruku, Unfree, Ancheta Wis, Andy, Peruvianllama, Andycjp, Prestonmarkstone, Chris j wood, Discospinster, Rich Farmbrough, Bender235, Sunborn, Oniongirl, CDN99, Feitclub, VBGFscJUn3, Tavdy79, Velho, Benbest, JFG, Kosher Fan, Marudubshinki, Sin-man, Ryoung122, Tlroche, Rjwilmsi, Koavf, Tawker, Nneonneo, Tintazul, Pinkville, Preslethe, Diza, No Swan So Fine, MoRsE, Wavelength, Jadon, Mikalra, RussBot, John Quincy Adding Machine, Chris Capoccia, Anomalocaris, NawlinWiki, Lsisson, Ragesoss, Xabian40409, Jcrook1987, SFGiants, Knotnic, Arad, Katieh5584, Trickstar, SmackBot, Kellen, Lcarsdata, Melchoir, Moshe Constantine Hassan Al-Silverburg, George Church, MaxSem, Azslande, AMK152, Nakon, Akriasas, Romanski, SS2005, Wtwilson3, Mgiganteus1, Green Giant, Xiaphias, InedibleHulk, Iridescent, Clarityfiend, Gavrilov, Joseph Solis in Australia, AStudent, PaddyM, Esn, Harold f, Mellery, Lavateraguy, BeenAroundAWhile, Redlock, Browserlong, Cydebot, NealIRC, Querencia, Robertinventor, Epbr123, Headbomb, Transhumanist, Jayron32, Mack2, Qwerty Binary, MER-C, Avaya1, Albany NY, Greensburger, Magioladitis, A4, TheAllSeeingEye, PenguinJockey, Ethanjohn, GirlForLife, A3nm, Bù hán er li, Roswell Crash Survivor, Fconaway, KeepItClean, Anonywiki, Jiu9, Student7, Dorftrottel, Useight, Thomas.W, Philip Trueman, A4bot, Sturunner, Pweltz, Aholladay, Whatiguana, Ecopetition, Sardonicone, Chill Factor Five, Stewart222, Joe3600, Flyer22 Reborn, KPH2293, Lisatwo, Fasthorse, Varanwal, Invertzoo, Luatha, ClueBot, Kumagoro-42, Fyyer, Mild Bill Hiccup, Timberframe, Sanjeev.singh3, Phenylalanine, John J Bulten, Cheakamus, Remember September 11, 2001, Sun Creator, Aitias, Ddog892, DumZiBoT, Hell Hawk, Ost316, Good Olfactory, Roentgenium111, Some jerk on the Internet, DOI bot, Non-dropframe, Krollbarf, RTG, Bernstein0275, PFSLAKES1, LinkFA-Bot, Terpfan20, Legobot, Yobot, Themfromspace, The Earwig, Lerichard, Leefeni de Karik, AnomieBOT, Andrewrp, Yachtsman1, Materialscientist, Citation bot, Flavius.Aettius, Alumnum, Locobot, Green06, Citation bot 1, AQUIMISMO, Jonesey95, Tom.Reding, Trappist the monk, Comet Tuttle, E.songhori, EmausBot, Cagrimertbakirci, Dcirovic, GermanJoe, Hkwatkins, Leoj83, DASHBotAV, ClueBot NG, Posywillos73, CarolineLG, Xagg, Apolo13, Lystopad, BG19bot, CitationCleanerBot, BattyBot, Cyberbot II, TheLionHearted, ChrisGualtieri, Ivan trus, Futurist110, Dexbot, Spindlerlabucr, Seacactus 13, Everything Is Numbers, I am One of Many, Ashorocetus, Anrnusna, Rbaughman7, Sandvich18, JohnBKeating, AwesoMan3000, Cassandra3001, NQ-Alt, Izkala, Quixotic Cleaner, Xavier troll, VRtrooper, WikiCnidaria, GreenC bot, Vaticidalprophet, Bender the Bot, Fick Lui and Anonymous: 216

- **Epigenetic clock** *Source:* https://en.wikipedia.org/wiki/Epigenetic_clock?oldid=768669371 *Contributors:* Xanzzibar, Rjwilmsi, Nihiltres, Bgwhite, Arthur Rubin, Chris the speller, Loadmaster, Cunningpal, Headbomb, Cgingold, CommonsDelinker, Doc James, William11405, Unbuttered Parsnip, Yobot, Worldbruce, AnomieBOT, Gap9551, Diannaa, Dewritech, Josve05a, Antiqueight, BG19bot, Stevetihi, JuniperRed, CitationCleanerBot, BattyBot, ChrisGualtieri, Dexbot, Dmitry Dzhagarov, Spyglasses, Seabuckthorn, Chaya5260, Charlotte Aryanne, Dan Eisenberg, XKidKemist, ACleaningPerson and Anonymous: 12

- **Senescence** *Source:* https://en.wikipedia.org/wiki/Senescence?oldid=768752651 *Contributors:* Bryan Derksen, Timo Honkasalo, SimonP, Azhyd, Heron, Netesq, Patrick, Infrogmation, Michael Hardy, Lexor, Adrianrf, Gabbe, Qaz, Gdvorsky, TakuyaMurata, (, Palfrey, Shino Baku, Schneelocke, Quizkajer, Timwi, Tedius Zanarukando, Selket, Samsara, Frazzydee, Robbot, P0lyglut, Witbrock, Wayland, Barbara Shack, Mcfaulr, Mintleaf~enwiki, Fudoreaper, IRelayer, Everyking, Jfdwolff, Dlamming, Pne, Neilc, MisterFusion, Dan aka jack, Beland, Socrtwo, Imjustmatthew, CALR, Rich Farmbrough, Drano, Bender235, Neurophyre, Ben Standeven, Aranel, Siggyllama, Nigelj, Shenme, Viriditas, Cayte, Scott Ritchie, La goutte de pluie, Eric Kvaalen, Iris lorain, Wdfarmer, Jost Riedel, Leoadec, RJII, TenOfAllTrades, TheCoffee, Georgia guy, LOL, StradivariusTV, Benbest, Canadian Paul, Dolfrog, GregorB, V8rik, ThrBigD, Rjwilmsi, Nightscream, Eyu100, Staecker, FlaBot, Vietbio~enwiki, HowardLeeHarkness, BMF81, King of Hearts, Chobot, Bgwhite, YurikBot, Wavelength, NTBot~enwiki, Reo On, Chris Capoccia, Emmanuelm, Hyperbole, Gaius Cornelius, CambridgeBayWeather, Shaddack, Voyevoda, Erielhonan, Epipelagic, DeadEyeArrow, GraemeL, Matt Heard, RupertMillard, SmackBot, Mitteldorf, KnowledgeOfSelf, TestPilot, Verne Equinox, Edgar181, Gilliam, Tyciol, Haenaroad, Kernigh, RDBrown, Droll, Deli nk, Somewildthingsgo, Ikiroid, Go for it!, JonHarder, Radagast83, Nakon, TedE, Richard001, Freemarket, Drphilharmonic, Virtualsim, Metamagician3000, G-Bot~enwiki, AThing, Anlace, Jidanni, Livemixlove, Mgiganteus1, Green Giant, Peterlewis, Ben Moore, Ckatz, Jon186, Succubus MacAstaroth, Elb2000, Gavrilov, Kaarel, Kcops7, Lord mortekai, CRGreathouse, Cogpsych, JamesAGreen, lokseng, Cydebot, Ppgardne, A876, Dominicanpapi82, Garik, NMChico24, Wikid77, Headbomb, Pjvpjv, Zachary, Transhumanist, KrakatoaKatie, WinBot, Silver seren, Alphachimpbot, Wayiran, Lfstevens, Skomorokh, Rothorpe, WolfmanSF, Lucyin, Pushnell, Upholder, GirlForLife, Benjamintchip, JaGa, WLU, Adriaan, Uriel8, Nono64, J.delanoy, Boghog, Tikiwont, Yoongkheong, Rod57, Jeepday, Mikael Häggström, Tarotcards, LittleHow, TomasBat, DadaNeem, Mikon, BrettAllen, Kniht, Funandtrvl, GIBBOUS3, Erwinser, Majoreditor, Triamus, Jim200, Afarorg, Spencersage, Guillaume2303, Ask123, Aholladay, Y tambe, Jermerc, Hedgehog33, Richwil, Enviroboy, Softlavender, Jeffw245, Doc James, Winchelsea, Georgesibbald, LeadSongDog, Yerpo, Ozkurede, Ausjay99, Tomahiv, Invertzoo, ClueBot, NickCT, Renzoy16, Billionaire142, Healthwise, Mild Bill Hiccup, Niceguyedc, Eldarion, Phenylalanine, Gnome de plume, Rhododendrites, Jonverve, PotentialDanger, XLinkBot, WordMachine, PatrickStar LaserPants, KarakasaObake, Facts707, Paddy Simcox, Addbot, DOI bot, Silvia3, Viper7882, Neodop, Sarah Lynne Nashif, Fluffernutter, Fli-Worker, Bernstein0275, LinkFA-Bot, Cozmic serpent, Thomas Bjørkan, Tassedethe, Bouncingball2, Ehrenkater, Jarble, Candidulus, Yobot, WikiDan61, Tiddlypeep, Der Zeitgeist, AnomieBOT, Materialscientist, Citation bot, G6cid, Flavius.Aettius, Capricorn42, Gap9551, Crzer07, Twirligig, John S. Peterson, Mathgod333, Imyoda69, Serenditious, FrescoBot, Paine Ellsworth, Melly2000, Himanshu2k7dude, Green06, Imdbbmmamf, Citation bot 1, Igor233, Pinethicket, Ohtavala, Fartherred, Trappist the monk, DARTH SIDIOUS 2, Matt-eee, EmausBot, Wakuteka, ZeniffMartineau, Lotez, EME44, Dcirovic, Allforrous, Pierlot, RockMagnetist, Hkwatkins, ClueBot NG, Excelsius, Wikigoodnews, IfYouDoIfYouDon't, Osterluzei, Frietjes, Dictabeard, Lshanahan, MerlIwBot, Helpful Pixie Bot, Curb Chain, BG19bot, Wp500, Stevetihi, Neuraxis, JuniperRed, Cameroncheng, Pablomoorhead, NotWith, BattyBot, Cyberbot II, Maxronnersjo, Manabeast333, GoldenSHK, Dexbot, Jackninja5, TwoTwoHello, Everything Is Numbers, Dmitry Dzhagarov, PC-XT, 19BrookStr, Yanivnizry, Sjrct, KeoRoo, Ulenspegel, YiFeiBot, Cytokinetics, Voodoogoogoo, Chaya5260, Captain Cornwall, Monkbot, Brantley.Holland, Notrubd, Mattster3517, Wasp32, Mayo Foundation, Accurate right, Wik5645, Readeraa1, GreenC bot, Wikisanchez, Tim280716 and Anonymous: 285

- **Shelterin** *Source:* https://en.wikipedia.org/wiki/Shelterin?oldid=763628195 *Contributors:* Benbest, Rjwilmsi, Bgwhite, Kaobear, Speciate, Jdaloner, Azim58, Trappist the monk, BG19bot, CitationCleanerBot, Biosas and Anonymous: 4

- **Hayflick limit** *Source:* https://en.wikipedia.org/wiki/Hayflick_limit?oldid=762087258 *Contributors:* William Avery, ShaunOfTheLive, Altenmann, Xanzzibar, Giftlite, Jfdwolff, Pascal666, Pgan002, DragonflySixtyseven, Thorwald, Percy, Discospinster, Cmdrjameson, R. S. Shaw, Richard Arthur Norton (1958- ), Mindmatrix, Benbest, Cosmicosmo, GregorB, Davidrust, Noit, BD2412, FreplySpang, Rjwilmsi, Tizio, Nneonneo, NeonMerlin, Diza, Sasuke Sarutobi, Ansell, Dysmorodrepanis~enwiki, Dtrebbien, Welsh, Mccready, Epipelagic, Marcelo-Silva, Leptictidium, 2over0, Edouard-lopez, JuJube, Sardanaphalus, SmackBot, ElTchanggo, Kirkaracha, Scray, Nahum Reduta, Drphilharmonic, JzG, Soap, Loadmaster, Smith609, Succubus MacAstaroth, Iridescent, Gavrilov, Banedon, Gregbard, Cydebot, OrbitSoldier, A876, Crum375, TimVickers, Tlabshier, Hoffmeier, Yurei-eggtart, Gabe1972, A3nm, Purslane, Miltnoda, Peter nann, Leonard Hayflick, Funandtrvl, Myles325a, Malljaja, Spiral5800, Richwil, SieBot, Phe-bot, Alexbrn, Sanya3, OKBot, Knownlike, Gustavocarra, MacGod, XLinkBot, Twitchbunny, Addbot, Mac Dreamstate, Debresser, SpBot, Lightbot, Legobot, Luckas-bot, Yobot, Nutriveg, Citation bot, R4nd0m U53R, Trilliumroots, Dac28, Paine Ellsworth, Citation bot 1, Citation bot 4, MolBioMan, RedBot, Recycled.jack, Bokeh.Senssei, Haljammy, Watcher0911, RjwilmsiBot, Ripchip Bot, Blackolives, Thubby12, Milotoor, WikitanvirBot, 9banDH, ClueBot NG, MusikAnimal, AvocatoBot, Aibrain, U4ealongan, YFdyh-bot, Futurist110, Дмитрий Дж., Everything Is Numbers, Reatlas, Alexwho314, TheFrog001, Ulenspegel, EtymAesthete, Anrnusna, Monkbot, Happy Attack Dog, Unician, TimothyBoland1, Azmistowski17, FadyP and Anonymous: 98

- **Apoptosis** *Source:* https://en.wikipedia.org/wiki/Apoptosis?oldid=767609558 *Contributors:* Magnus Manske, Ignaciovicario, Carey Evans, Derek Ross, LC~enwiki, Sodium, The Anome, Alex.tan, Andre Engels, Vanderesch, PierreAbbat, Ben-Zin~enwiki, Heron, Michael Hardy, Lexor, 168..., Ellywa, JWSchmidt, MichaK, Wikiborg, Wik, Steinsky, AHands, Furrykef, Jaimeglz, Populus, Quoth-22, Flockmeal, Acdbx, Robbot, Korath, R3m0t, Peak, Romanm, Chris Roy, Auric, Wikibot, Kent Wang, Diberri, Alan Liefting, Marc Venot, Giftlite, DocWatson42, Mintleaf~enwiki, Zuxy, Alison, Bensaccount, Chinasaur, Mboverload, Jds, Pne, Bobblewik, Neilc, Adenosine, Mporch, J. 'mach' wust, Knutux, Sam Hocevar, WpZurp, Sparky the Seventh Chaos, Brianhe, Rich Farmbrough, Guanabot, FT2, Jyp, Bender235, Onelamb, Pjf, Kwamikagami, Nickj, Jpgordon, Jonathan Drain, Bobo192, Billymac00, Arcadian, La goutte de pluie, Pschemp, HasharBot~enwiki, Alansohn, Thebeginning, Jared81, Tabor, Bootstoots, Helixblue, Tycho, Cburnett, Pauli133, Ceyockey, Feezo, Marcelo1229, Kelly Martin, TheIguana, Xmp, Benbest, EnSamulili, Flamingspinach, SDC, Palica, GFP~enwiki, Magister Mathematicae, Grammarbot, Phoenix-forgotten, Rjwilmsi, Biochemza, Brighterorange, Bhadani, Cassowary, FlaBot, Nihiltres, Kerowyn, Gurch, Jrtayloriv, Stevenfruitsmaak, Daev, Chobot, Ramorum, Sharkface217, Mhking, Antiuser, Bkhouser, YurikBot, Wavelength, DMahalko, Hede2000, Pi Delport, Epolk, Mastertypo, Hydrargyrum, JihemD, Gaius Cornelius, Shaddack, NawlinWiki, Aeusoes1, Snek01, Welsh, Wolbo, DRosenbach, Bob247, Leptictidium, Mike Dillon, Peoplez1k, Webster100, Kubra, Kungfuadam, Banus, GrinBot~enwiki, Zvika, SmackBot, Haza-w, Reedy, Ixtli, Eskimbot, P b1999, David.Throop, JoeKearney, Tyciol, Bluebot, RDBrown, Tito4000, Ben.c.roberts, Regford, Matthias koehler~enwiki, Snowmanradio, Nahum Reduta, Kingdon, Dcamp314, Drphilharmonic, Rockpocket, The Ungovernable Force, Esrever, Nishkid64, DO11.10, Jidanni, N3bulous, Jaganath, Shinryuu, Serephine, InedibleHulk, SandyGeorgia, Novangelis, Hu12, Tawkerbot2, Bioinformin, CmdrObot, Memetics, Lithium6, ShelfSkewed, Leujohn, Oroboros, Jeremykappasig, Sonare, AndrewHowse, Bevan lin, Eubanks718, Rhode Island Red, Adolphus79, Christk02, Wdspann, Orphu of io, Marcgillespie, Retroid, Hoobyjuice, Wikid77, Opabinia regalis, Qwyrxian, CopperKettle, Headbomb, Luigifan, Bio2mancer, Peter Znamenskiy, Escarbot, Stui, MoogleDan, Cajdd, Xolom, TimVickers, Exteray, Ftc68, Deflective, TAnthony, KEKPΩΨ, Magioladitis, Avjoska, ThoHug, WhatamIdoing, Ciar, Vivekanandbhardwaj, Robert M. Hunt, LorenzoB, Lenticel, Ligress, Djr5353, Nono64, Llamabread, Nbauman, Boghog, Theowl007, Yonidebot, RedPoptarts, Hodja Nasreddin, Dispenser, KylieTastic, Entropy, DorganBot, WinterSpw, IanRobertson72, Jeff G., Philip Trueman, A4bot, Rei-bot, Littlealien182, MuanN, Wiikipedian, Earthdirt, Dclyde, Messengercrow, TaintedCherub, Dirkbb, Venus14, Munozpinedo, Ceranthor, Twooars, Dawning, Fischer.sebastian, Doc James, Kosigrim, Sonnejw0, StAnselm, JerrySteal, Flyer22 Reborn, Sovbeos, Lightmouse, Kaklama, Sunrise, Mike2vil, Bogwhistle, Drgarden, Delphis wk, ClueBot, Deviator13, R000t, DevinCurrie, Botodo, Drmies, Mild Bill Hiccup, Highwind65, Jai20chand20, Opcassio~enwiki, Crjinlaca, Cookie28, Drsrisenthil, Rhododendrites, Arjayay, Questinthedesert, Lesarge, El bot de la dieta, Rfreed314, Pablufu, Ginbot86, Fiquei, XLinkBot, SilvonenBot, Daughter of Mimir, Snapperman2, Addbot, Seipjere, DOI bot, Neodop, Tanhabot, Nomad2u001, Low-frequency internal, Biotech Examiner, MrOllie, Dum da dump dum dum, Download, Chzz, Tide rolls, Saintjoep, Dreid1987, Luckas-bot, Yobot, AnomieBOT, Piano non troppo, RSXS, Materialscientist, Citation bot, Xqbot, Srich32977, Omnipaedista, Pginz, Turnvater Jahn, Misull, Howard McCay, Tobby72, Unomi, D'ohBot, Zh8n2, Citation bot 1, Belairrımkb, Sideamongst, Robinhaw, RedBot, Meaghan, My very best wishes, Jauhienij, TobeBot, Trappist the monk, Sylvwiki, 564dude, Onel5969, DASHBot, Orphan Wiki, Imflora, NotAnonymous0, Dcirovic, Werieth, BrandonEXPLOSION, Clementine2009, Access Denied, H3llBot, AbdulKareem92, Pierre Fauquenot, Rasinj, Egelberg, Vitagos, RockMagnetist, Llightex, Sven Manguard, Anita5192, Apoptosome, Petrb, ClueBot NG, CocuBot, Kupothechocobo, Chester Markel, Frietjes, Antibody123, Prf17, Helpful Pixie Bot, TheToxicologists, Curb Chain, Vigi.limi, Signaluser2, BG19bot, Abm8587, Drrajendrans, CatPath, PhnomPencil, AvocatoBot, Orangutans, Neuraxis, Dkrysko, Pattimurphy, NotWith, Syammohanm, Zedshort, EricEnfermero, ChrisGualtieri, Mediran, Melenc, Khazar2, Eottergonzalez, DotSlashSteven, Nmcca, MikeHBriggs, Vibitz, Dexbot, Everything Is Numbers, Bela Bartoszek, Corn cheese, Goodwinjinesh, Me, Myself, and I are Here, Dravijitmondal, AmericanLemming, JWNoctis, EtymAesthete, Bruinmed2017, AlmaMer, 7Sidz, Monkbot, Lisawisa, Nbzxy1993, Seraphthorn, Thenamestom, CV9933, Joh09518, Drsbreddi, KasparBot, Ltumanovskaya, Oatco, InternetArchiveBot, GreenC bot, OAbot, BQUB16Agomila, M.Majid Tanveer Jutt and Anonymous: 373

- **Programmed cell death** *Source:* https://en.wikipedia.org/wiki/Programmed_cell_death?oldid=762151963 *Contributors:* Josh Grosse, Brainsik, Lexor, Jaimeglz, Tophanana, Romanm, Pne, Williamb, Alexrexpvt, Rich Farmbrough, FT2, Rspeer, CanisRufus, Orlady, Arcadian, La goutte de pluie, Nihil~enwiki, Alansohn, Danhash, WadeSimMiser, Robertwharvey, Cyberman, Phoenix-forgotten, Rjwilmsi, Koavf, Biochemza, Jrtayloriv, Wavelength, Epipelagic, Jinkbl0t, Leptictidium, Arthur Rubin, PRehse, SmackBot, Ixtli, Tyciol, Kingdon, Dcamp314, Drphilharmonic, Dr.saptarshi, Makyen, Serephine, Michaelbusch, Alaibot, Headbomb, Dawnseeker2000, JAnDbot, Leuko, Ph.eyes, Albany NY, Magioladitis, Dekimasu, The cattr, MartinBot, Boghog, Mikael Häggström, Plindenbaum, DorganBot, AlnoktaBOT, Munozpinedo, MCTales, SieBot, Gknor, Louismaddox, ClueBot, Lysis rationale, Koumz, WikHead, Addbot, DOI bot, Andykinosis, Chzz, Ehrenkater, Lightbot, Xenobot, Xcrunnerjadon, Luckas-bot, Anypodetos, AnomieBOT, AndrooUK, Citation bot, J04n, FrescoBot, Jatlas, Citation bot 1, Jonesey95, Tom.Reding, Zbalmuth, Jesse V., John of Reading, Superboy12, Dcirovic, Het Sas Team, Wayne Slam, ClueBot NG, Peter James, Braincricket, YorrickDavis, Hz.tiang, CitationCleanerBot, DPL bot, MikeHBriggs, Everything Is Numbers, Me, Myself, and I are Here, Awdeakin, Ak247shay, FamAD123, Iztwoz, Ckorb89, Cribff, Robevans123, Monkbot, NeuroApoptosis, Pinjoeko, The Voidwalker, Bear-rings and Anonymous: 70

- **Cell division** *Source:* https://en.wikipedia.org/wiki/Cell_division?oldid=768379158 *Contributors:* Magnus Manske, Josh Grosse, Lexor, Ahoerstemeier, Nikki chan, CatherineMunro, JWSchmidt, AWhiteC, Marshman, Doug swisher, RadicalBender, Robbot, Paranoid, Arkuat, Dave6, Giftlite, Nunh-huh, Onco p53, GD~enwiki, G3pro, PFHLai, Mike Rosoft, Rich Farmbrough, Kooo, Vitamin b, Martpol, Bender235, Guettarda, Bobo192, Shenme, Reuben, Arcadian, La goutte de pluie, Jojit fb, Obradovic Goran, Nsaa, Jakew, Alansohn, Super-Magician, Joriki,

phanBot, Rrburke, Madman2001, Monotonehell, Dcteas17, Drphilharmonic, JzG, FrozenMan, Serephine, Ambuj.Saxena, IvanLanin, Twas Now, ChrisCork, Protiek, Dia^, CmdrObot, Svetachakrabarti, Cydebot, Ppgardne, Hwttdz, Crum375, SomethingCatchy, Thijs!bot, Headbomb, JustAGal, Oleykin, MoogleDan, TimVickers, Tlabshier, Darklilac, Pixelface, Lfstevens, Ozperp, MER-C, VoABot II, Decembermouse, Balloonguy, Destynova, Enquire, Rickard Vogelberg, AstarothCY, R'n'B, Pekaje, Fconaway, Boghog, Xris0, Rod57, ElinneaG, Sprunk, STBotD, Davrids, VolkovBot, Mirrordor, Mabidex, TXiKiBoT, Berichard, JCRansom, AlleborgoBot, Adr1liano, SieBot, Mikemoral, Noisy Hand, Bgordski, Henk Poley, Cmale, Plastikspork, Niceguyedc, Dylan620, Alexbot, Tundrill, Cowwaw, Bproman, Dana boomer, Sierrasciadmin, KarakasaObake, Noctibus, Dfoxvog, Kace7, Good Olfactory, Snapperman2, Addbot, Piz d'Es-Cha, DOI bot, Fyrael, Neodop, Freakonomicsfan, Bernstein0275, Deamon138, Bouncingball2, Tide rolls, Icons789, Saintjoep, שב, ررشک, Ben Ben, Luckas-bot, Yobot, Tiddlypeep, Theserialcomma, Legobot II, Vinodksingh1111, AnomieBOT, Superpoopoo, Materialscientist, Citation bot, Trilliumroots, LilHelpa, Flavius.Aettius, Srich32977, Makeswell, Kylelovesyou, FrescoBot, Citation bot 1, Citation bot 4, Jonesey95, RedBot, Bohalrantipol, Jeangabin, Alexorella, Trappist the monk, NickVertical, Autumnox, Time9, Theo10011, Amkilpatrick, Vitaliy0001, Feduchin, Watcher0911, RjwilmsiBot, Adarakdjian, TjBot, EmausBot, Wiki.Tango.Foxtrot, Lotez, Gnulinux, Dcirovic, TYelliot, Patch101, FeatherPluma, Roanotto, ClueBot NG, Pengortm, Widr, BG19bot, I7laseral, CitationCleanerBot, Dormantfreedom, Pndanilov, Emskorda, BattyBot, Germanbrother, Dexbot, Everything Is Numbers, Ivanjoksimovic, Alexwho314, BradYard, Lisamistowz, EtymAesthete, Anrrusna, AlmaMer, Pdavis9396, Maethes, Monkbot, MattLBeck, Ip1673, Nith117, Noora Al Batushi, Dan Eisenberg, Dr Scott Cohen, Mahfuzur rahman shourov, Jackehoe, Scottbradleycohen and Anonymous: 212

- **Telomerase reverse transcriptase** *Source:* https://en.wikipedia.org/wiki/Telomerase_reverse_transcriptase?oldid=765624080 *Contributors:* Stepp-Wulf, Rich Farmbrough, Arcadian, BD2412, Rjwilmsi, Vegaswikian, Hyperbole, Eleassar, SmackBot, Drphilharmonic, Cydebot, Was a bee, R'n'B, Boghog, Andrew Su, ProteinBoxBot, Svick, DumZiBoT, Addbot, DOI bot, Luckas-bot, Yobot, Citation bot, Citation bot 1, Yeast2Hybrid, Dcirovic, Ego White Tray, Teakt17, BG19bot, Everything Is Numbers, Splicevariant, EtymAesthete, Anrrusna, Monkbot, CV9933 and Anonymous: 7

- **Reverse transcriptase** *Source:* https://en.wikipedia.org/wiki/Reverse_transcriptase?oldid=768895161 *Contributors:* Magnus Manske, Paul Drye, WojPob, Graft, Ewen, Dwmyers, Gaurav, Suisui, JWSchmidt, Lancevortex, AhmadH, Thomasgl, Robbot, RedWolf, Peak, Jfdwolff, Jrdioko, Dullhunk, PDH, PFHLai, M1ss1ontomars2k4, Dr. Strangelove~enwiki, Arcadian, Physicistjedi, Pearle, Espoo, Loris, Rebroad, Richard Arthur Norton (1958- ), Rjwilmsi, Jivecat, FlaBot, Chobot, WriterHound, YurikBot, Alethiareg, Conscious, Splette, Rodasmith, Wimt, Morpheios Melas, Lt-wiki-bot, SmackBot, Espresso Addict, David Shear, Evenprime, Drphilharmonic, Mystaker1, Ohconfucius, Mgiganteus1, Serephine, Tawkerbot2, Fvasconcellos, Agathman, Cydebot, RelentlessRecusant, Master son, Carstensen, Wejstheman, Thijs!bot, Peter Znamenskiy, GAThrawn22, Escarbot, Lakster37, Tjmayerinsf, Smartse, Myanw, VoABot II, MartinBot, R'n'B, Boghog, Lantonov, Mikael Häggström, Alnokta, VolkovBot, TXiKiBoT, CarinaT, Richwil, SieBot, Fratrep, Svick, Iknowyourider, Ward20, Capitalismojo, Hariva, ClueBot, Doseiai2, DragonBot, Jed 20012, Veryhuman, Addbot, Echinoidea, Legobot, Luckas-bot, Yobot, Etektema-GMU, Aff123a, Citation bot, NajGMU, Dkabban-GMU, Skhanal-GMU, MauritsBot, Xqbot, Flower-gmu, Km2452-GMU, Eumeng.chong, FrescoBot, Jamesooders, Citation bot 1, Pinethicket, RedBot, MastiBot, Abuttlmao, SW3 5DL, Lb.at.wiki, RoadTrain, Reaper Eternal, Mrseanski, TuHan-Bot, Dcirovic, John Mackenzie Burke, Abergabe, WeigelaPen, Jtien512, Mikhail Ryazanov, ClueBot NG, Helpful Pixie Bot, Kalafati, Strike Eagle, Miguelferig, Ekennedt, Pratyya Ghosh, DarafshBot, Makecat-bot, Everything Is Numbers, Kaitanyk, Tentinator, Bilorv, Monkbot, MLGProGamer123, Narky Blert, Bender the Bot and Anonymous: 114

- **Chromosome** *Source:* https://en.wikipedia.org/wiki/Chromosome?oldid=768419605 *Contributors:* AxelBoldt, Magnus Manske, TwoOneTwo, NathanBeach, Bryan Derksen, The Anome, Malcolm Farmer, Josh Grosse, Youssefsan, Christian List, William Avery, AdamRetchless, Heron, Rsabbatini, Tox~enwiki, Youandme, Bernfarr, Edward, Michael Hardy, Lexor, Ixfd64, Miciah, Karada, Alfio, 168..., Ellywa, Ahoerstemeier, Arwel Parry, JWSchmidt, Darkwind, Julesd, Glenn, Llull, Andres, Tristanb, Grin, Cafemusique, Ptoniolo, RodC, Mkrose, Steinsky, Omegatron, Samsara, Topbanana, Raul654, GPHemsley, Phil Boswell, Donarreiskoffer, Robbot, Astronautics~enwiki, Fredrik, Altenmann, Peak, Sunray, Hadal, UtherSRG, Isopropyl, Xanzzibar, Tea2min, Giftlite, Tom harrison, Lupin, Ferkelparade, Everyking, FriedMilk, Guanaco, Pascal666, Alvestrand, Kandar, Erhudy, Chowbok, Gadfium, Pgan002, Ruy Lopez, J. 'mach' wust, Knutux, Quadell, Antandrus, G3pro, PDH, Bornslippy, Tail, JeffreyN, Neutrality, JohnArmagh, Adashiel, Iwilcox, Mike Rosoft, Venu62, Discospinster, Rich Farmbrough, Guanabot, Cacycle, Hippojazz, Lyndell, Mani1, Bender235, Billlion, RoyBoy, Dennis Brown, CDN99, Bobo192, Smalljim, R. S. Shaw, Arcadian, Giraffedata, Jerryseinfeld, Brainy J, Alansohn, JYolkowski, Arthena, Keenan Pepper, Andrewpmk, Wouterstomp, Calton, Curious1i, Plange, Snowolf, Velella, Wayne Schroeder, Cnickelfr~enwiki, ClockworkSoul, Esparkhu, Amorymeltzer, Sciurinæ, TheCoffee, Johntex, TriNotch, Adrian.benko, RyanGerbil10, Gmaxwell, Richard Arthur Norton (1958- ), Pinball22, Jitsang, Bkkbrad, Astrowob, Eleassar777, Hotshot977, Eras-mus, Kralizec!, MarcoTolo, GSlicer, Mandarax, Tslocum, RichardWeiss, Graham87, GoldRingChip, BD2412, Galwhaa, Avbetuw, Jclemens, Drbogdan, Rjwilmsi, Smoe, Venullian, TBHecht, Fred Bradstadt, Maurog, GregAsche, Sango123, Yamamoto Ichiro, Titoxd, FlaBot, Old Moonraker, Winhunter, Crazycomputers, RexNL, Gurch, Intgr, TeaDrinker, Daycd, Chobot, DVdm, Random user 39849958, Bgwhite, Bomb319, Gwernol, Siddhant, Cuahl, YurikBot, Wavelength, Kinneyboy90, Phantomsteve, RussBot, Postglock, Chris Capoccia, DanMS, SpuriousQ, RadioFan, Stephenb, Manop, Shell Kinney, CambridgeBayWeather, Eleassar, Ihope127, Wimt, Ugur Basak, NawlinWiki, EWS23, Dysmorodrepanis~enwiki, Wiki alf, Spike Wilbury, Cquan, Mindthief, Catamorphism, Adamrush, Lipothymia, Mysid, Bota47, Addps4cat, Marcelo-Silva, Leptictidium, Ajvphilp, Zzuuzz, Ninly, Closedmouth, Jwissick, Moogsi, Pb30, 2fort5r, Anclation~enwiki, Allens, Kungfuadam, Junglecat, Banus, Ghazer~enwiki, GrinBot~enwiki, Asterion, DVD R W, Tom Morris, Eog1916, SmackBot, Ariliand, Hazaw, Brianyoumans, Slashme, InverseHypercube, KnowledgeOfSelf, Unyoyega, C.Fred, Blue520, Bomac, Lankenau, AndreasJS, Matthuxtable, Delldot, BiT, Andrewkantor, Apers0n, Zephyris, Yamaguchi先生, Gilliam, Betacommand, Andy M. Wang, Luenlin, KaragouniS, Static Universe, Asclepius, Persian Poet Gal, RDBrown, Tito4000, MK8, Master of Puppets, MalafayaBot, SchfiftyThree, DHN-bot~enwiki, Darth Panda, Gracenotes, Akhtar Ali Khan, Salmar, E946, Can't sleep, clown will eat me, DéRahier, Yidisheryid, Abrahami, Khoikhoi, Aldaron, Nakon, TedE, Dreadstar, Drphilharmonic, DMacks, Jóna Þórunn, Vina-iwbot~enwiki, Clicketyclack, Blahm, Madeleine Price Ball, SashatoBot, ArglebargleIV, Potosino, Kuru, Heimstern, Shadowlynk, IronGargoyle, Mr. Vernon, Slakr, SimonATL, Mr Stephen, Waggers, BranStark, Ethelred Cyning, Iridescent, NEMT, Sir1, Andreas Rejbrand, Igoldste, Rnb, Tawkerbot2, George100, Chris55, DBooth, Will Pittenger, SkyWalker, Wikipeedio, Alexei Kouprianov, Ale jrb, Agathman, Pathh, Dgw, NickW557, Penbat, Tex, Future Perfect at Sunrise, Parslad, Michaelas10, Gogo Dodo, Anthonyhcole, Corpx, ST47, Tawkerbot4, Carstensen, Dynaflow, Ahmedbma, Chrislk02, JCO312, JodyB, חוקר, Thijs!bot, Epbr123, Opabinia regalis, Kablammo, Mojo Hand, Wompa99, Headbomb, Marek69, John254, Doyley, RFerreira, AgentPeppermint, AndrewKemendo, Dawnseeker2000, Mentifisto, David D., KrakatoaKatie, AntiVandalBot, Luna Santin, Shmumickel, QuiteUnusual, MrWood, Tangerines, DarkAudit, TimVickers, Zappernapper, Altamel, Sluzzelin, Zokie, DuncanHill, MER-C, Andonic, PhilKnight, Savant13, .anacond-

cess Denied, Scientistview, Donner60, Petrzak, Woodsrock, ClueBot NG, Horoporo, Rezabot, Widr, Helpful Pixie Bot, Uniquenick, AdventurousSquirrel, CitationCleanerBot, Blakethegundork, TuringMachine17, Suizei, Khazar2, Leilimosa, Kelvin13, EagerToddler39, Andrewrgross, Eyesnore, Everymorning, BruceBlaus, Sam Sailor, Powermelon, Bilorv, Monkbot, Acagastya, Jdvalencia, Medgirl131, Kaseypeesho, Pinnate foliage, Gladamas, KasparBot, Abidemi37, Bender the Bot, T.squad and Anonymous: 337

- **DNA** *Source:* https://en.wikipedia.org/wiki/DNA?oldid=768582387 *Contributors:* AxelBoldt, Magnus Manske, Peter Winnberg, Marj Tiefert, Lee Daniel Crocker, Eloquence, Mav, Bryan Derksen, Strubenstein, RK, Andre Engels, Fnielsen, Ted Longstaffe, LA2, Danny, Aldie, Unukorno, Toby Bartels, PierreAbbat, Ortolan88, Ben-Zin~enwiki, Anthere, Ellmist, Graft, Heron, Hephaestos, Someone else, Stevertigo, Spiff~enwiki, Lir, Erik Zachte, Lexor, Wwwwolf, Ixfd64, Fruge~enwiki, TakuyaMurata, (, Pde, Pcb21, Goatasaur, Egil, 168..., Looxix~enwiki, Ellywa, Mortene, Ahoerstemeier, Fcrick, Mac, Docu, Snoyes, CatherineMunro, JWSchmidt, Kingturtle, Glenn, Cyan, Poor Yorick, Kwekubo, Rotem Dan, Llull, Samuel~enwiki, Mxn, Raven in Orbit, Quickbeam, Hashar, Mulad, Crusadeonilliteracy, Adam Bishop, Timwi, Dcoetzee, Nohat, Wikiborg, Reddi, Lfh, Jwrosenzweig, Gutza, Rednblu, Wik, Zoicon5, Steinsky, Patrick0Moran, Tpbradbury, Tom Allen, Samsara, Thue, Bevo, Shizhao, Dbabbitt, Raul654, Gakrivas, Bcorr, Pakaran, Jerzy, Lumos3, Donarreiskoffer, Robbot, Astronautics~enwiki, Chris 73, Schutz, Vyasa, Peak, Stewartadcock, Sverdrup, Academic Challenger, Timrollpickering, Bkell, Factual, Moink, Hadal, Jstech, Anthony, Neckro, Pifactorial, Diberri, Cyrius, Dmn, Dina, Ancheta Wis, Giftlite, JamesMLane, Graeme Bartlett, DocWatson42, Nunh-huh, Kapow, Netoholic, Fastfission, MSGJ, Obli, Everyking, No Guru, Dratman, Curps, Michael Devore, Bensaccount, Guanaco, Jorge Stolfi, Pascal666, Horatio, Luigi30, Solipsist, Ojl, Avala, SWAdair, Bobblewik, Alan Au, Delta G, Wmahan, Stevietheman, Adenosine, Utcursch, Pgan002, Andycjp, Dullhunk, J.'mach' wust, CryptoDerk, Gazibara, Kums, Antandrus, Onco p53, MisfitToys, G3pro, PDH, Jossi, Rdsmith4, Mikko Paananen, OwenBlacker, Kevin B12, PFHLai, Magadan, Icairns, Troels Arvin, Figure, Asbestos, Neutrality, Golnazfotohabadi, JohnArmagh, Jh51681, Sonett72, Adashiel, Thorwald, Mike Rosoft, Alkivar, D6, Freakofnurture, Rdb, A-giau, William Pietri, ElTyrant, Rich Farmbrough, Guanabot, Ffirehorse, Cacycle, Qutezuce, Vsmith, EliasAlucard, Mgtoohey, Mjpieters, Zazou, Bender235, ESkog, Kbh3rd, Swid, Loren36, Danny B-), Brian0918, Charm, Ben Webber, Kwamikagami, Mwanner, Phoenix Hacker, Aude, Shanes, Susvolans, RoyBoy, EurekaLott, Andreww, Causa sui, Bobo192, Kghose, Infocidal, R. S. Shaw, Brim, Dungodung, Arcadian, Redquark, Timl, Tomgally, La goutte de pluie, Jojit fb, Malcolm rowe, Vanished user 19794758563875, Kierano, Hagerman, Bijee~enwiki, HasharBot~enwiki, Sam Burne James, Emoticon, Jumbuck, Danski14, Mithent, Gary, Anthony Appleyard, Chino, Borisblue, Atlant, SemperBlotto, Ricky81682, Loris, Benjah-bmm27, Wouterstomp, AzaToth, Yamla, Water Bottle, Echuck215, Seans Potato Business, Kocio, InShaneee, Hu, Malo, VladimirKorablin, Snowolf, PaePae, Wmitchell, Melaen, Schapel, BaronLarf, ClockworkSoul, Unconventional, KingTT, Knowledge Seeker, Cburnett, Evil Monkey, Cal 1234, Max Naylor, CloudNine, TenOfAllTrades, Sciurinæ, Inge-Lyubov, Lerdsuwa, LFaraone, Gene Nygaard, Alai, Mattbrundage, Netkinetic, Johntex, Kitch, Adrian.benko, RyanGerbil10, Falcorian, Tariqabjotu, Ashujo, DeAceShooter, Stemonitis, Natarajanganesan, Gmaxwell, Zanaq, Kelly Martin, OwenX, Woohookitty, Mindmatrix, Katyare, TigerShark, Yansa, Rocastelo, Carcharoth, Nitecrawler, Lincher, Colorajo, JeremyA, Duncan.france, MONGO, Nirajrm, Astrowob, Eleassar777, Bbatsell, GregorB, SCEhardt, TheAlphaWolf, Junes, Palica, Turnstep, Marudubshinki, Mandarax, Frostyservant, RichardWeiss, Yasha~enwiki, Graham87, Deltabeignet, Magister Mathematicae, 168..., 169, GoldRingChip, Buxtehude, BD2412, Patrick2480, FreplySpang, Lion Wilson, Yurik, RxS, Effeietsanders, Whoutz, Crzrussian, Drbogdan, Rjwilmsi, Biriwilg, Koavf, Virtualphtn, Collins.mc, Vary, JHMM13, Bruce1ee, Jmcc150, Salix alba, JonMoulton, Oblivious, DonSiano, Ligulem, Gjuggler, SeanMack, Bubba73, Brighterorange, Bhadani, Ucucha, AySz88, Sango123, Yamamoto Ichiro, Kevmitch, Wobble, Ravidreams, FlaBot, Tufflaw, RobertG, Latka, Nihiltres, Crazycomputers, Chanting Fox, Chill Pill Bill, RexNL, Takometer, Karrmann, Jrtayloriv, DevastatorIIC, Wikipedia Administration, ZScout370, Alphachimp, McDogm, Daycd, BMF81, GordonWatts, Smithbrenon, Mstroeck, King of Hearts, Chobot, Nauseam, Antilived, Bornhj, Bjwebb, Mhking, Bgwhite, Digitalme, Whosasking, The Rambling Man, Wavelength, Cathalgarvey, Sceptre, DarkAvenger, WAvegetarian, Jumbo Snails, Spaully, DNA EDIT WAR, Editing DNA, Anarchy on DNA, DNA is shyt, I hate DNA, Zell Miller's DNA, Trent Lott's DNA, RJC, Pigman, Bernie Sanders' DNA, Orrin Hatch's DNA, Bill Nelson's DNA, Tom Harkin's DNA, Chuck Grassley's DNA, Richard Durbin's DNA, Max Baucus' DNA, Splette, SpuriousQ, Stephenb, Wimt, GeeJo, Tavilis, Anomalocaris, Canadaduane, Alcides, Sentausa, Fabhcún, EWS23, Alzhaid~enwiki, Dysmorodrepanis~enwiki, Wiki alf, Chick Bowen, Jaxl, Tmueck, Psora, Taco325i, Yoninah, Djm1279, Ragesoss, Retired username, Sangwine, Robdurbar, Bert Macklin, PhilipO, Misza13, Grafikm fr, Fs, Tony1, Dbfirs, Lockesdonkey, Roy Brumback, DeadEyeArrow, Bota47, Biolinker, Cosmotron, Supspirit, Scope creep, Caerwine, Chris84~enwiki, Somoza, Wknight94, Leptictidium, WAS 4.250, FF2010, Alarob, Calaschysm, 2over0, Theodolite, BenBildstein, Theda, Spondoolicks, Hurricanehink, Jolt76, GraemeL, Acer, JoanneB, Heathhunnicutt, Zahiri, ArielGold, Curpsbot-unicodify, Michigan user, RunOrDie, Stuhacking, Kungfuadam, Banus, Airconswitch, Samuel Blanning, BiH, DVD R W, Jer ome, Luk, RichG, Blastwizard, Itub, Twilight Realm, A bit iffy, SmackBot, Eperotao, Spongebobsqpants, Tarret, Chodges, Slashme, Prodego, KnowledgeOfSelf, TestPilot, Royalguard11, Martin.Budden, Melchoir, Nitramrekcap, Shoy, Joconnol, Jacek Kendysz, Pkirlin, Delldot, Blondtraillite, Zephyris, Yamaguchi先生, Cool3, Aksi great, Peter Isotalo, Gilliam, Julian Diamond, Chaojoker, Jimwong, ERcheck, Carbon-16, Yaser al-Nabriss, Kazkaskazkasako, Izehar, Unint, Keegan, Persian Poet Gal, RDBrown, MidgleyDJ, Uthbrian, Adamstevenson, Diohcierekim's sock, DHN-bot~enwiki, Terraguy, Zven, Darth Panda, Firetrap9254, Gracenotes, Lightspeedchick, Mikker, Hgrosser, Zsinj, Can't sleep, clown will eat me, Jorvik, Danielkueh, Snowmanradio, OOODDD, Roadnottaken, TheKMan, Anita1988, GVnayR, Grover cleveland, Aldaron, NoIdeaNick, Spectrogram, Iapetus, Nakon, Savidan, Ne0Freedom, Jiddisch~enwiki, Rezecib, RandomP, Smokefoot, SteveHopson, Drphilharmonic, Portugue6927, Fagstein, DMacks, Kendrick7, Where, Jls043, Pilotguy, Madeleine Price Ball, SashatoBot, Nishkid64, Visium, Rory096, Zahid Abdassabur, Kuru, General Ization, Tazmaniacs, Kahlfin~enwiki, Epingchris, Statsone, Shadowlynk, AstroChemist, JoshuaZ, Edwy, Coredesat, Mgiganteus1, MichaelHa, IronGargoyle, Fernando S. Aldado~enwiki, The Man in Question, Loadmaster, Digger3000, Alexandremas~enwiki, Hvn0413, Mantissa128, Echo park00, Munita Prasad, Mr Stephen, Dicklyon, Serephine, 4u1e, SandyGeorgia, Sir.Loin, Midnightblueowl, Ryulong, Elb2000, Carlo.milanesi, Squirepants101, EggBurger, Autonova, Glen Hunt's DNA, Keith-264, PaulGS, SimonD, Vanished user, Paul venter, Ariel Pontes, Younusporteous, Paul Foxworthy, ScottBS, Esurnir, Adambiswanger1, Tawkerbot2, Alegoo92, Ouishoebean, Davidbspalding, Redneckjimmy, Fvasconcellos, Switchercat, JForget, RSido, Friendly Neighbour, Gatortpk, CmdrObot, Mattbr, Wafulz, Zarex, Lavateraguy, SupaStarGirl, Leevanjackson, CWY2190, GHe, STAN SWANSON, Mintman16, Outriggr (2006-2009), Cerberus lord, Logical2u, Nthornberry, Moorice~enwiki, RoddyYoung, GeoMor, Sopoforic, A D 13, WillowW, Bryan, Steel, YanWong~enwiki, Michaelas10, Gogo Dodo, Travelbird, Red Director, Trd300gt, Hughdbrown, Sloth monkey, Studerby, Shekharsuman, B, Tawkerbot4, Carstensen, Doug Weller, DumbBOT, Narayanese, ErrantX, Omicronpersei8, Daniel Olsen, Gimmetrow, Casliber, Sabbre, Konrad Foerstner, Thijs!bot, Bloger, Opabinia regalis, Ante Aikio, Russ47025, Headbomb, West Brom 4ever, NorwegianBlue, Tellyaddict, Peter K., GAThrawn22, EdJohnston, Eddycrai, Nick Number, S77914767, Sean William, Natalie Erin, Escarbot, David D., KrakatoaKatie, Cyclonenim, Ialsoagree, Mattjblythe, AntiVandalBot, Mr Bungle, Ty146Ty146, Majorly, Luna Santin, Jeka911, Why My Fleece?, Opelio, Jayron32, 17Drew, Pro crast in a tor, TimVickers, Isilanes, Darklilac, Davegrupp, Storkk, Sjollema, Lklundin, Figma, Canadian-Bacon, JAnDbot, Deflective, Leuko, MER-C, Plantsurfer, Janejellyroll, DigitalGhost, Db099221, Alpinu, Andonic, Hut 8.5, Tstrobaugh, Clivedelmonte, Yahel Guhan, Efbfweborg, Sangak, Magioladitis, Pigmietheclub, WolfmanSF, Pedro, Bongwarrior, Hb2019, Carlwev, NighthawkJ, Bhar100101, Hullaballoo Wolfowitz, Jlh29, CattleGirl, Docjames, Scincesociety, GODhack~enwiki, Tobogganoggin, ThoHug, Avicennasis, WhatamIdoing, BatteryIncluded, Torchiest, David Eppstein, Emw, Gurko, Doctor Faust, Mr Meow Meow, Mika293, TheRanger, Stolsvik, Squidonius, Patstuart, E. Wayne, NatureA16, PsyMar, Yobol, MartinBot, NochnoiDozor, Sjjupadhyay~enwiki, ShaunL, Jonrunles, Pierceno, Dudewheresmywallet, Motley Crue Rocks, Rettetast, Fishingpal99, Zouavman Le Zouave, R'n'B, Test100000, CommonsDelinker, Nono64, Armored Ear, Impamiizgraa, Hairchrm, Fconaway, Jarhed, WelshMatt, Grandegrandegrande, 3dscience, Cinnamon colbert, DrKay, CFCF, Trusilver, Nate1028, Spathaky, UBeR, Nbauman, Boghog, Shmee47, Maurice Carbonaro, Vandelizer, WarthogDemon, Tdadamemd, Andrew wilson, BikA06, Ellis O'Neill, TheChrisD, Bmtbomb, DJRafe, Tonyrenploki, Nemo bis, Dr d12, MrErku, Lhenslee, Kemyou, Pyrospirit, (jarbarf), WHeimbigner, Aquaplus, NewEnglandYankee, Antony-22, Nwbeeson, SmilesALot, ArazZeynili, Rwessel, Master dingley, Touch Of Light, MKoltnow, Aatomic1, Potatoswatter, Scoterican, Cornacchia123, FOTEMEH, Aj123456, WJBscribe, YOUR DNA, Mleefs7, Dr.Kerr, Cainer91, Treisijs, Vaernnond, Etanol~enwiki, Idioma-bot, Plonk2, Hammersoft, VolkovBot, Preston47, DrMicro, Seldon1, Midoriko, Coolawesome~enwiki, Mstislavl, Daniel987600, Hersfold, Orthologist, Lia Todua, AInoktaBOT, Fences and windows, The WikiWhippet, Loginbuddy, Philip Trueman, Director, TXiKiBoT, Oshwah, Muro de Aguas, Billmcgn189, CupOBeans, ElinorD, Qxz, Ocolon, Harianto, Sintaku, Michael H 34, Toninu, Chanora, Seb az86556, Iliu Kr., UnitedStatesian, Pristontale111, Chuck02, Wiae, Luuva, Johanvs, Hockey21dude, DJAX, Nighthawk380, Rajwiki123, Gilisa, Hannes Röst, Gwsrinme, Mentalmaniac07, MarvPaule, Flavaflav1005, Yomama9753, Madhero88, Amboo85, Ashnard, JamesMt1984, Usergreatpower, Synthebot, CephasE, Northfox, AlleborgoBot, Sly G, Timewatcher, Pediaknowledge, Isis07, EmxBot, Gustav von Humpelschmumpel, Kbrose, Safwan40, SieBot, Cubskrazy29, ShiftFn, Tiddly Tom, Graham Beards, BotMultichill, Sakkura, Gerakibot, Dawn Bard, Cwkmail, RJaguar3, Triwbe, Brockett, Execvator, Jerryobject, Keilana, Aaaxlp, Crowstar, Prestonmag, Hzh, Artoasis, Lightmouse, Nanettea, Escape Artist Swyer, Hairwheel, BenoniBot~enwiki, Cradlelover123, Diego Grez-Cañete, Vividonset, Juneythomas, Andrij Kursetsky, Taulant23, Bogwhistle, Randomblue, Paulinho28, Florentino floro, Lascorz, Fortuvoft, SaltyForth123, Stuart7m, Jimriz, ClueBot, GorillaWarfare, Artichoker, Wikievil666, Yikrazuul, Enthusiast01, TobyWilson1992, Niceguyedc, GoEThe, Llongland, DragonBot, Excirial, Alexbot, Pumpkingrower05, Gulmammad, ThreeDaysGraceFan101, Estirabot, Coinmanj, NuclearWarfare, Jotterbot, Medos2, Highfly3442, Jackrm, OekelWm, Heyheyhack, Br shadow, Ardyn, Muro Bot, Gaara san, Aitias, Jonverve, Versus22, Josq, Johnuniq, Wnt, Wkboonec, Psymier, Rishi.bedi, DumZiBoT, InternetMeme, Jetsetpainter, Mandyj61596, Simultaneous, SilvonenBot, Priscilla 95925, Lemchesvej, Addbot, Giftiger wunsch, DOI bot, Medessec, TutterMouse, OBloodyHell, Low-frequency internal, Keepweek~enwiki, NjardarBot, Bernstein0275, DFS454, LinkFA-Bot, Kisbesbot, 84user, Numbo3-bot, Tide rolls, Luckas Blade, Zorrobot, WikiDreamer Bot, Ettrig, Legobot, Drpickem, Luckas-bot, Yobot, Marcus.aerlous, Aquilla34, Nallimbot, KamikazeBot, LightFlare, AnomieBOT, Tryptofish, 1exec1, Jim1138, Galoubet, JWSurf, Mahmudmasri, Materialscientist, Citation bot, Bci2, ArthurBot, Quebec99, Xqbot, Sciencechick, Chaboura, Timir2, Kholdstare99, Khajidha, JWBE, P99am, Gap9551, GrouchoBot, Jhbdel, Ute in DC, Wizardist, RibotBOT, Wiki emma johnson, 7434be, Britand, GhalyBot, AustralianRupert, Metalindustrien, Hersfold tool account, Jack B108, Legobot III, FrescoBot, Paine Ellsworth, Dogposter, Rohitsuratekar, Jc3s5h, KerryO77, HJ Mitchell, Steve Quinn, Citation bot 1, SebastianHawes, Scarce, Javert, Chenopodiaceous, Redrose64, DrilBot, Pinethicket, I dream of horses, Elockid, HRoestBot, Jonesey95, Tom.Reding, Supreme Deliciousness, RedBot, Curehd, Jujutacular, DixonDBot, Jann, Dinamik-bot, Clarkcj12, Sgt. R.K. Blue, HayleyJohnson21, Earthandmoon, Jynto, Tbhotch, Jesse V., Minimac, Ugly Ketchup, DARTH SIDIOUS 2, Mean as custard, RjwilmsiBot, Cloakedyoshi, Salvio giuliano, Mandolinface, Toto Azéro, Mashin6, EmausBot, Orphan Wiki, Acather96, WikitanvirBot, Never give in, Jodon1971, Joeywallace9, GoingBatty, Black Yoshi, Winner 42, Dcirovic, K6ka, TeleComNasSprVen, Hhhippo, JSquish, Otterinfo, Empty Buffer, Dffgd, John Mackenzie Burke, SporkBot, AManWithNoPlan, Ocaasi, Thine Antique Pen, Hccc, JZuehlke, Brandmeister, Pkank, L Kensington, Perseus, Son of Zeus, Rr2wiki, Donner60, Dnacond, Nanodance, Scientific29, Puffin, Venkatarun95, Nofatlandshark, LikeLakers2, Davidartois, Woodsrock, Mikhail Ryazanov, Teaktl17, Will Beback Auto, ClueBot NG, Jnorton7558, Ds2207, Gareth Griffith-Jones, MelbourneStar, Prathfig, Kaushlendratripathi, Zynwyx, O.Koslowski, Widr, Argionember, Username 772, Theopolisme, Vogel2014, Helpful Pixie Bot, Iyentra Rasonica, Aaronstonestrom, Calabe1992, Gob Lofa, Bibcode Bot, Lowercase sigmabot, BG19bot, BOK602, Piguy101, Midnight Green, Astpurcell, JasonK33, Jazzlw, Min.neel, Snow Blizzard, Uluru345, Achowat, Djihinne1, Comfr, AndroidOS, Biosthmors, TuringMachine17, Stigmatella aurantiaca, Jimw338, Cyberbot II, ChrisGualtieri, Saxophilist, Khazar2, IjonTichyIjonTichy, Dexbot, Webclient101, Mogism, Oliverrichardson, Laxative Brownies, THFC1996, Fox2k11, Zziccardi, Hopefuldonor, Mshamza112, Phil2793, Fjozk, RandomLittleHelper, Cor Ferrum, Joeinwiki, ProDawg5, 9FireStar, Diegomanzana, Keiyashi, Dbzhero5000, IncredibleWondersYes1, FamAD123, Youngdro2, AmericanLemming, ManofQueens, EvergreenFir, Evolution and evolvability, Bever, BruceBlaus, Anrnusna, Meteor sandwich yum, Andrewmhhs, Jlmalcos, Abitslow, Csutric, Chaya5260, Zettek95, Mahusha, Giancarlobasile, Monkbot, Virion123, Owais Khursheed, Acagastya, Mr. 1100100, Entitymasterblaster, HMSLavender, Cyntiamaspian, KH-1, Bhootrina, Soldier of the Empire, JordiYiman, DiscantX, MicroPaLeo, Jensberzelius, Forscienceonly, Azealia911, Craftwerker, 3 of Diamonds, COOL ANKUR CHOUDHURY, ProprioMe OW, Unmaterial scientist, MANEVIL 187, Rrwanga17, Maddog9962002, Christianmorasco, Underlyingboss3, CAPTAIN RAJU, S281305, Nn9888, Ryanross123, JoshMuirWikipedia, Lemondoge, Swissinator, MoonmanMCD, Majora, I like doors, Matiar Chris Brown, Mirenoula, Sweatypalmsggg, Joshhassweatylankles, Mikemorris123456789, Nah172, TajBioinfo, MartinZ, Pro-Wiki21, Gantavya garg, PubSci, Lordmitchimus, CutiePie420, FormerBBC, Milkdudstruth, Alexander Davronov, Milkdudstruthback, Milkdudstruthagian, MariusOrion, I Study I Do, Harmonrasmu, Maria Cantana, InternetArchiveBot, Bender the Bot, Shagil Kannur, Heididoerr061, DennisPietras and Anonymous: 1051

- **DNA replication** Source: https://en.wikipedia.org/wiki/DNA_replication?oldid=768585663 Contributors: AxelBoldt, The Anome, Vanderesch, Graft, Heron, Lexor, Skysmith, Alfio, Ahoerstemeier, Habj, Samuel~enwiki, Fibonacci, Robbot, Baldhur, ZimZalaBim, Whiteniko, Hadal, Lupo, Oddharmonic, Giftlite, Bensaccount, FriedMilk, Adenosine, Pgan002, GD~enwiki, Jossi, Taka, Ary29, Neutrality, Julianonions, Imjustmatthew, Mike Rosoft, Rich Farmbrough, Guanabot, Vsmith, Xezbeth, Bender235, Kaisershatner, El C, RoyBoy, Adambro, Bobo192, Stesmo, Whosyourjudas, Smalljim, Cmdrjameson, Arcadian, Timl, Dennis Valeev, La goutte de pluie, Nhandler, Nsaa, Alansohn, Arthena, Wouterstomp, Stillnotelf, Snowolf, Ravenhull, Velella, ClockworkSoul, Tycho, Netkinetic, RyanGerbil10, Jackhynes, Benbest, Before My Ken, Fbv65edel, MONGO, LadyofHats, Plrk, Dysepsion, Snafflekid, Rjwilmsi, SMC, Thisismikesother, Alveolate, The wub, Mortice, FlaBot, Cless Alvein, RobertG, Nihiltres, GünniX, Alphachimp, BradBeattie, Chobot, Bgwhite, Whosasking, YurikBot, Borgx, WAvegetarian, Chris Capoccia, GLaDOS, SpuriousQ, Thiseye, Malcolma, Daniel Mietchen, Jpbowen, InvaderJim42, Raven4x4x, RUL3R, Wangi, Jhinman, Closedmouth, Maristoddard, Pb30, Xaxafrad, Svetlana Miljkovic~enwiki, Mfedder, JoanneB, CIreland, Rystic, SmackBot, Andthendougsaid, Ariliand, InverseHypercube, KnowledgeOfSelf, Royalguard11, Bomac, Stepa, Eskimbot, CapitalSasha, Bozartas, Edgar181, Blondtraillite, Zephyris, Ashermadan, Freddy S., Gilliam, DividedByNegativeZero, Skizzik, ERcheck, Andy M. Wang, Bluebot, Persian Poet Gal, JDCMAN, Tree Biting Conspiracy, Hichris, MalafayaBot, Celarnor, Zrulli, Bowlhover, Richard001, Drphilharmonic, Madeleine Price Ball, SashatoBot, Ser Amantio di Nicolao, Harryboyles, Scientizzle, Gobonobo, JulianBlow, Thegathering, Edwy, Steipe, Mgiganteus1, MichaelHa, PseudoSudo, A. Parrot, Stwalkerster, Noah Salzman, Wrlampe, Romanticcynic, TastyPoutine, Artman40, Sasata, Hu12, Dan Gluck, JayZ, Twas Now, Igoldste,

## 23.10.2 Images

- **File:Uracil_base_glycosidase.jpg** *Source:* https://upload.wikimedia.org/wikipedia/commons/7/7c/Uracil_base_glycosidase.jpg *License:* Public domain *Contributors:* Transferred from en.wikipedia to Commons. *Original artist:* TimVickers at English Wikipedia
- **File:Walter_sutton.jpg** *Source:* https://upload.wikimedia.org/wikipedia/commons/4/43/Walter_sutton.jpg *License:* Public domain *Contributors:* http://www.genomenewsnetwork.org/resources/timeline/1902_Boveri_Sutton.jpg *Original artist:* Unknown<a href='https://www.wikidata.org/wiki/Q4233718' title='wikidata:Q4233718'><img alt='wikidata:Q4233718' src='https://upload.wikimedia.org/wikipedia/commons/thumb/f/ff/Wikidata-logo.svg/20px-Wikidata-logo.svg.png' width='20' height='11' srcset='https://upload.wikimedia.org/wikipedia/commons/thumb/f/ff/Wikidata-logo.svg/30px-Wikidata-logo.svg.png 1.5x, https://upload.wikimedia.org/wikipedia/commons/thumb/f/ff/Wikidata-logo.svg/40px-Wikidata-logo.svg.png 2x' data-file-width='1050' data-file-height='590' /></a>
- **File:Wiki_letter_w_cropped.svg** *Source:* https://upload.wikimedia.org/wikipedia/commons/1/1c/Wiki_letter_w_cropped.svg *License:* CC-BY-SA-3.0 *Contributors:* This file was derived from Wiki letter w.svg: <a href='//commons.wikimedia.org/wiki/File:Wiki_letter_w.svg' class='image'><img alt='Wiki letter w.svg' src='https://upload.wikimedia.org/wikipedia/commons/thumb/6/6c/Wiki_letter_w.svg/50px-Wiki_letter_w.svg.png' width='50' height='50' srcset='https://upload.wikimedia.org/wikipedia/commons/thumb/6/6c/Wiki_letter_w.svg/75px-Wiki_letter_w.svg.png 1.5x, https://upload.wikimedia.org/wikipedia/commons/thumb/6/6c/Wiki_letter_w.svg/100px-Wiki_letter_w.svg.png 2x' data-file-width='44' data-file-height='44' /></a>
  *Original artist:* Derivative work by Thumperward
- **File:Wikiquote-logo.svg** *Source:* https://upload.wikimedia.org/wikipedia/commons/f/fa/Wikiquote-logo.svg *License:* Public domain *Contributors:* Own work *Original artist:* Rei-artur
- **File:Wiktionary-logo-v2.svg** *Source:* https://upload.wikimedia.org/wikipedia/commons/0/06/Wiktionary-logo-v2.svg *License:* CC BY-SA 4.0 *Contributors:* Own work *Original artist:* Dan Polansky based on work currently attributed to Wikimedia Foundation but originally created by Smurrayinchester
- **File:Working_principle_of_telomerase.png** *Source:* https://upload.wikimedia.org/wikipedia/commons/3/3b/Working_principle_of_telomerase.png *License:* CC BY-SA 3.0 *Contributors:* Own work *Original artist:* Fatma Uzbas
- **File:Zidovudine.svg** *Source:* https://upload.wikimedia.org/wikipedia/commons/3/32/Zidovudine.svg *License:* Public domain *Contributors:* Own work *Original artist:* Fvasconcellos

## 23.10.3 Content license

- Creative Commons Attribution-Share Alike 3.0

www.ingramcontent.com/pod-product-compliance
Lightning Source LLC
Chambersburg PA
CBHW081112170526
45165CB00008B/2429